P9-CKY-442

TECHNIQUES OF METALS RESEARCH

R. F. Bunshah, Editor

Volume I:	Techniques of Materials Preparation and Handling Parts 1 to 3
Volume II:	Techniques for the Direct Observation of Structure and Imperfections Parts 1 and 2
Volume IIA:	The Stereographic Projection and Its Applications (by O. Johari and G. Thomas)
Volume IIB:	Transmission Diffraction Electron Microscopy (by G. Thomas and W. L. Bell)
Volume IIC:	Handbook of Metallographic Techniques (by R. Gray, R. Crouse, and E. L. Long)
Volume III:	Modern Analytical Techniques for Metals and Alloys Parts 1 and 2
Volume IIIA:	Electrical Resistivity and Hall Effect Measurements (by R. Reich)
Volume IV:	Physicochemical Measurements in Metals Research (Editor: R. A. Rapp)
Volume V:	Measurement of Mechanical Properties
Volume VI:	Measurement of Physical Properties (Editor: E. Passaglia)
Volume VII:	Techniques Involving Extreme Environment, Nondestructive Techniques, Computer Methods in Metals Research, and Data Analysis

TECHNIQUES OF METALS RESEARCH

R. F. Bunshah, Editor
University of California
Los Angeles, California

TECHNIQUES OF METALS RESEARCH
Volume III, Part 1

Modern Analytical Techniques for Metals and Alloys

Part 1

Editor
R. F. BUNSHAH

Authors of Part 1

E. R. BLOSSER
D. A. BOLTZ
W. G. BOYLE
R. I. BYSTROFF
E. A. HAKKILA
J. E. HARRAR
W. H. HENRY
J. E. HOLLIDAY
J. A. HOWELL
S. KALLMANN
W. M. MALLETT
A. MIZUIKE

1970
INTERSCIENCE PUBLISHERS
a division of John Wiley & Sons, New York, London, Sydney, Toronto

The paper used in this book has pH of 6.5 or higher. It has been used because the best information now available indicates that this will contribute to its longevity.

1 2 3 4 5 6 7 8 9 10

Library of Congress Catalog Card Number 67-20260
SBN 471 12215 7
Printed in the United States of America

INTRODUCTION TO THE SERIES

Progress in metals research and all the related areas of materials research in the past two decades has been greatly affected by the introduction of new methods, techniques, and instruments, some of which are highly sophisticated. Much of the modern methodology of mathematics, physics, and chemistry has been called upon to refine the investigation of matter in the metallic state and in the solid state at large. Thus, a large number of many different types of techniques are currently used in metals research.

This has been the case not only for the fundamental and scientific exploration of the metallic state, but also for the techniques used to measure properties, and evaluate performance, and for applications in actual use.

In comparison to the immediate post World War II era (circa 1945), a much larger number of metals and their alloys are in use today. The much greater variety of their functions under normal and extreme conditions has led to the development of a vast body of knowledge and especially of highly specialized methods and techniques. As usual, information about techniques is more widely scattered than information on results.

The term "metals research" is used here in its broadest sense to include the "development" connotation, and it is not restricted to small-scale laboratory experiments. In fact, a successful pursuit of a research or development program nowadays often requires a thorough comprehensive consideration of all aspects of the problem, from preparation and handling of the primary purified metal through refining, consolidation, mechanical and heat treatment, sample machining, and preparation, to measurement of physical, chemical, and mechanical properties and chemical composition.

It seems therefore appropriate to put together a well-organized, integrated presentation of the total area of techniques of metals research comprising all of the aspects mentioned above.

One of the best ways to summarize a large body of information is a comprehensive, well-referenced review article. This series consists of review articles on the various techniques of metals research. The number of techniques is large and the rate of development of new techniques is rapid, which makes it possible that some newer techniques may have been omitted in this edition, although a conscientious effort has been made to make the coverage as complete as possible. Some topics which are very marginal to the concept of this series, e.g., ore mining and beneficiation, have been intentionally omitted. Similarly, only the newer techniques of chemical analysis, particularly those dealing with trace analysis, and analysis of interstitials are covered.

Several books have been published at irregular intervals in the past 20 years dealing with fragments of the area we are trying to describe. Our primary aim is to bring together a comprehensive collection of the techniques of metals research.

This could, of course, be done only as a collaborative effort. The result will be a series of self-consistent volumes which will as a whole, we hope, form a reliable guide to the multifarious problems which arise in every laboratory in which metals, metallic substances, and other similar materials are investigated. Each volume will deal with a broad subject area.

While it is not the aim to tell the expert in a particular method or discipline more than he knows already, the articles in this series will enable the graduate student, the researcher, or the supervisor with an adequate background in the physical sciences, metallurgy, or engineering to gain access to a technique with which he is unfamiliar and to enable him to judge its applicability to a special problem and, even more important, its limitations. The contributions are written in such a style that the reader should be able to use them immediately without further reference. Ample literature quotations will guide him to more detailed information.

It is not intended that the articles in this series be written in cookbook style, so that it would enable somebody to go straight to the laboratory and use the techniques he has been reading about.

Each article in the series is an entity in itself and the reader will therefore occasionally find overlap of subject matter between articles. However, it is the editor's philosophy that it is far more important to preserve the integrity and continuity of statements in an article than prune out a few pages here and there. The general organization of each article is to give the theoretical background necessary to understand the various techniques, a description of the technique or techniques, a discussion of their advantages and limitations, and a bibliography for further reading.

Within this general scope, the author has been given the freedom to develop his topic as he sees fit. Therefore, the relative weight given to each of these categories in a given article will depend on the subject matter being treated. Some articles dealing with new techniques are almost entirely descriptive, whereas others dealing with old, established techniques, like x-ray diffraction, are essentially organized, well-referenced summaries of groups of techniques which are dealt with in detail elsewhere.

It is rather difficult to organize and arrange the rather heterogeneous topics in a completely rational and systematic fashion. For example, the same topic could equally well be placed in more than one volume. I have chosen the following overall breakdown:

Volume I:	Techniques of Materials Preparation and Handling
Volume II:	Techniques for the Direct Observation of Structure and Imperfections
Volume IIA:	The Stereographic Projection and Its Applications
Volume IIB:	Transmission Diffraction Electron Microscopy
Volume IIC:	Handbook of Metallographic Techniques
Volume III:	Modern Analytical Techniques for Metals and Alloys
Volume IIIA:	Electrical Resistivity and Hall Effect Measurements
Volume IV:	Physicochemical Measurements in Metals Research
Volume V:	Measurement of Mechanical Properties
Volume VI:	Measurement of Physical Properties
Volume VII:	Techniques Involving Extreme Environment Nondestructive Techniques, Computer Methods in Metals Research, and Data Analysis

When a particular article becomes sufficiently large or a particular topic is unusually broad, it will be published as a separate part under its own cover. It is hoped that the comprehensive Index which will be published at the end of the series will help in locating any and all of the topics discussed.

In planning and preparing the work, the editor has enjoyed the assistance and counsel of the Editorial Advisory Board as well as many other colleagues. Their help is gratefully acknowledged herewith.

He is also very grateful to the Editors for Volumes IV and VI, Dr. Robert A. Rapp and Dr. E. Passaglia, respectively.

My sincere thanks are also due to the authors of the various articles in the series and to the Editorial and Production Departments of the publisher.

This is a first attempt to perform a very large task. The editor, therefore, invites the comments, suggestions, and criticism of the readers to improve these volumes in future editions.

R. F. Bunshah

Livermore, California

PREFACE TO VOLUME III

The topics covered in this volume deal with modern analytical techniques for metals and alloys. It is well understood that trace impurities can have a significant effect on the physical and mechanical properties of materials. The obvious examples are those of semiconductor materials, laser material, and the like, in which trace impurities produce dramatic changes in electrical conductivity and optical and other physical properties. Equally important, though perhaps not so well recognized, are the marked effects of trace impurities on mechanical properties; for example, removal of trace impurities from alloy steels results in a large and technically significant drop in the ductile-brittle impact transition temperature. Removal of trace impurities in stainless steels markedly improves their corrosion resistance! These observations open up virtually untouched areas for future development—namely high purity alloys.

The chemical characterization of a material requires a knowledge of the amounts of major constituents, the trace impurities, and the location of these impurities in the microstructure. The techniques covered in this volume address themselves to all of these aspects of the problem, with perhaps greater emphasis on those dealing with trace analysis and the location of impurities in the microstructure.

Most of the techniques discussed are modern and many of them are under active development; for example, mass spectroscopy, activation analysis, and soft X-ray spectroscopy. Thus in the coming years we should expect to see lower detection limits and higher precision from many of them. Considerable use is presently being made of combinations of techniques (e.g., microprobe analysis and electron microscopy) from which rapid developments are anticipated.

Standard techniques of wet chemical analysis have been omitted from this book; they are numerous and fully discussed elsewhere.

The reader's attention is directed toward two chapters in Volume 2 of this series which deal with related techniques. Chapter 6 on field ion microscopy describes a method for the location of impurities in the structure which is also a semiquantitative method for the analysis of such impurities. Chapter

20 on autoradiographic techniques describes methods of locating impurities in the microstructure and of measuring quantitatively the amounts of the major constituents in various parts of the specimen. The latter is particularly useful in studying such important phenomenon as diffusion and oxidation.

For the materials scientist the ideal state of chemical characterization of a material is a knowledge of the chemical composition of its major constituents, the amounts of trace impurities, and more important still the location and localized concentration of these impurities. The analytical techniques available today provide considerable assistance toward this end. We cannot say, however, that it has been reached. Progress toward this goal will be indispensable in the development, understanding, and utilization of new materials.

R. F. BUNSHAH

Los Angeles, California
August, 1969

Modern Analytical Techniques for Metals and Alloys

Volume III

CONTENTS

PART 1

1. Brief Survey of Various Analytical Techniques Used for Metals and Alloys. *W. M. Henry and E. R. Blosser,* Battelle Memorial Institute, Columbus Laboratory, Columbus, Ohio 1
2. Separation and Preconcentration Techniques. *A. Mizuike,* Nagoya University, Chikusa-ku Nagoya, Japan 25
3. Vacuum Fusion, Vacuum Extraction and Inert Gas Fusion Techniques. *Manley W. Mallett,* Neutron Physics Department, General Electric Company, St. Petersburg, Florida, and *S. Kallmann,* Ledoux and Company, Inc., Teaneck, New Jersey 69
4. Combustion Methods for Carbon and Sulfur. *Walter G. Boyle, Jr.,* Lawrence Radiation Laboratory, University of California, Livermore, California 115
5. Electroanalytical Techniques. *Jackson E. Harrar,* Lawrence Radiation Laboratory, University of California, Livermore, California 143
6. Analytical Flame Spectroscopy. *Roman I. Bystroff,* Lawrence Radiation Laboratory, Livermore, California 185
7. Spectrophotometry and Spectrofluorometry. *James A. Howell,* Western Michigan University, Kalamazoo, Michigan and *David F. Boltz,* Wayne State University, Detroit, Michigan 225
8. X-Ray Spectroscopic Methods. *E. A. Hakkila,* University of California, Los Alamos Scientific Laboratory, Los Alamos, New Mexico 275
9. Soft X-Ray Spectroscopy in Metals Research. *J. E. Holliday,* Edgar C. Bain Laboratory for Fundamental Research, United States Steel Corporation, Monroeville, Pennsylvania 325

Author Index, Volume 3, Part 1 3–1

Subject Index, Volume 3, Part 1 3–15

PART 2

10. Electron Probe Microanalysis. *J. Philibert,* IRSID, St. Germain-en-laye, France

11. Emission Spectroscopy Including Arc, Spark and Other Methods. *Victor G. Mossotti,* Materials Research Laboratory, University of Illinois, Urbana, Illinois

12. Mass Spectrometric Techniques. *Richard E. Honig,* RCA Laboratories, Princeton, New Jersey

13. Secondary Ion Emission Analysis. *J. Phillibert,* IRSID, St. Germain-en-laye, France

14. Elastic Nuclear Scattering Techniques. *Sylvan Rubin,* Stanford Research Instititute, Menlo Park, California

15. Activation Analysis—General Introduction. *Phillipe Albert,* C.N.R.S., Vitry-sur-seine, France

16. Activation Analysis with Thermal Neutrons. *Phillipe Albert,* C.N.R.S., Vitry-sur-seine, France

17. Activation Analysis with Fast Neutrons. *J. Laverlochere,* C.E.N., Grenoble, Grenoble, Isere, France

18. Activation Analysis with Gamma Rays and Charged Particles. *Ch. Engelmann,* C.E.N., Saclay, Gif-sur-yvette, France and *Ph. Albert,* C.N.R.S., Vitry-sur-seine, France

19. Use of Computers in Activation Analysis. *J. Laverlochere,* C.E.N., Grenoble, Grenoble, Isere, France.

20. General Conclusion on Methods of Activation Analysis. *Phillipe Albert,* C.N.R.S., Vitry-sur-seine, France

Author Index, Volume 3, Part 2

Subject Index, Volume 3, Part 2

Chapter 1

BRIEF SURVEY OF VARIOUS ANALYTICAL TECHNIQUES USED FOR METALS AND ALLOYS

W. M. HENRY and E. R. BLOSSER, Battelle Memorial Institute, Columbus Laboratories, Columbus, Ohio

I. Introduction 1
II. Trace Detection 7
A. Detectability and Sensitivity 7
B. Matrix Interferences 11
C. Effects of Various Parameters 12
D. Minute Samples 13
III. Precision 14
IV. Accuracy 18
V. Sample Form and Preparation 20
VI. Automation and Data Handling 21
VII. Unresolved Problems 22
References 23

I. INTRODUCTION

An introductory chapter on a subject as diverse as modern analytical techniques encounters considerable risk in encroaching on the ensuing discussions by authors of specific techniques. However, such a chapter does give the opportunity to present a broad, noncontroversial view of the overall field and can present considerations to enable the reader to interpret the subsequent writings with a searching but open mind in quest of an analytical technique for his own use.

The modern analyst is endowed with a vast array of techniques to characterize metals and alloys. Table 1 gives a generalized listing of the various techniques and parameters. The newer of these are mostly instrumental and many of these represent relatively large outlays of capital funds to acquire. Further expenditures are required to obtain experience in the operation of the instruments, to maintain them, and to replace component parts. Since

Table 1. Some identification techniques and parameters—generalized

Technique	Determined	Simultaneously[a]	Precision[b] (percent)	Destruction	Lower limit of detection	Amount of sample consumed or examined for best detection	Useful range	Approximate costs
Emission spectrography	Metallic elements	Yes	3–10	Yes	1–10 ppm	20 mg	Lower limit to 5%	\$25 to \$50 3–10 elements[c]
X-ray spectrography	Elements > atomic No. 11 (Na)	Yes	1–5	No	10–1,000 ppm	4 cm²	Lower limit to 50%	\$5 per element[c]
Spark mass spectrography	All elements	Yes		Partial	0.003–0.3 ppm	2–200 mg	Lower limit to 1%	
Photoplate readout			10–50					\$75 to \$200[c]
Electronic readout			1–10					\$75 to \$200[c]
Electron microprobe	Elements > atomic No. 5 (B) (plus elemental distribution)	Yes	1–5	No	100–1,000 ppm	1–200 μ²	Lower limit to major compound	\$75 to \$200
X-ray diffraction	Crystal structure phases, compounds	Yes	5–10	No	1–5%	1–10 mg	Lower limit to major compound	\$50 to \$200

Metallography	Visual and photographic	Yes	—	No	1–1,000 × magnification	> 1 μ^2	1 μ resolution	\$25 to \$150
Light microscopy	Visual and photographic	Yes	—	No	1–1,000 × magnification	> 1 μ^2	1 μ resolution	\$25 to \$100
Electron microscopy	Visual and photographic (usually by replica)	Yes	—	Partial	1,000–100,000 × magnification	10–100,000 Å	10 Å resolution	\$50 to \$300
Scanning electron microscopy	Visual and photographic (replica not required)	Yes	—	No	15–50,000 × magnification	150 Å to 1 cm^2	150–1,000 Å resolution	\$50 to \$300
Activation analysis	Most elements—some receiving chemical separation	No	1–10	Yes and No	0.001–100 ppm	grams	Lower limit to 5%	Too variable to estimate
Chemical	Most elements	No	1–5	Yes	10–100 ppm	grams	Lower limit to 50%	\$5 to \$35 per element
Chemical color, stain, spot test, etc.	Most elements	No	1–5	Yes	0.01–10 ppm	grams	Lower limit to 1%	\$5 to \$35 per element
Atomic absorption	Up to 50 elements	No	1–5	Yes	0.1–100 ppm in solution	1 ml	Lower limit to 10%	\$5 to \$10 per element[c]
Flame emission	Alkali and alkaline earth elements	No	1–5	Yes	0.1–10 ppm in solution	1 ml	Lower limit to 10%	\$5 to \$10 per element[c]

(continued)

Table 1 *(continued)*

Technique	Determined	Simultaneously[a]	Precision[b] (percent)	Destruction	Lower limit of detection	Amount of sample consumed or examined for best detection	Useful range	Approximate costs
Vacuum fusion	O, H, N	Yes	10	Yes	0.1–10 ppm	1 g	Lower limit to 1%	\$50 for O, H, and N
Carbon determination	C							
(gravimetric)		—	3–10	Yes	50–100 ppm	1 g	Lower limit to 20%	\$5 to \$10
(conductometric)	C	—	3–10	Yes	1–10 ppm	1 g	Lower limit to 0.1%	\$25
Nitrogen determination (Kjeldahl)	N	—	3–10	Yes	1–10 ppm	1 g	Lower limit to 5%	\$20
Sulfur determination (combustion)	S	—	3–10	Yes	50–100 ppm	1 g	Lower limit to 5%	\$5 to \$10
(Luke's method)	S	—	3–10	Yes	1–10 ppm	1 g	Lower limit to 0.1%	\$25

[a] With unit sample.
[b] With appropriate standard(s), accuracy and precision can be comparable.
[c] Standard preparation, if needed, is additional.

analytical technology is advancing so rapidly, still additional funds are needed to modify instruments so as to use them at their greatest potential and to ward off their obsolescence.

One of the major advantages provided by the use of modern instrumental techniques is the capability to obtain the desired characterization of a material with lowered manpower effort. Stated another way, the development and adoption of many instrumental techniques frequently represents man-hour and/or manpower savings and not characterization data totally unattainable, at least in part, by other means. As an example, the optical and X-ray emission techniques provide compositional data on 50–70 elements previously determined singly or in functional groups by classical chemical techniques. The solids spark-source mass spectrograph provides sub-ppm compositional data on elements present in a material. Previously, such determinations utilized separation and preconcentration steps followed by such techniques as optical or X-ray emission analyses, colorimetric, fluorimetric, or other chemical or instrumental methods. The electron microprobe permits analyses of minute areas previously determined at least in part by microscopy, microchemical, and chemical microscopic techniques. A second major advantage obtained through use of modern instrumental techniques is the capability to obtain characterization virtually impossible by other means, or at least impractical from an economic viewpoint. Such is the case for the spark-source mass spectrograph in determining all elements, including the interstitials, on a small sliver of metal. Such also is the case of an ion bombardment-source mass spectrometer used to determine the composition of monolayer surfaces and an electron microprobe used to profile compositional differences across a grain boundary with high accuracy.

Today's analyst, gifted with numerous analytical techniques available which lessen his manpower effort and increase his analytical capabilities, stands ready in most cases to meet the increasing needs for compositional characterization demanded in a rapidly advancing materials technology. Although presently he may be using only certain of the many analytical tools available, he should be aware of the advantages and disadvantages of all. Thus, he can measure his present techniques and practices against the potentialities offered by other techniques and intelligently select replacement or augmentation when economics of his present analytical work or changes dictate a change.

Intelligent awareness of the potentialities of the newer analytical techniques and decisions on their selection are not always easy to come by. Very often similar characterization data can be provided by the use of any of several techniques, and true evaluation of all of the potentialities, advantages, and disadvantages of the several techniques can be hidden in the understandable enthusiasms of the manufacturers or analysts using a given technique. Thus, one encounters such statements as the following which appeared in the analytical literature: "Any analytical chemist who is not making use of the techniques of . . . stands perilously near the brink of anachronism," or "Probably no other analytical method is so useful or so rapid as . . . for making qualitative and quantitative determinations of the elements in a sample over a wide range of composition, regardless of the crystal structure, chemical combinations, valence state or of the nature of the sample whether it be a massive solid, a thin film, a powder or a liquid." The specific techniques named in the above quotes were omitted because similar statements are seen for nearly all the newer techniques and there is no point in distinguishing these particular techniques over the others.

The number of modern analytical techniques are too many to cover adequately in an introductory chapter. This introduction is intended to point out certain salient features of general interest which may be factors in evaluating a given technique or several techniques described in the ensuing chapters.

II. TRACE DETECTION

The capability to detect minute concentrations is an important aspect of analytical characterization methods. In certain cases the need for ultra detection can apply to only one or a few selected elements and in other cases "across the board" ultra detection may be required. Many materials and particularly pure metals now can be and are being characterized far more completely than in the past, when designations such as "3 nines" or "6 nines"—i.e., 99.9 or 99.9999—were common. These latter designations usually were based only on emission spectrographic data which did not include the everpresent interstitial elements nor did it provide very good detectability on several metallic elements.

A. Detectability and Sensitivity

The analyst often must provide data not only on the elements detected in a sample but also on the concentrations of other elements possibly present but not detected by the techniques used for the analyses. Presenting complete information is quite time-consuming and often needless for a given analysis. Tabular data can be used to provide generalizations on the detection limits attainable for all elements determinable by a given technique. Such generalizations can be a form listing all elements in the periodic table with space provision for filling in the concentrations of impurities found and the detectability limits for those elements not found present. The data of course should be revised when a particular technique is changed or when the equipment on which the original tabular data were obtained is modified or replaced. Rapid advances in analytical technology and equipment necessitates frequent such revisions of detectability data.

A tabulation weight of detectabilities attainable by various techniques is given in Table 2. This table lists the minimum weight, to the nearest power of 10, of an element detectable in favorable cases as compiled from data based on published work. In actual practice many of these limits will not be realized owing to interferences, blanks and nonideal samples. To convert these weight limits to percentages or parts per million it would be necessary to know the weight of sample that can be analyzed in a given run. It is often possible to preconcentrate specific elements from a sample, thus gaining a magnitude or more in percentage detectability. Parenthetically, note that most chemical techniques are in themselves concentration techniques that isolate the sought element from its matrix, and that certain specialized optical emission techniques (e.g., use of "carriers") also effect a separation during the analysis. Many manufacturers provide current detectability lists obtainable with their instruments. Other lists of detectabilities are contained in volumes on trace–characterization and materials–purification practices.[1–4] When used knowledgeably these tabulations enable a laboratory to compare its data with that of others, to select a particular technique for application to a given problem, and to evaluate the potential offered by the addition of techniques other than those presently being used on a given or new laboratory work load.

Intelligent use of data given on detectability attainable by a given technique and particularly the use of such data to intercompare detectability by several techniques requires that several variables concerning such data be examined. With what confidence limit were the data compiled? ASTM E-2 defines detection limit as "A stated limiting value which designates the lowest concentration or mass that can be estimated or determined with confidence and which is specific to the analytical procedure used. Note: Unless otherwise stated the detection limit is assumed to have a confidence level of about one standard deviation."[5] Others suggest up to four times the standard deviation as the proper confidence level to be used. Morrison and Skogerboe give a good discussion of the varying criteria used in determining detection limits and summarize their discussion by stating that whatever level of confidence is chosen, it and the method used for testing it should be stated when reporting detection limits.[6] Variations of detection limits attainable by a given technique can vary a magnitude or more depending on the criteria chosen for the confidence in the detection.

Another factor affecting the utility of data given on detectability is the equipment with which it was established. Activation methods will vary

Table 2. Approximate detectabilities attainable by various techniques (Grams in powers of 10[a])

Element	Optical emission spectrography	Chemical	Flame (atomic absorption and emission)	Gamma and neutron activation	Spark-source mass spectrography	Thermal (combustion and vacuum fusion)	Electron microprobe
H	—	—	—	−1	−11	−6	—
He	—	—	—	—	−11	—	—
Li	−10	—	−12	−10	−11	—	—
Be	−10	−7	−9	−1	−11	—	—
B	−8	−6	—	—	−11	—	−13
C	—	−6	—	−8	−11	−6	−13
N	—	−6	—	−8	−11	−5	−12
O	—	—	—	−8	−11	−5	−13
F	—	−7	—	−7	−11	—	−14
Ne	—	—	—	−7	−11	—	—
Na	−9	—	−10	−9	−11	—	−14
Mg	−9	−5	−9	−7	−11	—	−14
Al	−9	−6	−7	−9	−11	—	−14
Si	−8	−5	−7	−7	−11	—	−14
P	−6	−6	—	−8	−11	—	−14
S	—	−5	—	−7	−11	—	−14
Cl	—	−5	—	−9	−11	—	−14
Ar	—	—	—	−9	−11	—	—
K	−8	—	−9	−8	−11	—	−13
Ca	−9	−4	−9	−6	−11	—	−14
Sc	−8	−4	—	−10	−11	—	−14
Ti	−8	−5	−7	−7	−11	—	−14
V	−8	−6	−7	−10	−11	—	−14
Cr	−8	−6	−8	−8	−11	—	−13
Mn	−9	−6	−8	−10	−11	—	−13
Fe	−9	−6	−7	−7	−11	—	−13
Co	−8	−6	−8	−10	−11	—	−13
Ni	−8	−6	−8	−7	−11	—	−13
Cu	−9	−6	−8	−9	−11	—	−13
Zn	−7	−5	−8	−8	−10	—	−13
Ga	−8	−7	−7	−9	−10	—	−13

(continued)

Table 2 (*continued*)

Element	Optical emission spectrography	Chemical	Flame (atomic absorption and emission)	Gamma and neutron activation	Spark-source mass spectrography	Thermal (combustion and vacuum fusion)	Electron microprobe
Ge	−8	−7	−6	−9	−10	—	−13
As	−7	−7	−6	−10	−10	—	−13
Se	—	−5	−6	−8	−10	—	−13
Br	—	−5	—	−9	−10	—	−14
Kr	—	—	—	−9	−10	—	—
Rb	−8	—	−9	−8	−10	—	−13
Sr	−8	—	−9	−8	−10	—	−13
Y	−8	—	−7	−9	−10	—	−13
Zr	−8	−5	−5	−6	−10	—	−13
Nb	−7	−6	−6	−9	−10	—	−13
Mo	−9	−7	−8	−7	−10	—	−13
Ru	−8	−5	−7	−7	−10	—	−13
Rh	−8	—	—	−10	−10	—	−13
Pd	−8	—	—	−9	−10	—	−13
Ag	−9	−6	−8	−10	−10	—	−13
Cd	−7	−6	−8	−7	−9	—	−13
In	−8	−5	−8	−11	−10	—	−13
Sn	−8	−6	−7	−7	−9	—	−13
Sb	−8	−6	−7	−9	−10	—	−13
Te	−7	−5	−7	−8	−9	—	−13
I	—	−5	—	−9	−10	—	−13
Xe	—	—	—	−7	−10	—	—
Cs	−7	—	−8	−7	−10	—	−13
Ba	−8	−4	−8	−7	−10	—	−13
La	−7	−4	−6	−9	−10	—	−13
Ce	−6	−4	−5	−7	−10	—	−13
Pr	−6	−4	−6	−10	−10	—	−13
Nd	−6	−4	−6	−8	−9	—	−13
Sm	−7	−4	—	−10	−9	—	−13
Eu	−8	−4	−9	−11	−10	—	−13
Gd	−7	−4	−6	−9	−9	—	−13
Tb	−7	−4	−6	−10	−10	—	−13

(*continued*)

Table 2 *(continued)*

Element	Optical emission spectrography	Chemical	Flame (atomic absorption and emission)	Gamma and neutron activation	Spark-source mass spectrography	Thermal (combustion and vacuum fusion)	Electron microprobe
Dy	−7	−4	−7	−11	−9	—	−13
Ho	−7	−4	—	−10	−10	—	−13
Er	−7	−4	−7	−8	−9	—	−13
Tm	−8	−4	−7	−10	−10	—	−13
Yb	−9	−4	−8	−10	−9	—	−13
Lu	−8	−4	−7	−10	−10	—	−13
Hf	−7	−5	−5	−9	−9	—	−13
Ta	−7	−5	−5	−9	−10	—	−13
W	−6	−6	−6	−9	−9	—	−13
Re	−6	−5	−6	−10	−10	—	−13
Os	−8	−4	−5	−9	−9	—	−13
Ir	−7	−4	−6	−10	−10	—	−13
Pt	−8	−6	−6	−8	−9	—	−13
Au	−8	−6	−7	−10	−10	—	−13
Hg	−7	−7	−7	−9	−9	—	−13
Tl	−8	−5	−7	−7	−10	—	−13
Pb	−8	−7	−8	−5	−9	—	−13
Bi	−7	−5	−8	−7	−10	—	−13
Th	−6	−6	−4	−9	−10	—	−13
U	−7	−9	−5	−9	−10	—	−13

[a] Example: −11 means approximate detectability is 10^{-11} g.

considerably in the detection limits attainable on given elements depending on the irradiation facility used, i.e., the type and energy of irradiation and the quality of the counting equipment. Despite this it is not infrequent to see a list of the detection limits attained for many elements entitled simply "activation analysis." Dectectabilities attainable by atomic absorption change considerably for certain elements dependent on the light source and the excitation gas mixture employed.

Differences in interpretation as to what is meant by the terms detection limit or detectability and sensitivity limits or sensitivity can cause confusion not only among analysts but among users of analytical data. Presently there is a trend to limit the use of the terms sensitivity limits and sensitivity to the increase in signal response corresponding to an increase in the concentration

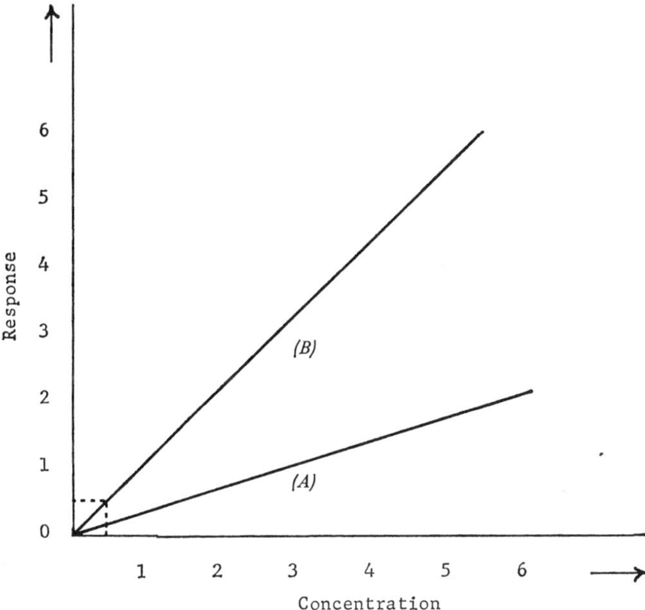

Fig. 1. Low- and high- sensitivity curves: *(A)* low sensitivity; *(B)* high sensitivity.

of the element being determined. As such, sensitivity is slope dependent. As illustrated in Figure 1, Curve A shows a low change in signal for a corresponding change in concentration. This low signal change precludes as fine a distinction in concentration variation as would be obtained if Curve B were the signal response. However, in the area of the dotted lines, near zero concentration and near zero signal response, the two curves are quite similar. In this area the detection and the sensitivity of a given analytical technique are nearly synonomous.

B. Matrix Interferences

Further confusion arises when detectabilities of various techniques are compared without regard for special cases of matrix and impurity interferences and limitations imposed by the amount of sample available or by its physical and chemical form. Matrix factors can affect tremendously the determination of low amounts of one element in another by a given analytical technique. The analyst is familiar with many such problems as the determination by emission spectrography of traces of arsenic in cadmium, the determination of elements adjacent in the periodic table by X-ray fluorescence, the determination of aluminum in iron by spark-source mass spectrography, etc. Sample form also can affect the values for the lowest concentrations of elements which can be detected by various techniques.

To resolve the ambiguity of elemental detectability arising from a matrix interference, the terms "absolute detection limit," "absolute detectability," and "absolute sensitivity" have been employed. The terms absolute detectability or absolute sensitivity often are used to designate the minimum concentration or amount of an element which can be determined under the most favorable conditions.

C. Effects of Various Parameters

Practically speaking, detection limits for a given element can vary considerably, depending on the analytical technique used, the matrix material, the physical form and state of the material analyzed, the presence of interfering elements, and the equipment used as well as the analyst's skill. Useful data on the detectability attained for a given element in a given matrix should include information on the method by which the data were derived and the confidence limit of the method. When this information is presented with a description of the technique and the instrument, if any, used, detectability terms are highly useful and informative. Commonly such terms are given in percent, parts per million, parts per billion, or weight. Incomplete information can lead to confusion, particularly on the part of persons not closely associated with analytical term usage and its occasional vagaries. For example, the electron-microprobe technique commonly "looks at" or requires a sample size of less than 10 μ^3. Detection of 100 ppm of an element in a 10 μ^3 volume represents a detection limit of about 10^{-14} g—a very good detection limit but of course obtained only in optimum cases and only in the specific area examined.

As can be seen from the foregoing discussion, there are at least five specific problems involving detectability, each of which requires that detectabilities be stated in terms relevant to the problem. First, there is the isolated, minute sample in which only the major elements can be determined. Here, absolute detectabilities stated in amount (grams) are most meaningful. Second, there is the nonisolated minute sample in a very much larger matrix—an inclusion, for example, in a steel sample. For this analysis detectabilities for the elements of interest in that matrix are desired, as well as a knowledge of the ability of the technique to confine the analysis to the specific area of interest. Third, there is a very thin film distributed more or less uniformly on a matrix surface. Absolute detectability plus detectability in the presence of the matrix should be known for this problem; the interference from the matrix becomes less important as the film thickness and uniformity increase, or as the depth of sampling (penetration) decreases. Fourth, there are the uniformly distributed trace elements in a sizeable sample. Detectabilities for these trace elements in the matrix are the proper way to express the data for this problem. Fifth, there is a small sample with trace impurities. The same data are required as in case four, plus assurance of the ability of the chosen technique to utilize efficiently the small sample.

Note that in each case mentioned above it is necessary to know either the absolute detectability or the detectability in the matrix, or both, to judge the applicability of a technique in a specific instance. Thus, to be comprehensive, a table of detectabilities should list these data for every element in every matrix for every technique. Clearly this is an almost impossible task.

D. Minute Samples

In respect to small samples it is interesting to review Benedetti-Pichler's projections on the minimum quantities of material required for analysis.[7] In his very informative volume on the identification of materials he projects that about 1000 atoms of a solid or liquid are an ultimate quantity for any identification. This corresponds to a mass of about 3×10^{-19} g for a material of average molecular weight. He states that a safety factor of 1000 to 10000 should be imposed on this 3×10^{-19} g figure to ensure that sufficient mass is available to provide the customarily accepted values of density, refraction, optical rotation, absorption of radiation, etc., necessary for identification. This results in an estimated minimum theoretical mass requirement of about 10^{-15} to 10^{-16} g.

A projection of the capability to identify extremely small particles, however, must reflect recognition of the fact that a considerable differential can exist between what can be learned from analysis of a known ideally suited material isolated from any substrate or interferences versus analysis of an unknown material perhaps hidden in a substrate together with many interfering materials. This differential can represent many magnitudes with regard to the size and number of particles required to obtain definitive information or in regard to the amount and quality of information that can be obtained on a particle of a given size.

Thus, while Benedetti-Pichler projects that a minimum theoretical mass of 10^{-15}–10^{-16} g or about 0.1 μ^3 is necessary for identification, a more practical definition of a minimum mass points to about 5×10^{-12} g or 1 μ^3 for an element of average density. If the problems of locating the particle of interest in a substrate, isolating it, and placing it in the identifying radiation beam within the detector are considered, it seems more practical to consider that a 10 micron-sized particle equal to 5×10^{-9} g or 1×10^3 μ^3 for a material of average density would be required in most cases for a fairly complete identification. The electron microprobe is ideal in many cases for use in analyzing small particles since the electron beam can be confined to an area of about 1–10 μ, and, as mentioned earlier, detectabilities as low as 10^{-14} g can be obtained in the sample volume observed. However, the electron microprobe does not provide complete identification of all components. The unequivocal identification of extremely small particles, 1–10 μ in size, nearly always would require more than a single analytical technique and in most cases would necessitate the application of a number of techniques, initially using those of a nondestructive nature.

III. PRECISION

Selection, use, and evaluation of analytical characterization methods must include considerations of the precision attainable by the methods. A fairly wide range of criteria is involved in such considerations. Such decisions are not always in the province of the analyst but can be set by the production department, the materials research worker, the purchasing department, or others for whom the analyses are made. Generally, in most metals and alloy analyses at least two precision levels are required, in addition to those associated with semiquantitative and qualitative work. A high level of precision is demanded when a product is being made or purchased with the price of the product depending on the assay result of the product or in production where close tolerance compositional specifications must be met. High precision levels often are required also in materials development work in which knowledge of exact alloy and impurity contents may be necessary to understand the properties of the material. A lower level of precision frequently is satisfactory to meet quality control demands or to meet the less stringent analytical data requirements of materials engineers.

In both cases the analyst should be able to provide data on the precision-he is attaining for the work. This necessitates a measurement of the repros ducibility of the analytical technique used. It must be kept in mind that precise results are not necessarily accurate results, because constant errors can exist in a technique and yet not affect its reproducibility. Thus, a calculation of precision should not be misinterpreted as being a calculation of accuracy. Measures of precision can be obtained relatively simply. However, it is not unusual for analysts to have little or no statistical proof for their estimated precision and to merely interpret their data intuitively. Many thorough discussions on techniques for calculating analytical precision appear in the chemical literature and these should be referred to in order to ensure that proper measuring conditions for testing precision are being met.[8–11,16]

To obtain precision measurements one must obtain a number of values, preferably a considerable number. Precision ordinarily is determined by calculating the standard deviation, s, as

$$s = [\Sigma(X - m)^2/(n - 1)]^{1/2} \tag{1}$$

where X is the value obtained, m is the average value, and n is the number of determinations of X. When the standard deviation is approximately the same over a wide range of concentrations, precision data can be expressed conveniently in percent by using the coefficient of variation as:

$$\text{coefficient of variation} = \frac{\text{standard deviation}}{\text{mean}} \times 100$$

Confidence limits should be given when expressing the standard deviation, s, for a given technique since the standard deviation can vary considerably when the number of repeat measurements, n, is small. Confidence limit or level is the percentage of observations expected to fall within a stated range and is related to the standard deviation of the observations by the equation

$$X - m = st \tag{2}$$

where x, m, and s have the same meanings as above and t is a tabulated constant whose magnitude is dependent on the number of observations and the confidence level required.

Suppose the analyst desires to estimate what maximum error will occur in 95% of his determinations. For example, assume the true value for an element is 10 units, and the analyst obtains 12, 10, 11, 8, and 9 units in 5 determinations. His *average* value is accurate, his standard deviation is 1.6, his maximum error at the 95% confidence level is 4.4 but his maximum observed error is only 2. Alternatively, the question can be asked, "To what confidence level can a maximum error of 2 be expected?" Based on the above example, the tabulated table of t values gives about 70% as the confidence level.

In some analytical techniques it is possible to make successive, repetitive observations of the blank and the sample. When working near the detection limit, these paired observations can be treated statistically in a manner similar to that described above to establish the confidence level at which the sample signal is real and not random noise and error.

Once the analyst has a measure of the precision of data obtained by a given technique he then is in a position to estimate its accuracy. Further he is in a position to evaluate the quality of his data with those obtained by other techniques and so is able to retain or change his operation as new developments occur or as his analytical demands change.

Table 3 lists the precision and sensitivity of various analytical techniques as given in Reference 15.

Table 3. Precision and sensitivity of analytical techniques (from Materials Advisory Board Report MAB-229M on Characterization of Materials)

Technique	Applications	Sensitivity	Average detectability in grams in powers of 10[a]	Precision
Wet chemistry—titrimetry	Major and minor phase—concentration; also impurities	$10^{-2}M$ in solution		0.01%
		$10^{-5}M$ in solution		0.1%
		10^{-6}–$10^{-7}M$ in solution	−4	
Wet chemistry—gravimetry	Major and minor phase—concentration	1000 ppm/ g sample	−3	0.01%
Wet chemistry	Major phase—valence			0.01%
Coulometry	Major phase—concentration			0.001–0.005%
Mössbauer spectroscopy	Major phase—valence also impurities—valence	Down to 0.1%–0.0001% depending upon density of matrix		Semiquantitative now
Electron probe microanalysis	Homogeneity of major phase minor phases	Down to 0.1% over a 1–5μ scan diameter	−13	0.5%
Emission spectroscopy	Impurities (survey)	0.1–100 ppm	−8	5–10%
Spark-source mass spectrometry	Impurities (survey)	0.01–0.1 ppm	−10	Semiquantitative now
Atomic absorption	Impurities	0.005–0.1 ppm solution	−7	5–10%
		0.1–10 ppm in solution		1–5%

Flame emission	Impurities	0.002–0.1 ppm in solution	−9	5–10%
		0.1–10 ppm in solution		1–5%
Spectrophotometry	Impurities	0.005–0.1 ppm in solution	−5	5–10%
Polarography	Impurities	All in solution:		
		0.1–1 ppm		2–10%
		10–100 ppm		0.1–0.02%
		0.005 ppm	−6	20%
		0.001 ppm (with anodic stripping preconcentration)	−8	5–10%
Neutron activation	Impurities	0.001–0.01 ppm	−9	2–10%
Vacuum fusion—mass spectrometry	Impurities: O_2, N_2, H_2	0.07 ppm	−6	20%
		100 ppm		5%
X-ray fluorescence spectrometry	Major and minor constituents	20–200 ppm generally; 0.1 at best	−6	0.1%

[a] Example: −11 means elemental detectabilities average 10^{-11} g.

IV. ACCURACY

Inevitably, and properly, any discussion of analysis involves accuracy. By accuracy is meant the degree to which an obtained answer matches the true situation, as differentiated from precision, which is a measure of the repeatability of an answer. The degree of accuracy for a given problem must be evaluated from many aspects, including cost, technical feasibility and capability, time, and, above all, realistic requirements.

Realistic requirements are stressed because there is much misunderstanding about true needs. One can cite instances where plus or minus 0.01% at the 50% level is important, but they are usually either stoichiometry problems or isotopic assays. At the other extreme, it sometimes suffices to know what element is the major constituent and to ignore everything else. Somewhere between these limits fall most of the analytical requirements. Materials produced for commercial use (as distinguished from materials to be studied in research) are usually sold on a specification basis with limits between which certain elements, compounds, or properties will lie. To debate whether these limits are justified is pointless; the analyst is obliged to produce answers with sufficient accuracy to say that the material is or is not within the specifications.

For most of the important elements in most of the commercial materials, techniques have been developed, as will be described in the following chapters, which presumably satisfy the imposed requirements. But, as the so-called advanced materials—exotic alloys, refractory metals, solid-state materials, and refractory oxides, to name a few—are becoming increasingly common, these proven techniques are stretched to the limit, and as purity has been increased, accuracy has tended to suffer.

The materials engineer has every right to expect good accuracy at the percent level, but at present he cannot get this accuracy in the ppm range. For example, stainless steel can be analyzed routinely by several methods for the chromium, nickel, etc., with an accuracy better than plus or minus 1% of the amount present. But the same accuracy for trace elements—the same elements in a highly refined bar of iron—is unattainable today. And is it really necessary? Except for the solid state field, these ultrapure materials are not commercial items. In many instances information of real value to the researcher can be quite crude by the usual standards of accuracy. Error factors of two or more in an answer may still show the effects of a purification step or an unsuspected contamination. This is not to imply that accuracy is unimportant; rather, that accuracy must be viewed in relation to the other parameters of an analysis.

Accuracy can be established most conveniently by the use of standards.

Standards in turn imply that their composition is accurately known. The National Bureau of Standards, equivalent organizations in other countries, and several industrial companies issue standards of the more common commercial metals and alloys. The elements certified in these have been carefully checked, usually by several competent laboratories and often by different techniques. In evaluating the performance of a laboratory or a new technique, these standards must be used with care. Homogeneity on a scale comparable with the amount used for the analysis must be known. Certain standards are certified homogeneous only for a given weight of sample, or only in a specified portion of a solid piece. It would be folly to use such a standard to calibrate a microtechnique such as the electron microprobe analyzer unless sufficient samples were taken to ensure that the *average* analysis was truly representative.

Lacking suitable primary standards, most analysts resort to a variety of standardization techniques. One of the best is the cross check, especially if the element in question can be determined by methods based on two or more completely different physical and/or chemical properties of that element. For example, if in a given material optical emission spectroscopy, classical wet chemistry, neutron activation, and atomic absorption give essentially the same answer for an element, that element could be considered standardized. Less desirable, but still better than no standardization, is the use of several laboratories employing the same technique. Here the possible bias of the technique must be evaluated.

Synthetic standards are commonly used if the physical properties of the sample can be duplicated or changed. Solutions are the best example because standards can be prepared, from "pure" metals, which should closely duplicate a solution of the sample. Dry mixed or fused standards are slightly less satisfactory because the possibility exists of inhomogeneity and residual physical differences. Other useful techniques are "spiking" a sample with a known amount of an element or comparing the sample with a standard of a different but similar matrix.

Quite obviously the requirements for standardization vary. A laboratory connected with a production operation can and should expend a considerable effort to ensure that its standards are both accurate and suitable for routine or automated use. This means, in many cases, solid metal standards for emission spectrometry or X-ray fluorescence. The standards themselves, of course, will previously have been standardized by other methods. Research laboratories, on the other hand, rarely are concerned solely with one or a few materials. Rather, they are required to analyze a wide variety of samples, for many of which no primary standards exist. It would be impractical to prepare well-standardized solid standards for each different sample. Instead, one or more alternative procedures such as described above should be used.

Once a method of standardization has been selected, the precision and accuracy of the resulting determination can be assessed. Many authors have treated this question statistically, and the reader may consult these references for detailed mathematical techniques.[8–11] Basically, it consists of making many determinations and finding the difference of each from the true value. The differences are then entered in the standard deviation equation 1 given above, where m is now the true value, not the mean. The deviation so found is a measure of the "average" accuracy of the determinations because the sign of the difference is lost by squaring. The absolute accuracy can be estimated by separate computations of plus and minus standard deviations, by plotting histograms of the differences, or by other mathematical techniques described in the references.

V. SAMPLE FORM AND PREPARATION

One of the considerations in the choice of an analytical technique is the amount and form of the sample available. As the amount decreases, the suitable techniques from which one can choose become fewer. At the many-grams level, any applicable method can be used; at the gram level, wet chemical techniques for more than a few elements will probably not be suitable; milligram and lesser amounts must be analyzed by the more sensitive techniques such as emission and mass spectrography. Very small samples, isolated in a larger matrix, are best handled by a probe technique such as the laser–emission spectrograph or preferably by the electron microprobe.

Samples which are not solid metals (plating solutions, for example, or slags from a melting operation) can be converted to a convenient form by dissolution, drying, or grinding. Sometimes it is possible to use the sample with little or no pretreatment as in the case of the emission spectrographic analysis of lubricating oils for metal pickup.

If the sample is not expendable, nondestructive analyses are indicated. These would include X-ray fluorescence and diffraction, electron microprobe, and activation. Even these techniques usually require that the sample dimensions be between definite limits. The same is generally true for the "seminondestructive" techniques including spark-source mass and emission spectrography.

If pretreatment is required before the analysis can proceed, the effects of this step must be considered. For example, drilling chips taken from a stainless steel sample in all probability would not be contaminated to any significant degree by the drill wear. But an ultrapure titanium sample might pick up more iron from the drill than was present in the original sample. Equally serious contamination can result from the use of ball mills, cutoff wheels, files, milling cutters, wire nippers, and the like.

Techniques which examine primarily surfaces, such as point-to-plane optical emission spectroscopy, mass spectrography, X-ray fluorescence, and the probes, are especially sensitive to surface contamination. Sometimes it is this contamination or surface coating which is of analytical interest, not the bulk. In these cases little or no prior cleaning is possible and one must be certain that the surface is not accidentally contaminated, as by fingers, at any time. If, instead, the bulk is of interest, then the surface must be cleaned. The usual treatments are abrasion, chemical etching, and ion bombardment. Each has its advantages and limitations.

In every instance a close liasion between the analyst and the materials engineer is vital, no less before a sample is actually taken than during any subsequent step of the analysis. *If at all possible*, the engineer should plan for samples adequate and suitable for the analytical technique to be used. This may mean casting separate pins or buttons of a melt, preparing a larger batch of an experimental alloy than had been planned, or perhaps even redesigning the experiment so that his accuracy requirements will be in an attainable range. If the specifications of the analyst are met, it is far more likely that he in turn can meet the needs of the engineer.

VI. AUTOMATION AND DATA HANDLING

High manpower costs, plus the increasing demand for speed and reliability, have led to the replacement of man by machine in analytical chemistry just as in almost every other part of our society. Instruments have long been used to reduce the man–hours required for a given determination; in the past 20 years automatic data readout has become commonplace with the direct-reading emission spectrometer. In the last few years the wide availability of computers has prompted several persons to use computers for data reduction.[12,13] Particularly in a mass spectrographic analysis, where many hundred bits of data are normally obtained on a single plate, the manual computation of all this information would be prohibitive. Computer programs have been written which accept microphotometric readings, i.e., percent transmission or absorbance and line position, and from these calculate concentrations. Even semiquantitative data on a routine basis have been obtained in this way using a remote link to a time-sharing computer.[14] Another technique which can use computer data reduction is activation, where overlapping decay curves can be resolved by solving a series of exponential equations.

Automation is found in several techniques, especially X-ray fluorescence, emission spectroscopy, and wet chemistry. In the latter, commercial equipment such as robot or automatic analyzers are available which can perform a series of operations on a sample solution and read out the amount of the sought element by absorption spectrophotometer or flame spectrophotometer. These automated instruments have been used extensively in clinical laboratories and now are finding places in metallurgical analytical programs.

Before either automation or computer data reduction are employed, several questions should be answered. Is the work load of the laboratory sufficiently heavy and repetitive to justify the expense of automation and/or computer data processing? Are there known to be no unexpected variations in the samples which might be interpreted falsely by the automatic equipment? Does the accuracy gained by computerization of complex data justify the cost of renting the equipment? Is manual data reduction really the limiting factor in either accuracy or speed, or do the limitations lie further back in the analytical train? In short, automation and computerization can offer real savings in time and gains in data interpretation if their place in the total analytical program is properly evaluated. If not, they may tend to tie the laboratory to a fixed, and no longer optimum, program because of the investment required to make necessary conversions.

VII. UNRESOLVED PROBLEMS

From the number, breadth, and variety of techniques as described in subsequent chapters it would appear that analytical capability is ahead of other fields in meeting the demands of rapid developments in metals and alloy technology. In certain instances this is the case, but many problems remain in providing needed analytical characterization. The ability to provide exact stoichiometric analyses of semiconductor alloys is well beyond the accuracies presently available by analytical techniques. The capability to determine the compositions of surfaces and thin electrical films, clearly differentiated from their attendant atmospheric and bulk interface compositions, is at the best only partially fulfilled. The accuracies presently obtainable in determining the highly important elements carbon, oxygen, nitrogen, and hydrogen in metals often are very poor when these elements are present at low parts-per-million concentrations.

Many other examples could be cited where present analytical technology is inadequate to provide data now needed in materials research studies. A serious drawback which exists in many laboratories can be termed contamination-noise-level limitations. Inherent capabilities to analyze materials are blocked by background impurities in the laboratory atmosphere, by contamination from laboratory hardware, laboratory floors, walls, and benches, and particularly by contamination from analytical reagents. The need for clean operational conditions must be recognized and met in laboratory construction to provide facilities to minimize contaminant noise levels.

A serious need also exists for more primary metal standards tailored for specific analytical techniques. It has been found in many cases that standards suitable for optical or X-ray emission work are not suitable for spark-mass spectrographic work which utilizes a much smaller sample area. The preparation and the documentation of the compositions of a wide variety of metal standards are challenging tasks. However, the use of primary standards still provides the most useful way a laboratory can standardize its data.

References

1. W. W. Meinke and B. F. Schribner, Eds., *Trace Characterization, Chemical and Physical*, (Natl. Bur. Std. Monograph 100) (1967).
2. G. H. Morrison, Ed., *Trace Analysis: Physical Methods*, Interscience, New York, 1965.
3. *Purification of Materials*, New York Academy of Sciences, Art. 1, **137**, 1 (1966).
4. M. S. Brooks and J. K. Kennedy, Eds., *Ultrapurification of Semiconductor Materials*, MacMillan, 1962.
5. *Methods for Emission Spectrochemical Analysis*, American Society For Testing and Materials, 4th ed., (1964).
6. G. H. Morrison, Ed., *Trace Analysis: Physical Methods*, Interscience, New York, Chap. 1, (1965).
7. A. A. Benedetti-Pichler, *Identification of Materials*, Academic, New York, 1964.
8. W. J. Youden, in *Treatise On Analytical Chemistry*, Part I, Kolthoff and Elving, Eds., Interscience, New York, 1959, Chapter 3.
9. H. A. Laitenen, *Chemical Analysis*, McGraw-Hill, New York, 1960.
10. W. J. Youden, *Statistical Methods for Chemists*, Wiley, New York, 1951.
11. A. B. Alder, *Anal. Chem.*, **36**, 9 (1964), 25A-34A.
12. B. F. Schribner, Ed., *Activities of the National Bureau of Standards Spectrochemical Analytical Section*, Technical Note 401, July 1965 through June 1966 (September 30, 1966) p. 62.
13. J. C. Franklin and E. B. Griffin, "Spark-Source Mass Spectrographic Analysis of Steels and Nickel Alloys", AEC Contract W-7405 eng 26, Report Y-1543 (Chemistry), Union Carbide Corporation, Nuclear Division, Oak Ridge Y-12 Plant (July 6, 1966).
14. P. R. Kennicott, General Electric Company, Research Laboratory, Schenectady, New York. (Private communication, Picker Nuclear MS-7 Users' Conference, April, 1967.)
15. "Characterization of Materials," Materials Advisory Board Report MAB 229-M, National Academy of Sciences, March, 1967.
16. C. Eisenhart, *Science*, **160**, 120 (1968).

Chapter 2

SEPARATION AND PRECONCENTRATION TECHNIQUES

ATSUSHI MIZUIKE, Nagoya University, Chikusa-ku, Nagoya, Japan

I. Introduction 26
II. General Considerations 27
A. Recovery 27
B. Separation Factor 28
C. Contamination Hazards 28
1. Contamination due to Reagents 29
2. Contamination due to Apparatus 29
3. Airborne Contamination 31
4. Other Sources of Contamination 33
III. Volatilization Methods 33
A. Distillation from Solutions 33
B. Volatilization at High Temperatures 36
IV. Liquid–Liquid Extraction Methods 37
A. Techniques 41
B. Applications 42
V. Selective Dissolution Methods 46
VI. Precipitation Methods 47
A. Precipitation of Ordinary Amounts of Elements 47
B. Collection of Trace Elements on Gathering Precipitates 48
VII. Electrodeposition Methods 50
A. Electrodeposition on a Mercury Cathode 50
1. Electrodeposition of the Desired Trace Elements 52
2. Electrodeposition of Matrix or Interfering Elements 53
B. Electrodeposition on Other Electrodes 55
VIII. Ion-Exchange Methods 55
A. Techniques 56
B. Effect of the Composition of the Solution 59
C. Applications 60
IX. Miscellaneous Methods 62
References 63

I. INTRODUCTION

Separations and preconcentrations in analytical chemistry are considered as very useful techniques in association with both modern instrumental methods of analysis and classical methods such as gravimetric and volumetric analyses. In quantitative analyses, either instrumental or wet chemical, coexisting elements in the sample often interfere with the determination of a desired element in various ways. Even if there is no interference from coexisting elements, it is generally impossible to directly determine trace elements in the low parts per million (ppm), parts per billion (ppb), or lower concentration range in samples such as metals and alloys, because the signals obtained in instrumental methods of determination diminish to the noise level with decrease in concentration of elements. Although improvements in selectivity and sensitivity by the use of modern instrumental techniques, specific organic reagents, or masking agents have overcome these difficulties in many cases, analysts still have to use separation or preconcentration techniques in many analytical procedures to remove interfering or matrix elements prior to the determination and thus obtain greater accuracy or sensitivity.

Since, in instrumental techniques, the same concentration of an element gives different signal strengths in different matrices, standards used for calibration are generally required to duplicate both the chemical and physical nature of the sample as much as possible. Hence, preparation of standards is a very difficult task, especially when solid samples are analyzed for trace constituents without decomposition. This problem, however, becomes much less serious after separation. Also, heterogeneous distribution or segregation of elements in a sample may be homogenized during the separation step.

On the other hand, separations and preconcentrations are, in general, time consuming and tedious, and subject to large loss and contamination hazards (see Section II). Therefore, these steps should be omitted from analytical procedures whenever possible.

A number of separation techniques based on physical, chemical, and mechanical principles can be applied to analytical chemistry.[1] In this chapter, separation and preconcentration methods which are especially useful in analysis of metals, alloys, and related materials are discussed from the practical viewpoint. After discussion of general problems, six most widely used separation and preconcentration techniques, i.e., volatilization, liquid–liquid extraction, selective dissolution, precipitation, electrodeposition, and ion exchange, will be described in detail with several typical or interesting examples mainly selected from the preconcentration of low ppm or ppb level impurities in various metals and alloys. Some other methods of separation useful for metals analysis are briefly mentioned in the final section.

II. GENERAL CONSIDERATIONS

In analytical chemistry, selection of separation and preconcentration techniques depends mainly on (*1*) nature and size of the sample; (*2*) desired elements and their concentrations; (*3*) number of the desired elements to be separated simultaneously—single element or multielement (group) separation; (*4*) method of determination following the separation; (*5*) accuracy and precision required; (*6*) number of samples; and (*7*) speed required. There are several criteria used in the evaluation of separation and preconcentration techniques. These include (*a*) recovery of the desired elements; (*b*) separation factors for undesired constituents with respect to the desired elements; (*c*) contamination hazards; (*d*) complexity of the technique; (*e*) time required for a separation; and (*f*) cost. Since the first three are of primary importance, they are discussed further in detail.

A. Recovery

The recovery or yield of a desired element is defined as the ratio of the quantities of the element after and before the separation, which is ordinarily expressed in percent. There exist several possible sources of loss of desired elements during separation and related steps, which result in low recovery. These include evaporation of desired elements and formation of insoluble residues containing the desired elements during decomposition of the sample prior to the separation, imperfect separations, careless manipulations, and adsorption of the desired elements on the surfaces of the apparatus. Loss of trace elements is notorious in extremely dilute solutions. In such solutions, the solute often shows anomalous behavior such as strong adsorption on the wall of the vessel and formation of radiocolloids in carrier-free radioactive solutions. Therefore, generally speaking, the smaller the amount of the elements, the more the danger of losses.

The recovery, loss, and entire behavior of an element through the separation process can be investigated very conveniently by the use of radioactive tracers. A radioactive isotope of the element in the same chemical form is added to the sample before the separation, and the behavior of the element is followed by sensitive, rapid, and selective radioactivity measurements. Even continuous measurements can be made easily. The tracer technique is especially invaluable to trace element separations, because contamination hazards can be ignored in this technique.

Although a recovery of 100% is required in most cases, a known and constant recovery between 90 and 100% is sometimes acceptable for the determination, especially in trace element analysis. When the recovery fluctuates very much each time, the radiometric correction method or the isotope dilution method (by the use of either radioactive or stable isotope) should be applied. A limitation of the application of tracers to separation techniques is the difficulty in introducing tracers into solid samples. Radioactive labeling of elements in solid samples is required when the tracer technique is applied to investigate the recovery and loss of elements during decomposition of solids and separations involving volatilization from solids and selective dissolution of solids. Radioactivation techniques are sometimes useful for this purpose.

Conversely, for the separation of radionuclides, milligram (sometimes microgram) quantities of their stable isotopes are usually added as carriers prior to the separation. This technique enables one not only to improve recoveries of the radionuclides but also to correct measured radioactivities by yields of the carriers determined by gravimetric, photometric, or other methods. Carrier-free separations, i.e., radiochemical separations without isotopic carriers, are also used. A series of monographs[2] presents very useful information in radiochemical separations and related fields.

B. Separation Factor

The separation factor ($S_{B/A}$) for an undesired constituent B with respect to a desired element A is defined as follows:

$$S_{B/A} = (Q_B/Q_A)/(Q_B{}^0/Q_A{}^0) \tag{1}$$

where $Q_A{}^0$ = the quantity of A in the sample; $Q_B{}^0$ = the quantity of B in the sample; Q_A = the quantity of A after the separation; Q_B = the quantity of B after the separation. The reciprocal of $S_{B/A}$ is called the concentration factor or the enrichment factor for A, when A is the trace element and B is the matrix.

When percent level elements in the sample are separated from coexisting elements for gravimetric or volumetric analysis, separation factors required are usually about 10^{-4} or larger. However, in the separation of trace elements from the matrix, separation factors of less than 10^{-4} are often required, although larger values are allowed in some cases.

C. Contamination Hazards

During the separation and related steps, various unfavorable foreign substances including the same elements as those to be determined and other interfering elements are introduced into the sample from various external sources. This phenomenon, which is simply called "contamination," is one of the most troublesome problems in chemical analysis, especially in trace element analysis. Contamination sources include reagents, vessels, and other apparatus used for the separations, the atmosphere of the laboratory, etc. Since some kinds of contamination are not reproducible, it is very difficult to estimate them accurately. Usually a "blank run" is carried through all the steps of the separation without the sample to estimate the contamination. However, this method is quite unsatisfactory. The smaller the amounts of elements the greater will be the effect of contamination. Therefore, in trace analysis, it is necessary to reduce contamination as much as possible to obtain reliable analytical rsults.

1. CONTAMINATION DUE TO REAGENTS

Various impurities present in water and reagents used in trace analysis become serious sources of contamination. Although very pure analytical reagents are now commercially available, and some reagents are easily purified or prepared in laboratories, they still contain detectable amounts of impurities. Low ppb of metallic impurities are almost always found even in very carefully purified water or acids. However, it is not difficult to reduce concentrations to the parts per trillion (ppt) range for a limited number of particularly undesirable impurities. Reagents may be contaminated during storage by the material of which the container is made. Precautions and means to minimize contamination due to reagents are as follows: (*1*) selection of the purest commercial reagents; (*2*) purification or preparation of high purity reagents in laboratories (see Table 1); (*3*) storage of reagents in clean proper containers in clean laboratories; (*4*) avoidance of prolonged storage of reagents; and (*5*) use of minimum amounts of reagents.

2. CONTAMINATION DUE TO APPARATUS

Under certain conditions, the surfaces of vessels and other apparatus which come in direct contact with the samples are attacked to some extent and contaminate the sample. On the other hand, strong adsorption of elements on the surfaces of the apparatus causes contamination of later samples from previous samples. The latter phenomenon is called "memory." Generally speaking, high temperature, high pressure, and longer time of contact increase the corrosion of the apparatus and the adsorption of elements on the surface. Some of the enamels used for graduation marks in pipets[3] and foreign matter strongly adhered on the surfaces of new glass or plastic ware may cause serious contamination.

Table 1. Selected laboratory methods for purification and preparation of high purity reagents

Reagent	Method	Ref.
Water	Repeated distillation	
	Ion exchange	
	Ion exchange followed by distillation	
HCl	Distillation of constant boiling mixture	
	Isothermal distillation	a
	Dissolution of anhydrous HCl gas in pure water	b
	Percolation through a Cl-form strong-base anion exchange resin column	
HNO_3	Repeated distillation of 65% HNO_3	
H_2SO_4	Repeated distillation	
$HClO_4$	Repeated distillation	
HF	Distillation	c,d
	Dissolution of HF gas in pure water	e
	Isothermal distillation	f
NH_4OH	Distillation	
	Isothermal distillation	a
	Dissolution of anhydrous NH_3 gas in pure water	
NaOH	Conversion of NaCl (purified by extraction with oxine and dithizone) with an OH-form anion-exchange resin	b
KOH	Conversion of KCl (purified by extraction with oxine and dithizone) with an OH-form anion-exchange resin	b
Organic solvents	Distillation	
	Back-extraction	
Salts	Shaking a neutral aqueous solution with dithizone–carbon tetrachloride or oxine–chloroform	
	Mercury cathode electrolysis	g

[a] H. Irving and J. J. Cox, *Analyst*, **83**, 526 (1958).
[b] R. E. Thiers, in *Methods of Biochemical Analysis*, D. Glick, Ed., Vol. 5, Interscience, New York, 1957, p. 273.
[c] R. Wickbold, *Z. Anal. Chem.*, **171**, 81 (1959).
[d] W. T. Rees, *Analyst*, **87**, 202 (1962).
[e] H. Stegemann, *Z. Anal. Chem.*, **154**, 267 (1957).
[f] W. Kwestroo and J. Visser, *Analyst*, **90**, 297 (1965).
[g] L. Meites, *Anal. Chem.*, **27**, 416 (1955).

Since contamination due to apparatus is quite difficult to estimate because of its poor reproducibility, much attention has to be paid to the selection of the apparatus materials, the cleaning and storage of the apparatus, and the history of the apparatus. Main considerations in the selection of the materials are: chemical composition, chemical resistivity, thermal resistivity, and price. Materials most commonly used for vessels include borosilicate glass, Vycor, fused silica, porcelain, platinum, polyethylene, and Teflon. It must be remembered that these commercial materials often contain various impurities in the low ppm level. Impurities in polyethylene[4,5] Teflon[5] and fused silica[6] are reported. Materials used for mortars and pestles include agate, mullite, hardened steel, vitreous alumina, tungsten carbide, and boron carbide because of their great hardness.

With regard to cleaning of glass, silica, and plastic ware, the use of a one-to-one mixture of concentrated sulfuric acid and nitric acid is recommended[4] instead of conventional chromic aicd cleaning solution, because the latter causes strong retention of chromium.[7–9] Platinum ware is cleaned by sodium pyrosulfate fusion followed by soaking in 6*M* hydrochloric acid. To avoid airborne contamination, various covers are used during storage of the apparatus. In trace analysis, it is also advisable to wash the vessels immediately before use with a chloroform solution of dithizone for heavy metals determinations and with steam for the Kjeldahl distillation of ammonia, respectively.

3. AIRBORNE CONTAMINATION

The atmosphere in laboratories may contain various kinds of harmful gases, vapors, aerosols, and dust particles which cause contamination of the sample. To minimize airborne contamination, the following points should be considered: (*1*) design of the proper laboratories for trace and microanalysis—selection of surface materials and ventilation systems are most important; (*2*) good housekeeping; (*3*) removal from the laboratory of all reagents and other substances which have any possibility of contamination; (*4*) covering beakers or other vessels with watch glasses, etc.; and (*5*) use of special apparatus for evaporation and drying of the sample (see Fig. 1).

Airborne lead contamination was reduced by using the evaporation chamber shown in Figure 1*c* in a clean laboratory, where lead-laden dusts and aerosols were removed from the air by electrostatic precipitation and filtration.[10] The experiment was carried out in Pasadena, California. The results summarized in Table 2 indicate that lead contamination deposited in an open beaker in the ordinary laboratory exceeds that from the evaporation chamber in the clean laboratory by a factor of 20 or 30.

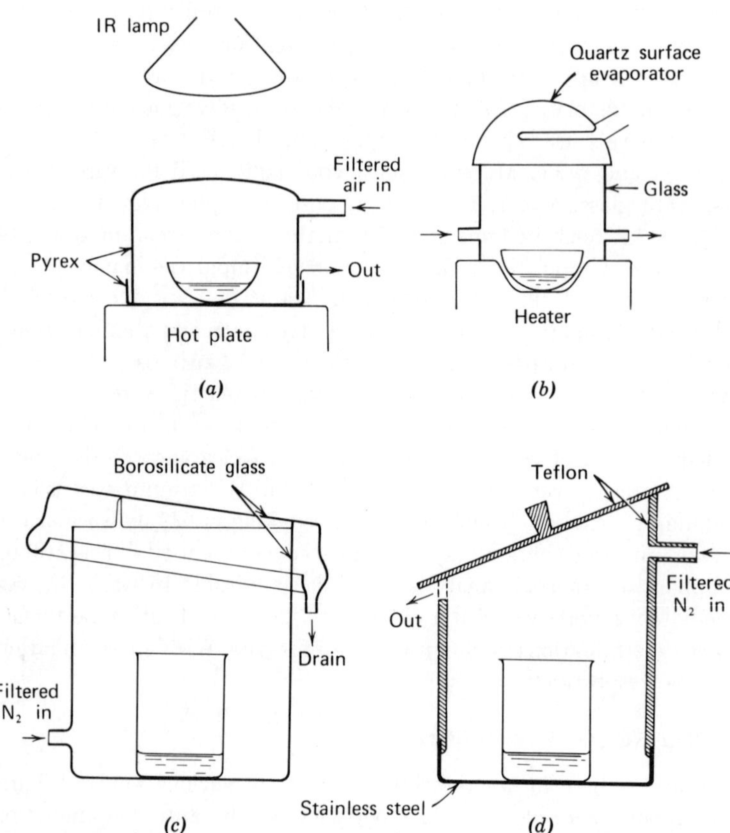

Fig. 1. Evaporation chambers. (*a*) See reference 4 and 11. (*b*) See reference 6. (*c*) and (*d*) See reference 10.

Table 2. Lead contamination from air[a,b]

Laboratory	Evaporation chamber[c]	Time, days	Amounts of lead, μg
Ordinary	Not used	8	4.07, 2.32
Ordinary	Used	8	1.13
Clean[d]	Not used	8	0.44
Clean[d]	Used	8	0.18, 0.13
Clean[d]	Used	1	0.02

[a] From ref. 10.
[b] Five hundred ml of $6M$ HCl in a Teflon beaker was evaporated to dryness.
[c] See Figure 1*c*.
[d] The air was cleaned by electrostatic precipitation and filtration.

4. OTHER SOURCES OF CONTAMINATION

There are some other sources of contamination not discussed previously. Careless manipulations, of course, cause serious contamination. Touching the surfaces of the sample or the apparatus (which contacts directly with the sample) with fingers is very dangerous, because the sample is contaminated with appreciable amounts of sodium, chlorine, organic matter, etc. The insides of air ovens, muffle furnaces, and fume hoods should be carefully checked to avoid contamination.

For further details of contamination problems, see articles by Thiers.[4,11]

III. VOLATILIZATION METHODS

Volatilization, one of the oldest methods of separation in chemistry, has a wide applicability in modern analytical chemistry. Either desired or undesired elements are volatilized from liquid or solid samples at appropriate temperatures, which range from room temperature to 2000°C or even higher. Most essential to this technique is a great difference in volatility between the elements to be volatilized and those to remain in the residue. Therefore, both elements are often converted to other more favorable chemical forms before or during the volatilization step.

There are several sources of loss of the desired elements during the separation by volatilization. When the desired elements are volatilized, losses may occur due to (*1*) adsorption of the volatilized elements on the walls of the apparatus; (*2*) imperfect conversion of the desired elements to volatile forms; and (*3*) slow diffusion rate of the desired elements through solid or molten samples. In the volatilization of relatively large amounts of undesired elements, losses often occur due to (*1*) partial volatilization of the desired elements; (*2*) mechanical entrainment of tiny liquid or solid particles containing the desired elements in the escaping vapor; (*3*) adsorption of the desired elements on the walls of the vessels; and (*4*) formation of acid-insoluble residues containing the desired elements. These losses as well as various kinds of contamination are especially serious when trace elements are separated from matrix elements. The proper selection of experimental conditions such as temperature, material of the vessels, and reagents added to the sample, minimizes these losses and contaminations. Generally speaking, the lower the temperature, the less the danger of loss and contamination due to the vessels used.

A. Distillation from Solutions

From appropriate solutions, a number of elements can be distilled[12–14] as various volatile compounds as shown in Table 3.

Table 3. Distillation of the elements from solutions

Element	Volatile form	Solution
B	$B(OCH_3)_3$	CH_3OH
N	NH_3	Alkaline solution
F	H_2SiF_6	H_2SO_4, $HClO_4$
Si	SiF_4	HF–H_2SO_4, HF–$HClO_4$(–HNO_3)[a]
S	H_2S	HCl, HCl–HI–H_3PO_2
Cr	CrO_2Cl_2	HCl–$HClO_4$
Ge	$GeCl_4$	HCl–HNO_3–$HClO_4$
As	AsH_3, $AsCl_3$, $AsBr_5$	HCl(–H_2SO_4),[a] HBr–$HClO_4$, HBr–H_2SO_4
Se	$SeBr_4$	HBr(–H_2SO_4),[a] HBr–Br_2
Br	Br_2	H_2CrO_4–H_2SO_4
Ru	RuO_4	HCl–$HClO_4$, HBr–$HClO_4$, $HClO_4$–H_2SO_4
Sn	$SnCl_4$, $SnBr_4$	HCl–HBr(–Br_2),[a] HCl–$HClO_4$, HBr–$HClO_4$, HBr–H_2SO_4
Sb	$SbCl_3$, $SbBr_3$	HCl–H_3PO_4, HBr–$HClO_4$, HBr–H_2SO_4
Re	Re_2O_7	HCl–$HClO_4$, HBr–$HClO_4$, HBr–H_2SO_4, H_2SO_4
Os	OsO_4	$HClO_4$, HNO_3
Hg	$HgCl_2$	H_2SO_4–HCl_{gas}

[a] With or without a reagent in parentheses.

These techniques are widely used in analysis of metals, alloys, and related materials for trace impurities. First, several examples are selected from volatilization of the desired trace elements. As little as 0.1 ppm of boron in metals such as beryllium, zirconium, thorium, uranium, nickel, and Zircaloy is separated by distillation as methyl borate from methanol solutions and determined spectrophotometrically.[15–17] The Kjeldahl method, in which nitrogen is distilled as ammonia from alkaline solutions, is regarded as the most reliable one for the determination of N in various metals.[18] Combined with sensitive spectrophotometric methods, the lower limit of determination is about 1 ppm. Low ppm of silicon in uranium and plutonium is distilled as silicon tetrafluoride from a perchloric–hydrofluoric–nitric acid solution for spectrophotometric determination.[19] A few ppm of sulfur in nickel and other metals is separated by distillation as hydrogen sulfide from a hydrochloric or a hydrochloric–hydriodic–hypophosphorous acid solution prior to volumetric or spectrophotometric determinations.[20,21] As little as 0.1 ppm of bromine in uranium fluorides and oxides is separated by distillation as elemental bromine from a chromic and sulfuric acid solution and determined by spectrophotometry.[22]

There are several examples of removal of matrix elements by distillation. Various metallic impurities such as Pb, Cd, Fe, In, Cu, Ni, Tl, Bi, Zn, P, and Al down to 0.01 ppm in high purity silicon and silica are preconcentrated by volatilization of silicon tetrafluoride from hydrofluoric–nitric–perchloric acid or hydrofluoric–sulfuric acid solutions, and then determined by polarography, spectrophotometry, or fluorimetry.[23–25] Platinum or Teflon vessels are used in this case. For the emission spectrographic determination of 0.1–30 ppm of Fe, Ni, Al, Mn, Ti, V, Mg, and Cu in high purity chromium, the matrix is evaporated as chromyl chloride from a hydrochloric–perchloric acid solution.[26] As, B, and Sn are partially or completely volatile at 200–220°C, the temperature at which chromium is evolved. As low as 0.01 ppm of Ga, In, P, Sb, Cu, Ag, Be, Fe, Mg, Mn, Al, Bi, Cd, Ni, and Pb in germanium or germanium dioxide are concentrated by distillation of germanium tetrachloride from hydrochloric–nitric–perchloric acid solutions, and determined by spectrophotometry or emission spectrography.[27–29] In the determination of impurities in gallium arsenide, arsenic is volatilized in the presence of hydrochloric acid, carbon tetrachloride, and bromine, followed by removal of gallium by liquid–liquid extraction and emission spectrography[30]; 0.1–0.01 ppm of Si, Mg, Mn, Cr, Ni, Bi, Al, V, Zr, Co, Cd, and Be can be detected. Almost the same procedure is applied to the determination of trace impurities in arsenic.[31] As little as 0.1 ppm of Cu, Pb, Ni, Co, and Te in selenium dioxide are separated by volatilization of selenium from concentrated sulfuric acid solution for polarographic determination.[32] Also, in the polarographic determination of Cu, Cd, Tl, Pb, Te, and Fe in high purity selenium, the matrix is removed by distillation as selenium bromide from hydrobromic acid solution.[33] The lower limits of detection are 0.0015–0.2 ppm. Evaporation of selenium is also applied to the spectrophotometric determination of Te[34] and the flame photometric determination of alkalies, alkaline earths, and Ga[35] in selenium in the fractional ppm range. Fractional ppm of iron in tin are determined spectrophotometrically after removal of tin as bromide from a hydrochloric acid–hydrobromic acid–bromine solution.[36] As little as 1 ppm of lead in high purity bismuth is concentrated by volatilization of bismuth as bromide for spectrophotometry.[37]

Evaporation of water, volatile acids, or organic solvents from dilute solutions is often required in trace analysis, because dilute solutions of the desired trace elements are generally obtained after the application of various separation methods such as solvent extraction, ion exchange, etc. Although the sample solution is usually simply heated in a vessel made of glass, silica, platinum, etc., it is sometimes advisable to add suitable reagents to retain the desired trace elements as nonvolatile compounds in the residue. Addition of reagents may also facilitate the perfect redissolution of the residue by formation of soluble salts of the desired trace elements or by carrier action of the added salts. As low as 1 ppb of metallic impurities such as Mg, Mn, Cr, Bi, Al, Ni, Mo, Be, In, Zn, Ti, Zr, Co, Fe, Pb, and Ga in high purity hydrochloric, nitric, hydrofluoric, and acetic acids are concentrated in the sulfate residue by evaporation with a small amount of sulfuric acid, and determined by emission spectrography.[38]

B. Volatilization at High Temperatures

Dry techniques of volatilization, i.e., volatilizations at high temperatures, are also used in metals analysis. H, C, N, O, and S are extracted as gases from solid or molten metals at high temperatures ranging from several hundred to 2000°C.[18,39–41] The gases are collected and then determined by gas analytical techniques. Hydrogen is completely extracted from a solid metal when the sample is heated in vacuum up to the temperature at which the diffusion velocity of hydrogen is appreciably fast. H, O, and N are simultaneously extracted into vacuum as molecular hydrogen, carbon monoxide, and molecular nitrogen, respectively, when the metal is melted in a graphite crucible by means of high frequency heating. This technique, which is called vacuum fusion, is regarded as the most reliable and sensitive one for oxygen determination in various metals. Inert gas fusion, a modification of this technique, uses an inert gas atmosphere such as argon instead of vacuum. Oxygen and nitrogen in various metals are also extracted by a dc carbon arc discharge in a static helium atmosphere and determined by gas chromatography.[42,43] Oxygen can be extracted as water vapor by heating the metal in hydrogen, in hydrogen sulfide,[44] or in hydrogen fluoride, and as sulfur dioxide by heating with sulfur vapor.[45] Hydrogen, carbon, and sulfur are volatilized as water, carbon dioxide, and sulfur oxides, respectively, when the metal is heated in oxygen. Carbon is also extracted as carbon disulfide from metallic silicon and germanium by treating with sulfur vapor at about 1000°C.[46] The lower limits of determination are about 1–10 ppm in most of the methods described above.

A number of metallic and nonmetallic impurities in refractory materials such as oxides of uranium, aluminum, thorium, zirconium, and beryllium are evaporated at 1500–2000°C from the sample in a small electrically heated crucible, and condensed on a water-cooled metallic or graphite rod placed above the crucible. The rod is then used as an electrode for emission spectrography.[47,48] Sensitivities of 0.02–20 ppm are obtained in this technique. The volatilization of the desired elements at high temperatures is also applied to the preconcentration of traces of Tl, Be, Zn, and Cd in various metals and minerals for emission spectrography,[49–53] and to the separation of lead in rocks and meteorites.[54]

Since many metal and nonmetal elements form relatively volatile chlorides, the sample is treated with chlorine gas at elevated temperatures to remove the matrix elements by volatilization. This technique is successfully applied to the flame photometric determinations of fractional ppm of Li, Na, K, and Ca in phosphorus, arsenic, and antimony,[55] and of low ppm of Ca in zirconium and its alloys.[56] Also, stable oxides such as silica and alumina in iron and steels are separated by the chlorination–volatilization technique,[57] which is applied to the sample directly or to the residue obtained by electrolytic dissolution of the sample discussed in Section V.

IV. LIQUID–LIQUID EXTRACTION METHODS

In recent years, liquid–liquid extraction[6,13,58–64] has been extensively used in metals analysis, especially for trace determinations, because of its selectivity, simplicity, rapidity, wide applicability, and relatively small contamination hazards. The method is based on the distribution of an element between water and an essentially immiscible organic solvent such as ether, chloroform, and carbon tetrachloride. The formation of an uncharged chemical species is essential for extracting an element from an aqueous solution into an organic solvent. Under proper conditions, most elements form extractable chemical species by simple coordination, heteropoly acids formation, chelation, ion association, and combinations of some of the above. Various organic chelating reagents, such as dithizone, oxine, cupferron, sodium diethyldithiocarbamate, thenoyltrifluoroacetone, and dimethylglyoxime, as well as organic solvents themselves (e.g., ether, acetylacetone) play very important roles in the formation of the extractable chemical species of the metals of interest. Liquid ion exchangers[65,66] are also useful in extraction. Examples of metal extraction systems are given in Table 4.[6]

In liquid–liquid extraction, the distribution ratio, i.e., the ratio of the total concentrations of an element in the two liquid phases after equilibrium, is a main concern. The larger the distribution ratio of the desired element, and the smaller that of the undesired one, or vice versa, the better separation can be achieved. Since the distribution ratio of an element is determined by various chemical interactions such as dissociation, polymerization, etc. in the aqueous and organic phases as well as distribution of an extractable species between the two phases, the proper choice of an extraction system, solvent, reagent concentration, pH or acidity of the aqueous phase, and the valence states of the elements is most important to achieve better separations. In halide, nitrate, and thiocyanate extraction systems, the addition of salting-out agents such as inorganic aluminum, calcium, or ammonium salts to the aqueous phase enhances the extraction of metals into the organic phase.

Table 4. Liquid–liquid extraction systems

Extraction system	Aqueous phase	Organic phase	Elements extracted into the organic phase
Fluoride	20*M* HF solution	Ethyl ether	Nb, Ta, Re
Chloride	6*M* HCl solution	Ethyl ether	Fe, Ga, Ge, As, Mo, Sb, Au, Tl
Bromide	4.5–5*M* HBr solution	Ethyl ether	Fe, Ga, As, In, Sn, Sb, Au, Tl
Iodide	6.9*M* HI solution	Ethyl ether	As, Cd, Sb, Au, Hg, Tl
Nitrate	8*M* HNO_3 solution	Ethyl ether	Ce, Au, U, Np
Thiocyanate	Thiocyanate solution	Ethyl ether	Be, Sc, Ti, Fe, Co, Zn, Ga, Nb, Mo, Ru, Rh, In, Sn, W, Re, Os, Au
Oxine	Aqueous solution	Chloroform solution of oxine	Be, Al, Ca, Sc, Ti, V, Mn, Fe, Co, Ni, Cu, Zn, Ga, Sr, Y, Zr, Nb, Mo, Pd, Cd, In, Sn, Sb, Ce, Nd, Er, W, Hg, Tl, Pb, Bi, Th, Pa, U, Pu
Cupferron	Aqueous solution containing cupferron	Chloroform, ether	Al, Ti, V, Fe, Co, Cu, Mo, Sn, Sb, Ce, Th, Pa, U
Dithizone	Aqueous solution	Chloroform or carbon tetrachloride solution of dithizone	Mn, Fe, Co, Ni, Cu, Zn, Pd, Ag, Cd, In, Sn, Te, Pt, Au, Hg, Tl, Pb, Bi, Po
Diethyldithiocarbamate	Aqueous solution containing sodium diethyldithiocarbamate	Chloroform, carbon tetrachloride, ethyl acetate	Ti, V, Cr, Mn, Fe, Co, Ni, Cu, Zn, Ga, As, Se, Mo, Pd, Ag, Cd, In, Sn, Sb, Te, W, Au, Hg, Tl, Pb, Bi, U

For the extraction of a metal element with a weakly acidic chelating agent such as dithizone and cupferron from an aqeuous solution into an organic solvent, the distribution ratio, D, is calculated as follows. The main equilibria included in this system are:

1. Distribution of the chelating agent (HR) between the two phases.

$$HR_{aq} \rightleftharpoons HR_{org} \qquad K_R = [HR]_0/[HR] \tag{2}$$

2. Dissociation of HR in the aqueous phase.

$$HR \rightleftharpoons H^+ + R^- \qquad K_a = [H^+][R^-]/[HR] \tag{3}$$

3. Formation of the extractable chelate (MR_n) in the aqueous phase.

$$M^{n+} + nR^- \rightleftharpoons MR_n \qquad K_f = [MR_n]/[M^{n+}][R^-]^n \tag{4}$$

4. Distribution of MR_n between the two phases.

$$MR_{n,aq} \rightleftharpoons MR_{n,org} \qquad K_x = [MR_n]_0/[MR_n] \tag{5}$$

Assume the metal chelate, MR_n, is the only metal-containing species in the organic phase and the metal ion, M^{n+}, essentially the only metal-containing species in the aqueous phase. Then, from the equations 2–5, the distribution ratio, D, which is independent of initial concentration of the metal, is expressed as follows:

$$D = \frac{[MR_n]_0}{[M^{n+}]} = \frac{K_f K_a{}^n K_x}{K_R{}^n} \cdot \frac{[HR]_0{}^n}{[H^+]^n} \tag{6}$$

These theoretical considerations agree with experimental results.

However, in many extraction systems, chemical interactions in both phases are more complex, and the distribution ratio sometimes depends on initial concentration of the metal. The relationships between the percent extraction, E, and pH or acidity of the aqueous phase are shown in Figures 2 and 3. The percent extraction, i.e., the percentage of the extracted amount to the total amount, is related to D as follows:

$$E = 100D/(D + V/V_0) \tag{7}$$

where V and V_0 represent volumes of the aqueous and organic phases, respectively.

In most extraction systems, a number of metals are extracted simultaneously into the organic phase under certain conditions. Although this is often favorable as a group separation technique for a multielement analysis (e.g., emission spectrography), it is also necessary to separate the extractable elements from each other. The proper adjustment of pH or acidity of the aqueous phase and the valence states of the elements and the use of masking agents improve the selectivity of the extraction methods. Masking agents prevent the extraction of certain elements into the organic phase by forming strong water-soluble (usually negatively charged) complexes of the elements. Most commonly used masking agents include cyanide, tartrate,

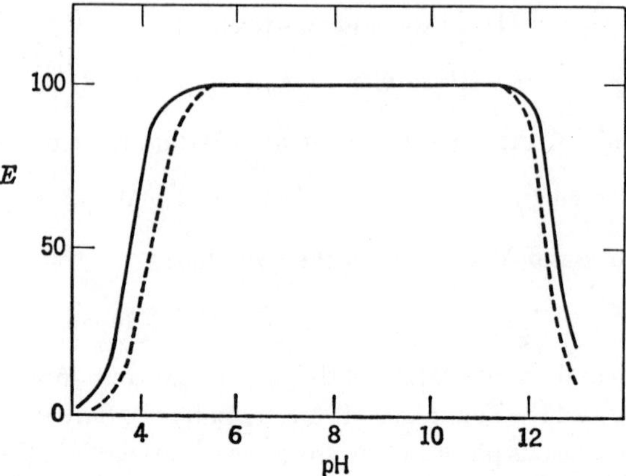

Fig. 2. Extraction of lead from sodium perchlorate solution with 0.002% (w/v) dithizone in carbon tetrachloride (———) and with 0.001% (w/v) dithizone (- - -). Lead taken = 62.5 μg. $V = V_0 = 25$ ml. The rapid decrease in *E* above pH ~ 11.5 is due to the formation of biplumbite ion. [O. B. Mathre and E. B. Sandell, *Talante*, **11**, 295 (1964)]

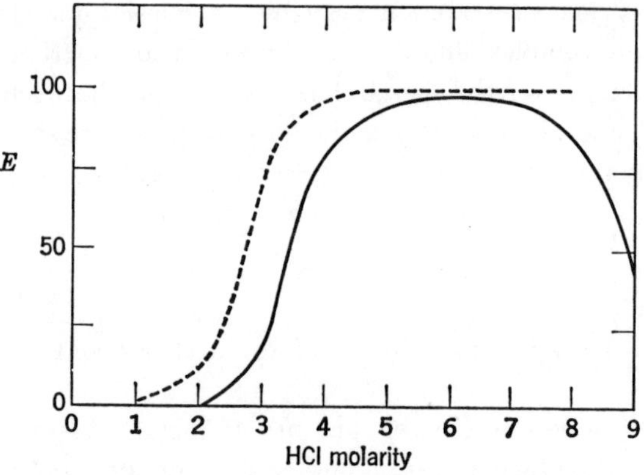

Fig. 3. Extraction of Fe(III) from hydrochloric acid solution with ethyl ether (———) and with methyl isobutyl ketone (- - -). See reference 6.

citrate, fluoride, and ethylenediaminetetraacetate (EDTA). Since these masking agents are weak bases, they work effectively only in weakly acidic or basic media. In the dithizone extraction, the use of masking agents and the adjustment of pH improve the selectivity as shown in Table 5[58] (cf. Table 4).

Table 5. Use of masking agents in dithizone reactions

Solution	Masking agents used	Metals reacting
Basic solution	Cyanide	Pb, Sn, Tl, Bi
Slightly acid solution	Cyanide	Pd, Hg, Ag, Cu
Dilute acid solution	Thiocyanate	Hg, Au, Cu
Dilute acid solution	Thiocyanate plus cyanide	Hg, Cu
Dilute acid solution	Bromide or iodide	Pd, Au, Cu
Dilute acid solution	EDTA	Ag, Hg
Slightly acid solution (pH 5) (Carbon tetrachloride solution of dithizone)	Thiosulfate	Pd, Sn, Zn (Cd, Co, Ni)[a]
Slightly acid solution (pH 4–5)	Thiosulfate plus cyanide	Sn, Zn
Basic solution	Citrate and tartrate	Usually do not interfere with extraction of reacting metals

[a]Large amounts of metals in parentheses react slightly.

A. Techniques

The technique most widely used in analytical chemistry is a simple batch extraction. Usually, an aqueous sample solution and an immiscible organic solvent are thoroughly contacted in a pear-shaped separatory funnel by shaking it by hand or with a mechanical shaker, and after establishment of equilibrium the two layers are settled and then separated. For working with micro or ultramicro samples, various kinds of apparatus have been proposed.[67,68] Sometimes vigorous shaking must be avoided because of the formation of an emulsion which makes the separation of the two phases difficult. If the distribution ratio is not large enough, the batch extractions are repeated twice or more with portions of fresh solvent, and the organic phases are then combined. Nearly complete extraction, i.e., a recovery of more than 99%, of the element of interest is thus obtained.

When the distribution ratio is small so that a large number of batch extractions are required, the continuous extraction technique[58] may be conveniently used. Also, when the distribution ratios of the elements to be separated from each other are of the same order of magnitude, the discontinuous countercurrent distribution extraction technique[69] should be applied for the separation.

After the extraction, the desired elements in the organic phase can be determined directly by spectrophotometric, flame photometric, atomic absorption spectrometric, or radiochemical procedures in many cases. Extraction with a small volume of organic solvent is a good concentration method. In addition, organic solvents often enhance sensitivity in flame photometry and atomic absorption spectrometry. In some cases, however, removal of the organic solvent, namely stripping, is required prior to the following separation or determination steps. Two methods are used: one is the evaporation of the organic solvent followed by the destruction of residual organic complexes with oxidizing acids, and the other is the back-extraction of the desired elements into an aqueous solution containing acids or other reagents under conditions whereby extractable complexes are destroyed. The separation of the extracted elements from each other is often effected in the back-extraction.

Sometimes the organic phase is washed prior to stripping or other manipulations by shaking with an aqueous solution of an appropriate composition. This operation, which is called backwashing, is useful to remove small amounts of the undesired elements (extracted together with the desired elements into the organic phase) without an appreciable loss of the desired elements. The separation factors are thus improved.

In ordinary extraction techniques, an excess of a chelating agent is used to extract the desired elements as much as possible by a single batch operation. However, a substoichiometric amount of a chelating reagent (e.g., dithizone and cupferron) is successfully used in the isotope-dilution method to extract a constant and extremely small amount of a metal element (e.g., Zn, Hg, Cu, and Fe) for specific activity measurements. High sensitivity as well as better selectivity are attained by this technique.[70,71] The lower limits of detection are 10^{-8} to 10^{-10} g/ml. The same technique is also useful in radiochemical separations for activation analysis.[72]

B. Applications

Separation by liquid–liquid extraction is especially useful in trace analysis. Under proper experimental conditions, the extraction of extremely small amounts of the desired elements into the organic phase is carried out almost perfectly leaving large amounts of undesired elements in the aqueous phase. Conversely, large amounts of undesired elements are extracted into the organic phase without any loss of extremely small amounts of the desired elements remaining in the aqueous phase. These techniques may be successfully applied even to carrier-free separations of radioactive nuclides. However, it must be kept in mind that trace amounts of elements are apt to be lost by adsorption on the wall of the separatory funnel, the third solid phase,

Table 6. Liquid–liquid extraction in analysis of metals and related materials for trace impurities

Matrix	Impurities	Concentration	Extraction system	Method of determination	Ref.
		A. Extraction of Trace Impurities			
Cadmium	Tl	Low ppm	Chloride	Polarography	a
Indium	Tl, Fe	0.01 ppm	Chloride	Polarography	b
Aluminum	Fe, Ga	Low ppm	Chloride	Spectrophotometry	c,d
Copper	Au	Low ppm	Chloride-nitrate	Spectrophotometry	e
Nickel	Co, Fe	0.1 ppm	Thiocyanate	Spectrophotometry	f,g
Uranium	B	5 ppb	Tetraphenylarsonium chloride	Emission spectrography	h
Thorium compounds	Cd	0.01 ppm	Liquid anion exchanger	Polarography	i
Zinc	Cd	0.5 ppm	Liquid anion exchanger	Spectrophotometry	j
Beryllium	Li	Low ppm	Dipivaloylmethane	Spectrophotometry	k
Germanium	As	0.01 ppm	Diethylammonium diethyldithiocarbamate	Gutzeit test and spectrophotometry	l,m,n
Silicon	As	Low ppm	Diethylammonium diethyldithiocarbamate	Spectrophotometry	l
Germanium	As, Bi	0.01 ppm	Diethylammonium diethyldithiocarbamate	Emission spectrography	o
Aluminum, titanium, and zirconium	26 elements[a]	0.1 ppm	Ammonium pyrrolidine dithiocarbamate and dithizone	Emission spectrography	p
Selenium	30 elements[b]	0.1 ppm	Oxine and dithizone	Emission spectrography	q
Uranium compounds	Cu, Hg, Ag	Low ppm	Dithizone	Spectrophotometry	r
Tungsten	Cu, Ni, Pb, Zn, Co	Low ppm	Dithizone	X-ray fluorescence	s
Tungsten	Ni	Fractional ppm	Dimethylglyoxime	Spectrophotometry	t

Table 6. *continued*

Matrix	Impurities	Concentration	Extraction system	Method of determination	Ref.
B. Extraction of Matrix Elements					
Iron	Mn, Cu, Ni, Cr, V, Co, Al, Pb, Ti, Zr, Ag, As, Bi	Low ppm	Chloride	Emission spectrography	u,v
Iron	Pb, Bi	0.1 ppm	Chloride	Cathode ray polarography	w
Gallium	17 elements[c]	0.005–1 ppm	Chloride	Emission spectrography	x,y
Gallium	Zn	0.04 ppm	Chloride	Spectrophotometry	z
Gallium arsenide	12 elements[d]	0.01 ppm	Chloride	Emission spectrography	aa
Uranium	Mn	Low ppm	Tributyl phosphate	Spectrophotometry	bb
Indium	Cu, Cd, Pb, Bi, Zn	0.01 ppm	Bromide	Polarography	b
Indium	28 elements[e]	0.1–1 ppm	Bromide	Emission spectrography	cc
Gold	Bi, Cd, Cu, Fe, Ni, Pb, Zn	0.1 ppm	Bromide	Polarography	dd
Mercury	Fe, Cu, Ni, Mn, Pb, Cd, Zn, Bi	0.01–1 ppm	Iodide or iodide-bromide	Spectrophotometry and voltammetry	ee
Bismuth	Fe, Cu, Pb, Ni, Zn, As	Low ppm	Iodide	Spectrophotometry, polarography, and voltammetry	ff

Footnotes:

[a] V, Cr, Mn, Fe, Co, Ni, Cu, Zn, Ga, As, Se, Mo, Pd, Ag, Cd, In, Sn, Sb, Te, Pt, Au, Hg, Tl, Pb, Bi, and U.

[b] Al, Sc, Ti, V, Mn, Fe, Co, Ni, Cu, Zn, Ga, Y, Zr, Mo, Pd, Ag, Cd, In, Sn, Sb, La, Hf, Pt, Au, Hg, Tl, Pb, Bi, Th, and U.

[c] Mn, Cr, Bi, Ni, Mo, Be, Ti, Zr, Cd, V, Zn, Co, Mg, Ca, Cu, Pb, and Al.

[d] Si, Mg, Mn, Cr, Ni, Bi, Al, V, Zr, Co, Cd, and Be.

[e] Li, Na, K, Rb, Cs, Be, Mg, Ca, Sr, Ba, Al, Sc, Y, Eu, Yb, Ti, Zr, V, Cr, Mn, Co, Ni, Rh, Pd, Pt, Cd, Pb, and Bi.

References:

a. R. Carson, *Analyst*, **83**, 472 (1958).
b. F. A. Pohl and W. Bonsels, *Z. Anal. Chem.*, **161**, 108 (1958).
c. H. Jackson and D. S. Phillips, *Analyst*, **87**, 712 (1962).
d. G. Parissakis and P. B. Issopoulos, *Mikrochim. Acta*, **1965**, 28.
e. S. Hirano, A. Mizuike, and K. Yamada, *Japan Analyst*, **9**, 164 (1960).
f. E. Jackwerth, *Z. Anal. Chem.*, **206**, 335 (1964).
g. E. Jackwerth, and E. -L. Schneider, *Z. Anal. Chem.*, **207**, 188 (1965).
h. S. Hirano, H. Kamada, and T. Nishiya, *J. Chem. Soc. Japan, Ind. Chem. Sect.*, **62**, 622 (1959).
i. S. Hirano, A. Mizuike, and Y. Iida, *Japan Analyst*, **11**, 1127 (1962).
j. J. R. Knapp, R. E. Van Aman, and J. H. Kanzelmeyer, *Anal. Chem.*, **34**, 1374 (1962).
k. R. F. Apple and J. C. White, *Talanta*, **13**, 43 (1966).
l. C. L. Luke and M. E. Campbell, *Anal. Chem.*, **25**, 1588 (1953).
m. S. T. Payne, *Analyst*, **77**, 278 (1952).
n. E. W. Fowler, *Analyst*, **88**, 380 (1963).
o. T. J. Veleker, *Anal. Chem.* **34**, 87 (1962).
p. O. G. Koch, *Mikrochim. Acta*, **1958**, 92, 151, 347.
q. O. G. Koch, *Mikrochim. Acta*, **1958**, 402.
r. J. Mareček and E. Singer, *Z. Anal. Chem.*, **203**, 336 (1964).
s. G. L. Hubbard and T. E. Green, *Anal. Chem.*, **38**, 428 (1966).
t. T. E. Green, *Anal. Chem.*, **37**, 1595 (1965).
u. R. E. Heffelfinger, D. L. Chase, G. W. P. Rengstorff, and W. M. Henry, *Anal. Chem.*, **30**, 112 (1958).
v. P. Répás, I. Sajó, and E. Gegus, *Z. Anal. Chem.*, **207**, 263 (1965).
w. R. C. Rooney, *Analyst*, **83**, 83 (1958).
x. E. B. Owens, *Appl. Spectry.*, **13**, 105 (1959).
y. J. H. Oldfield and E. P. Bridge, *Analyst*, **86**, 267 (1961).
z. D. Monnier and G. Prod'hom, *Anal. Chim., Acta*, **31**, 101 (1964).
aa. J. H. Oldfield and D. L. Mack, *Analyst*, **87**, 778 (1962).
bb. K. Motojima, H. Hashitani, and T. Imahashi, *Anal. Chem.*, **34**, 571 (1962).
cc. J. F. Duke, in *Ultrapurification of Semiconductor Materials*, M. S. Brooks, and J. K. Kennedy, Eds., Macmillan, New York, 1962, p. 356.
dd. F. A. Pohl and W. Bonsels, *Mikrochim. Acta*, **1961**, 314.
ee. E. Jackwerth, *Z. Anal. Chem.*, **202**, 81 (1964); **206**, 269 (1964).
ff. E. Jackwerth, *Z. Anal. Chem.*, **211**, 254 (1965).

or the liquid–liquid interface. Formation of an emulsion is very harmful in the separation. Proper selection of experimental conditions and careful manipulations is essential to minimize loss and contamination and to reduce separation factors. Several examples of the application of liquid–liquid extraction to the separation and determination of low ppm or ppb level impurities in metals and related materials are given in Table 6. Extraction of trace elements in nonaqueous liquid samples with aqueous solutions or other immiscible liquids is sometimes carried out. For example, low ppm or ppb amounts of phosphorus and boron in silicon tetrachloride are extracted with concentrated sulfuric acid and quinalizarin-sulfuric acid, respectively, and then determined by spectrophotometry.[73,74]

V. SELECTIVE DISSOLUTION METHODS

The selective dissolution of the desired trace elements with appropriate solvents from solids can be a very useful means of separation in trace analysis. In this technique, the original sample is usually converted to another favorable chemical or physical state before or during the separation step. Some examples in metals analysis follow. For the spectrophotometric determination of about 1 ppm of B in sodium metal, the sample is first converted to sodium chloride and the B is selectively dissolved in 95% ethanol.[75] Two very interesting methods are proposed for the preconcentration of low ppb of B in silicon. In Morrison and Rupp's method, the sample is dissolved in sodium hydroxide solution, and the sodium ion in the solution is then removed by electrolysis through a cation exchange resin membrane until the pH of the solution is lowered to 7–8. The electrolysis cell used is shown in Figure 4. The solution containing borate and crystallized silica is then evaporated, and 1 ppb to 1 ppm of B is extracted with water and determined by emission spectrography[76] or the mass spectrometry–isotope dilution method.[77] In the method proposed by Luke and Flaschen, silicon is converted to crystalline silica by heating the sample with small amounts of

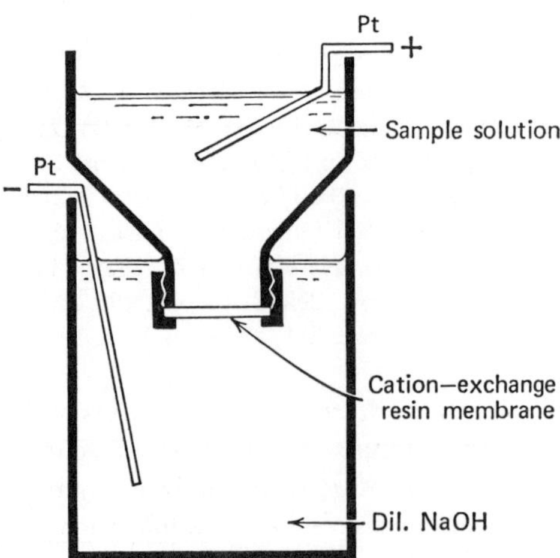

Fig. 4. Polyethylene electrolysis cell.

0.5% sodium hydroxide solution in an autoclave at 350°C and 350 atm. Boron present in the liquid phase is then determined by spectrophotometry[78] down to 0.02 ppm or by fluorimetry.[79]

It is generally difficult to dissolve the matrix elements selectively leaving small amounts of the residue containing the desired trace elements without any loss of the latter elements. However, this technique is very useful in the separation of oxide inclusions in iron, steel, and other metals,[57] where sulfuric or nitric acid, and a methanol solution of iodine or bromine are used as solvents to dissolve the matrix. An anodic dissolution is also applied to this technique.

VI. PRECIPITATION METHODS

Precipitation,[80,81] one of the most classical methods of separation, has several distinct disadvantages which make the separation imperfect. These include (*1*) solubility of a precipitate, (*2*) supersaturation of the solution, (*3*) coprecipitation, and (*4*) postprecipitation. Coprecipitation means the precipitation of a compound in conjunction with other compounds by mechanisms such as isomorphous mixed crystal formation, anomalous mixed crystal formation, adsorption, occlusion, and inclusion. When the precipitations of two solid phases take place in a consecutive manner with a varying time interval between the formation of the two phases, this phenomenon is called postprecipitation. However, if proper techniques are employed, precipitation is successfully applied to separations in trace analysis, in radiochemical analysis, as well as in precise gravimetric analysis. The use of masking agents improves the selectivity of the precipitation separations. The separation of the precipitate from the mother liquor is generally carried out by filtration or sedimentation usually facilitated by the use of a centrifuge. Materials used for a filter such as paper and glass frits are sometimes serious sources of loss (due to adsorption) and contamination.

A. Precipitation of Ordinary Amounts of Elements

This technique has been much investigated by many workers over the years mainly for gravimetric analysis. In ordinary separations, yield, purity, and filterability of the precipitate are main concerns. A satisfactory separation is effected by the proper selection of a precipitant, precipitation conditions, and a washing solution, and the use of the reprecipitation technique. Precipitation from homogeneous solution,[82] the best method of slow precipitation, is especially useful to improve the purity and filterability of the precipitate.

Matrix elements can be removed by precipitation for the determination of trace impurities in metals, although application of this technique to the ppb level seems to be difficult because of loss due to coprecipitation and contamination due to the use of relatively large amounts of precipitants. Some examples follow. For the spectrophotometric determination of as little as 0.5 ppm of bismuth in lead, the matrix element is precipitated as chloride.[83] Precipitation of nickel as hexamminoperchlorate after conversion of cobalt to a stable cobaltic ammine is applied to the spectrophotometric determination of low ppm of cobalt in nickel.[84] As little as 0.1 ppm of boron in silicon, germanium, and germanium dioxide is separated by precipitation of sodium silicate or germanate with methanol followed by distillation of boron, and determined by spectrophotometry.[85] In the spectrophotometric determination of traces of Fe and Pb in high purity copper metal and analytical reagent grade cupric salts, the copper is preliminarily separated by precipitation as cuprous thiocyanate from homogeneous solution.[86] No Fe or Pb is lost during this precipitation, whereas the conventional precipitation of cuprous thiocyanate leads to erratic results.

B. Collection of Trace Elements on Gathering Precipitates (Carriers)

When the elements of interest are present in solution in extremely low concentrations such as 1 μg/ml or less, it is very difficult to precipitate them by the ordinary technique. Even if the precipitate has an extremely low solubility, supersaturation and colloidal formation prevail in very dilute solutions. Therefore, trace elements are generally collected on a small amount of another precipitate, which is called a collector, carrier, or gathering precipitate, by mechanisms such as mixed crystal formation, adsorption, and occlusion. This technique can be applied even to sub-ppb level trace elements in various solutions containing large amounts of coexisting elements. It is also useful in carrier-free separations of radionuclides.

Ordinarily a milligram quantity of another element, i.e., a carrier element, which forms a collector with a precipitant is added to the solution beforehand. Organic reagents which form water-insoluble complexes with various metals are successfully used as precipitants with or without carrier elements. Addition of a relatively large excess of a water-insoluble organic precipitant to an aqueous sample solution causes the crystallization of the precipitant itself, which coprecipitates the trace elements of interest even without a carrier element. Fractional precipitation is useful when the desired trace elements form less soluble precipitates than does the matrix element with a common precipitant. In favorable cases, all of the desired trace elements are coprecipitated with small part of the matrix element, the latter acting as a carrier element. Sometimes a separately prepared precipitate is added to a sample solution to collect trace elements. In a modification of this technique, trace elements are collected on a disk of filter paper impregnated with a collector by passing the sample solution through the paper.[87]

In the selection of a collector, the following should be especially considered: (*1*) yield of the desired element; (*2*) selectivity; (*3*) ease of separation of the precipitate from the mother liquor; and (*4*) interference of the collector in the later separation or determination steps. A number of collectors have been reported.[13,88] These include halides, sulfides, hydroxides, phosphates, sulfates, thiocyanates, and elementary substances, as well as various organic compounds. Some examples are shown in Table 7.

Table 7. Collection of trace elements on gathering precipitates

Gathering precipitate	Elements collected
CuS	Zn, Ga, Mo, Tc, Rh, Pd, Ag, Cd, In, Sn, Sb, Pt, Au, Hg, Tl, Pb, Bi
HgS	Zn, Ga, Ge, Ag, Cd, In, Tl, Pb, Bi
PbS	Cu, Pd, Au, Tl
CdS	Fe, Cu, Hg
$Al(OH)_3$	Be, Ti, V, Cr, Fe, Co, Ni, Zn, Ga, Ge, Zr, Nb, Mo, Ru, Rh, Sn, La, Eu, Hf, W, Ir, Pt, Bi, U
$Fe(OH)_3$	Mg, Al, Ti, V, Cr, Mn, Co, Ni, Zn, Ge, As, Se, Zr, Mo, Ru, Rh, Cd, Sn, Te, W, Ir, Pt, Tl, Bi, Th, U
$Mn(OH)_4$	Al, Cr, Fe, Mo, Sn, Sb, Au, Tl, Bi, Th, Pa
ZnO	Ti, V, Cr, Mn, Fe, Co, Cu, Zr, Nb, Mo, In, Hf, Ta, W
Te	Pd, Ag, Pt, Au, Hg
Cu oxinate	Mg, Al, Ca, Mn, Fe, Cu, Zn, Cd, Hg
Cu cupferride	Ti, V, Fe, Zr, Nb, Mo, Sn, Hf, Ta, W, Bi

Applications of this technique in metals analysis for trace impurities are given below. Rare earths in beryllium, uranium, zirconium, titanium metals, alloys, and oxides are separated by coprecipitation with calcium and magnesium fluorides by using yttrium as a carrier element, and determined by emission spectrography down to 0.1 ppm.[89] As little as 0.2–1 ppm of Se and Te in copper and lead are separated by coprecipitation with arsenic, and determined spectrophotometrically.[90] Low ppm of Sb and Tl in lead are separated by coprecipitation with manganese dioxide and determined by spectrophotometry.[91] Cupferride and sulfide precipitations with copper as a carrier element are applied to the preconcentration of ppm or fractional ppm level impurities in indium for emission spectrographic determination.[92]

Sn, Ti, Zr, Hf, V, Nb, Ta, Mo, W, Fe, and Bi are coprecipitated as cupferrides, whereas Mo, Pd, Sb, Bi, Au, Hg, and Cd as sulfides.

Boron in silicon tetrachloride is coprecipitated by partial hydrolysis of silicon tetrachloride by addition of a dilute aqueous methyl cyanide solution, and after evaporation of silicon tetrachloride the boron in the residue is determined by emission spectrography down to 1 ppb.[93] Lead in telluric acid is coprecipitated by partial precipitation of tellurium at pH 1.6–2.1 with hydrazine, and determined spectrophotometrically down to 0.1 ppm.[94]

VII. ELECTRODEPOSITION METHODS

Deposition of heavy metal and halogen ions on liquid or solid electrodes such as mercury, platinum, silver, etc. from solutions by electrolysis[95–98] offers a very useful means of separation in metals analysis. Over 30 elements are deposited from various electrolytes on the cathode or sometimes on the anode as metals, amalgams, oxides, salts, etc. Elements such as As and Sb are separated by evolution as gases at the electrodes. The electrodeposition behavior of an element depends greatly on the composition of the electrolyte, material and shape of the electrodes, the type of electrolysis cell, and various other experimental conditions. Controlled potential electrolysis, where the electrode potential is externally kept constant during the electrolysis with respect to a reference electrode such as a calomel electrode, enables one to separate various metal ions having different deposition potentials from each other. The internal electrolysis technique is sometimes used for this purpose. The constant current or the constant applied voltage technique is also useful in many separations. Even in these cases, the cathode potential is kept nearly constant in the presence of so-called oxidation–reduction (redox) buffers such as large amounts of U(IV)–U(III), Ti(IV)–Ti(III), V(III)–V(II), or H(I)–H(0) couples present in electrolyte.[95]

Perfect electrodeposition of the element of interest is generally required. In stripping analysis,[99] however, the desired element is partially deposited on an electrode such as a tiny hanging-drop mercury electrode from a sample solution, and the stripping process, i.e., electrolytic redissolution of the element, is measured by voltammetry or coulometry. Higher sensitivities are obtained in this technique.

A. Electrodeposition on a Mercury Cathode

A number of heavy metals can be deposited on a mercury cathode even from acidic aqueous solution because of the high overpotential of hydrogen on mercury. The behavior of various elements in mercury cathode electrolysis[100,101] in dilute sulfuric acid solution is summarized in Table 8.[102]

Although dilute sulfuric or perchloric acid solutions are most frequently used as electrolytes, various other electrolytes containing nitric acid, hydrochloric acid, phosphoric acid, hydrofluoric acid, acetic acid, oxalic acid, citric acid, tartaric acid, buffer solutions, and anodic depolarizers such as hydrazine can also be employed.

Table 8. Mercury cathode electrolysis in ca. 0.3*N* sulfuric acid solution

Quantitatively deposited in the cathode	Cr, Fe, Co, Ni, Cu, Zn, Ga, Ge, Mo, Tc, Rh, Pd, Ag, Cd, In, Sn, Re, Ir, Pt, Au, Hg, Tl, Bi, Po
Quantitatively separated from the electrolyte, but not quantitatively deposited in the cathode	As, Se, Te, Os, Pb
Incompletely separated from the electrolyte	Mn, Ru, Sb, La, Nd

In mercury cathode electrolysis, the optimum cathode potential is easily known from polarographic data. Although controlled potential electrolysis is sometimes required for a separation, most mercury cathode electrolyses are carried out by the constant current or the constant voltage technique, often with simultaneous evolution of hydrogen. The latter techniques are simpler and sometimes result in a higher rate of deposition of the elements.[103]

Various electrolysis cells are reported for mercury cathode electrolysis, including those for micro or ultramicro sample solutions.[68] The larger the ratio of the surface area of the mercury to the volume of the electrolyte, the faster the deposition rate of the element. A mercury pool or sometimes a solid electrode covered with mercury is used as the cathode, and platinum, platinum–iridium (10%) alloy, or sometimes silver or lead used as the anode in the shape of straight wire, spiral, gauze, or plate. Selection of the material and shape of the anode is important from the standpoint of anodic oxidation of the electrolyte, anodic deposition of metals, and dissolution of the anode material. Even a platinum anode dissolves in the electrolyte (dilute sulfuric acid) and then deposits on the cathode.[104] Although the amount is very small, the platinum thus deposited sometimes interferes with the later determination.[104,105] Cooling and stirring of the electrolyte and the mercury, renewal of the mercury during electrolysis, and removal of the mercury and the electrolyte after completion of electrolysis are other considerations in design and choice of the cells.

1. ELECTRODEPOSITION OF THE DESIRED TRACE ELEMENTS

The electrolysis cell shown in Figure 5 is conveniently used for this purpose. The desired trace elements can be deposited in the mercury cathode within a few hours, leaving the matrix elements in the electrolyte. After completion of the electrolysis, the resulting dilute amalgam is taken out of the cell, and the desired trace elements are separated from the mercury by the following two methods.

One is the evaporation of mercury at about 350°C in a stream of nitrogen. A few micrograms or less of metals such as Co, Fe, Ag, and Au in a few milliliters of dilute amalgam can be perfectly recovered in the residue, although there is some danger of loss for metals with a low melting point such as cadmium. The other technique is the anodic stripping, in which electrolytic dissolution of the desired trace elements from the amalgam is effected by using the amalgam as the anode controlled at a potential a little (0.1–0.5 V) lower than the dissolution potential of mercury in solutions such as 0.1–0.5N potassium sulfate or potassium chloride solution. Since a dilute amalgam of less than $10^{-6}M$ has the same potential as that of pure

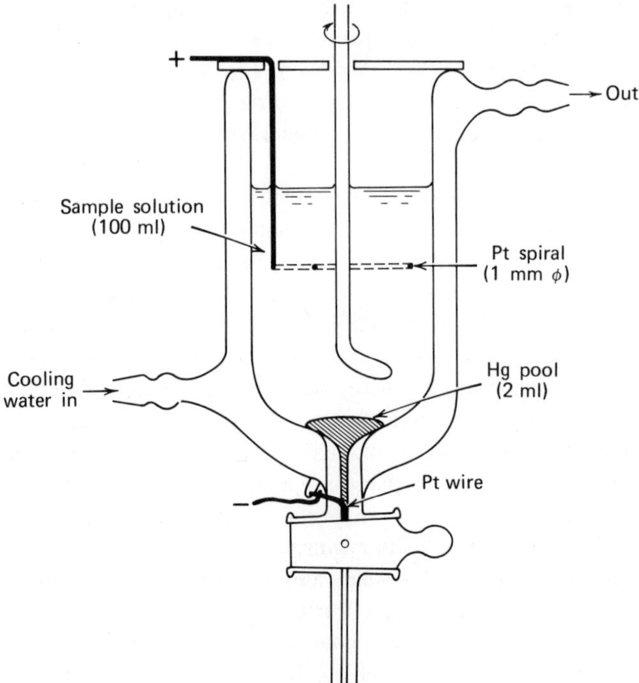

Fig. 5. Mercury cathode electrolysis cell.

mercury,[106] it is estimated that approximately 0.06 μg of Zn, 0.1 μg of Cd, 0.2 μg of Pb, or 0.06 μg of Cu remains in 1 ml of mercury after anodic stripping.[107] The desired trace elements in dilute amalgam may also be separated by treatment with acids or other appropriate solutions. It is, however, very difficult to dissolve the former completely without introduction of appreciable amounts of mercury into the solution.

Several examples in metals analysis follow. Low ppm of Cd, Co, Cu, Fe, Ni, Pb, Zn, and Bi in uranium and its compounds are deposited from dilute sulfuric or perchloric acid solutions.[105,108,109] After evaporation of mercury, the trace elements are determined by spectrophotometry or polarography. Cobalt in titanium and zirconium,[110] Fe, Ni, Co, Cu, and Cd in high purity magnesium[111] and aluminum[112] are also deposited from sulfuric acid solutions. By using the evaporation technique for the removal of mercury and spectrophotometric or polarographic determinations, the lower limits of detection are approximately 0.1 ppm. In the above procedures, the constant current technique is used, with total current of 1–2 A, current density of 0.1–1 A/cm^2 (at the cathode), applied voltage of 7–20 V, and electrolyte volume of about 100 ml.

A controlled potential electrolysis is applied to the separation of low ppm of Pb and Cd in zinc-base alloy.[113] Pb, Cd, as well as Cu are deposited in a mercury cathode controlled at −0.9 V versus the saturated calomel electrode (SCE) from hydrochloric acid solution containing hydrazine as anodic depolarizer. Anodic stripping of Pb and Cd from the amalgam is then carried out at −0.35 V versus the SCE in 0.1*N* potassium chloride solution containing hydrazine, and both elements are determined polarographically. Copper remains in the amalgam.

Noble metals such as Ag, Au, and Pt deposit spontaneously on mercury from appropriate solutions without application of an external potential. After addition of 0.5–2 ml of mercury, 50–500 ml of a sample solution is stirred by a magnetic stirrer. A few tenths of a microgram or less of noble metals are completely collected in a number of small mercury globules (1–4 mm in diameter) within half an hour to a few hours, leaving base metals such as Cu in the solution. Mercury is then separated by the volatilization technique. Silver in copper and iron[114] and Au in copper[115] are separated by this technique from ammoniacal solutions. The lower limits of detection are 0.01–1 ppm using spectrophotometry.

2. ELECTRODEPOSITION OF THE MATRIX OR INTERFERING ELEMENTS

The cell shown in Figure 6 is useful for removal by electrodeposition of the matrix or interfering elements from a sample solution. The magnetic field[116] stirs the mercury–electrolyte interface vigorously, removes deposited ferro-

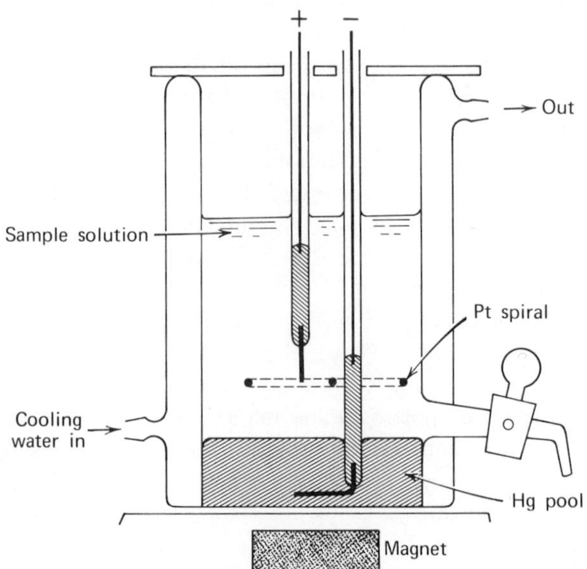

Fig. 6. Mercury cathode electrolysis cell for removal of major constituents.

magnetic metals continuously from the mercury surface, and accelerates the deposition rate of heavy metals. Electrolysis is often carried out without application of magnetic field; mechanical stirring devices are ordinarily used in this case.

This technique is widely used for the separation of trace or small amounts of B, Al, Ti, W, V, Ca, Mg, rare earth elements, etc. in iron, steel, nickel, and other metals and alloys. For example, low ppm of rare earths in stainless steel are separated from the matrix elements and determined by emission spectrography.[117] Chirnside et al.[118] proposed a novel procedure for the separation of low ppm of B in nickel prior to the spectrophotometric determination. The sample piece itself is used as the anode in a mercury cathode electrolysis with $0.02N$ sulfuric acid as the electrolyte; thus the dissolution of the sample and the deposition of the matrix element take place simultaneously. There is no danger of contamination from the anode material. In addition, contamination due to impurities in sulfuric acid is further reduced than the ordinary technique, where larger amounts of the acid are required for the dissolution of the metal. The same technique is applied to the polarographic determination of Al down to 0.3 ppm in iron by using $0.5N$ perchloric acid as the electrolyte.[119]

Controlled potential electrolysis with a mercury cathode is employed in the separation of as little as 1 ppm of Ni and Zn in copper and its compounds.[120] Copper is deposited in a mercury cathode controlled at −0.85 V versus the SCE from ammonium hydroxide–ammonium chloride buffer solution containing hydrazine in a nitrogen atmosphere. After removal of about 99.99% of the copper, Ni and Zn remaining in the electrolyte are determined polarographically. When Ni exists in large amounts compared with Zn, the former is also removed by electrolysis at −1.20 V versus the SCE prior to the polarographic determination of Zn. A similar technique is applied to the separation of low ppm of Zn in cadmium.[95]

B. Electrodeposition on Other Electrodes

A number of metals including Au, Hg, Ag, Cu, Bi, Sb, Sn, Pb, Cd, Zn, Ni, and Co are deposited on a platinum electrode from various solutions. Although this technique is mainly used in electrogravimetric analysis over the years, it is also applicable for the separation in trace analysis. Halide ions are deposited on a silver anode. High purity carbonaceous materials such as pyrolytic graphite and glassy carbon are sometimes useful as electrodes. The desired trace elements are deposited on such an electrode, which is then irradiated in a nuclear reactor. Because of low induced radioactivity and low absorption cross section of the electrode material, activation analysis combined with gamma spectrometry can be carried out conveniently.[121,122]

VIII. ION-EXCHANGE METHODS

Ion exchange[123–130] occupies a very important position in analytical separations because of its simplicity, selectivity, and wide scope. Briefly, this method is based on the distribution of an element between a solution and a solid ion exchanger. A solid ion exchanger used in analytical separations is generally an insoluble but permeable organic or inorganic high polymer with fixed positive or negative charges. To maintain electrical neutrality, there always exist inside this material mobile ions of opposite charge, which can be replaced nearly reversibly by other ions of the same charge when it is immersed in a solution. The ion exchanger is called a cation or anion exchanger, depending on the charge of the mobile ions.

Ion-exchange resins, i.e., synthetic organic high polymers with charged functional groups, are most commonly used in the shape of small beads or membranes. Sulfonated styrene-divinylbenzene copolymers (Dowex 50, Amberlite IR-120, etc.) and styrene-divinylbenzene copolymers into which quaternary ammonium groups are introduced (Dowex 1, Amberlite IRA-400, etc.) are cation and anion exchangers which correspond to strong acids and bases, respectively. There are also weak-acid or weak-base resins, which are occasionally used in analytical separations. Ion-exchange resins with special chelating groups[131] have higher selectivity. Ion exchange paper and cellulose-base ion exchangers are sometimes used. Synthetic inorganic ion exchangers[132,133] such as zirconium phosphate, zirconium tungstate, zirconium molybdate, hydrous oxides of zirconium(IV), thorium(IV), titanium(IV), and tin(IV) are especially useful in separations of highly radioactive materials (because of their resistivity to radiation) as well as in separations of alkali and alkaline earth elements (because of their excellent selectivity).

When a solid cation exchanger with mobile hydrogen ions (called an H-form cation exchanger) is immersed in an aqueous solution of sodium chloride, the following essentially reversible reaction takes place:

$$\underset{\text{Solid phase}}{R^-{-}H^+} + \underset{\text{Solution phase}}{Na^+ + Cl^-} \rightleftharpoons \underset{\text{Solid phase}}{R^-{-}Na^+} + \underset{\text{Solution phase}}{H^+ + Cl^-}$$

The mass action expression is ordinarily used for this equilibrium. However, the great change in activity coefficients limits its practical applicability. If the total amount of the sodium ion is very small compared with the exchange capacity (number of ion-exchangeable sites expressed in milliequivalents) of the ion exchanger, the distribution coefficient, D, defined as below is independent of the initial sodium concentration.

$$D = \frac{\text{Amount of Na/g of dry resin}}{\text{Amount of Na/ml of solution}} \tag{8}$$

The same discussion is applied to anion exchange or ion exchange involving multivalent or complex ions.

A. Techniques

The most widely used technique is a column operation. In general, it consists of three steps: adsorption, washing, and elution. In the adsorption step, a sample solution is introduced into the top of a column packed with solid ion-exchanger beads. The ions of the same electrical charge as the mobile ion of the exchanger adsorb on the upper part of the column by ion exchange, whereas the oppositely charged ions as well as uncharged chemical species ordinarily do not adsorb on the ion exchanger and pass through the column. In the washing step, a washing solution of an appropriate composition is passed through the column to remove the unadsorbed matter remaining in the interstitial volume of the column without loss of the adsorbed ions. Thus the separation is effected during the above two steps. In the elution step, suitable eluents containing acids, alkalies, complexing agents, etc., are passed through the column to desorb the adsorbed ions from the column.

The chromatographic elution, where the adsorbed ions are eluted from the column successively due to differences in affinity to the ion exchanger, is effected as shown in Figure 7.

If a trace amount of an element adsorbed as a narrow band on the extreme top of the column is eluted with an appropriate eluent, the elution curve approximates the Gaussian normal curve, and the equations 9–11 are derived from "plate theory." In this theory, one imagines that the column consists of a large number of "theoretical plates," within each of which the distribution of the trace element between the solution and the exchanger is considered to be in equilibrium.

$$V = I + DM \tag{9}$$

$$\Delta V/V = (2/N)^{1/2} \tag{10}$$

$$C_{\text{max}} = (N/2\pi)^{1/2}\, W/V \tag{11}$$

where V = volume of effluent at which the elution band has maximum concentration (ml); ΔV = half the band width at $C = C_{\text{max}}/e$ (ml); C_{max} = maximum concentration of the trace element in effluent; I = interstitial volume in the column (ml); M = total amount of ion exchanger in the column (g); D = distribution coefficient of the trace element; N = number of theoretical plates in the column; W = total amount of the trace element. The number of theoretical plates in a given column depends on the flow rate, the temperature, the eluent, and the ion being eluted. When a large amount of an element is adsorbed on the column and then eluted, an asymmetrical curve with a diffuse trailing edge is obtained.

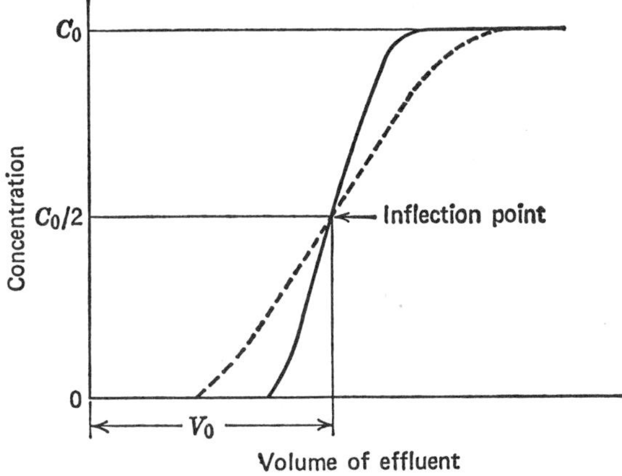

Fig. 7. Chromatographic elution curves.

In general, decrease in particle size of the ion exchanger, increase in column length, and decrease in flow rate increase the number of theoretical plates of the column, which results in better separation (see equations 9–11 and Figures 7 and 8). On the other hand, shorter column length and faster flow rate are preferred from the viewpoint of time required for a separation. With extremely small particle size, the resistance to flow becomes very large so that it is often necessary to use an inconvenient pressurizing device to get the desired flow rate. The optimum operational condition differs in each case. For most separations in metals analysis, particle size of 50–200 mesh, column length of several to several tens of centimeters, column diameter of 1–2 cm, and flow rate of 0.5–2 ml/cm²-min are used.

The adsorption of ions on an ion exchanger is also carried out simply by immersing the exchanger in a solution. This technique, which is called a batch operation, is occasionally used when the distribution coefficient of the element is very large. Stirring is often applied to attain the equilibrium quickly.

Desorption of the desired elements from the ion exchanger is sometimes omitted. For example, after adsorption of the desired elements on an ion exchanger (beads or membrane), the latter is directly employed in X-ray spectroscopy[134–136] and activation analysis.[137,138] Also, ashing of ion-exchange resins[138,139] and dissolution of inorganic ion exchangers[140] are used. The ashing is especially useful for metals such as Au and Pt, whose perfect

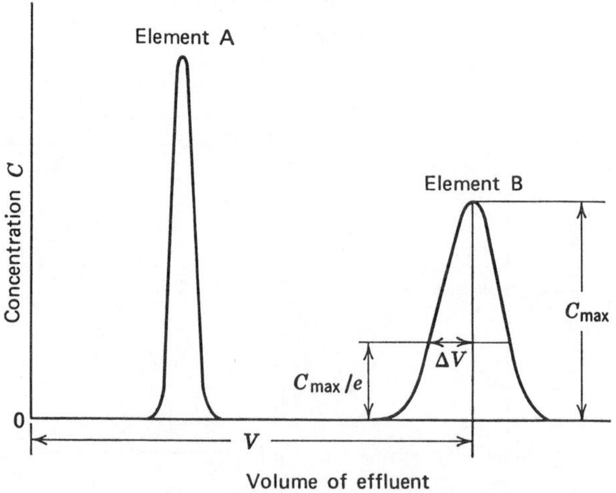

Fig. 8. Breakthrough curves. C_0:concentration in the original solution; $V_0 = I + DM$; (———) column of many theoretical plates; (– – –) column of fewer theoretical plates.

desorptions from the resin are very difficult with commonly available reagents.

The substoichiometric technique described in Section IV-A is also used in the ion-exchange separation.[72,141]

With regard to loss and contamination in ion-exchange separations, the following should be kept in mind. First, some elements adsorb on an ion exchanger so tightly that they are desorbed very slowly. This phenomenon causes not only loss of the desired elements being separated but also contamination of later samples. Second, ion exchangers sometimes decompose and contaminate the sample solution. For example, unidentified organic substances are detected in effluents from anion-exchange resin columns, which give harmful effects on a polarographic determination. Careful washing of a new ion exchanger with water, acids, alkalies, or complexing agents prior to use is very important to remove soluble inorganic and organic impurities in the exchanger as much as possible. Third, elution of the desired elements is greatly influenced by the column and the operational conditions as described previously. This sometimes becomes a serious source of loss of the desired elements. Careful attention must be paid to assure the reproducibility of all experimental conditions, especially when a chromatographic elution is carried out.

B. Effect of the Composition of the Solution

The ion-exchange behavior of an element depends greatly on the composition of the solution as well as the nature of the ion exchanger. Use of complexing agents, highly concentrated solutions, and aqueous organic mixed solvents often markedly improve the selectivity of ion-exchange separations. A number of papers have been published in these fields.[123–130]

The use of metal complexes, especially negatively charged complex ions, extends greatly the applicability of ion-exchange separation of metals. Most metals exist as cations in ordinary aqueous solutions, and there are relatively small differences in their adsorbability on cation exchangers, so that their separation from each other is limited. By formation of negatively charged complex ions or uncharged metal complexes, the possibilities for separation are increased. For example, in hydrochloric acid solution (0–12*M*),[142] almost all metals other than alkali and alkaline earth metals, Al, Y, rare earths, Ac, Th, and Ni form negatively charged chloride complexes and adsorb on an anion-exchange resin. Similarly, from hydrofluoric acid solutions (1–24*M*), elements such as Be, B, Al, P, Sc, Ti, Fe, As, Zr, Nb, Mo, Pd, Sn, Sb, Te, Hf, Ta, W, Re, Pt, Au, Hg, and U are adsorbed on an anion-exchange resin.[143] Various metals including Mo, Tc, Pd, Re, Au, Hg, Bi, Th, Pa, U, Np, and Pu are adsorbed on an anion-exchange resin from nitric acid solutions (0.1–14M).[144] Also, anion exchange of a number of elements in nitric–hydrofluoric acid mixtures was studied.[145] From dioxane–hydrochloric or –nitric acid mixtures alkali and alkaline earth elements are adsorbed on an anion-exchange resin at higher dioxane concentrations.[146]

Complexing agents which form negatively charged, uncharged, or positively charged complexes are also useful in cation-exchange separation of metals. Selective adsorption as well as selective or rapid elution of metal ions are achieved by the use of such complexing agents.

It is very interesting that anomalous adsorption of metals often occurs in concentrated solutions. For example, many metals including Ca, Sr, Sc, Y, rare earths, Ti, Zr, Hf, Th, Mo, W, U, Fe, Ga, and Bi adsorb strongly on a cation-exchange resin from concentrated hydrochloric or perchloric acid solutions beyond expectation.[147]

C. Applications

A number of analytical separations are successfully carried out by using ion exchange. These include the separation of rare earth elements from each other, carrier-free separation of radioactive nuclides, and many other difficult separations. The ion-exchange separation of trace impurities in metals from the matrix elements is discussed below.

In the case, where the desired trace elements are adsorbed on an ion exchanger and the matrix elements are not, one can introduce a large amount of a sample solution into a small column having a low exchange capacity without fear of breakthrough of the desired elements. The adsorbed trace elements are then eluted from the column with a small amount of an appropriate eluent. Thus, concentration of the desired trace elements is effected easily and rapidly. If necessary, a chromatographic elution of the trace elements can be carried out. Although ordinarily the ion exchanger is not saturated with the trace elements, the adsorption front of the trace elements gradually moves downwards during the adsorption step, and finally the trace elements break through the column as shown in Figure 8. The value $(V_0 - I)/D$ gives roughly the required amount of the ion exchanger, M, where V_0 is the total volume of the sample solution, I the interstitial volume of the column, and D the distribution coefficient of the trace element. Since the adsorption front also moves during the washing step, it is necessary to use a little longer column. When a chromatographic elution of the trace elements is required, a much longer column has to be used.

For the separation of trace metallic impurities in high purity metals and compounds, anion exchanges of chloride complexes of metals are very useful. Thus, Cu, Cd, Zn, Co, and Fe in high purity aluminum and magnesium are adsorbed on a Cl-form strong base anion-exchange resin from hydrochloric acid solution, followed by elution with hydrochloric acid and nitric acid for spectrophotometric and polarographic determinations.[112,148] The lower limits of detection are about 0.1 ppm. The same technique is also applied to the spectrophotometric determination of ppb or ppm amounts of cobalt in nickel and its compounds[149] and polarographic determination of cadmium in thorium compounds down to 0.01 ppm.[150] Gold in copper is separated by adsorption of Au on a Cl-form anion-exchange resin from hydrochloric–nitric acid solution.[138] The resin is then irradiated in a nuclear reactor, and low ppb of Au is determined by activation analysis. Alternatively, spectrophotometric determination is used after ashing of the resin, the detection limit being about 0.1 ppm in this modification. Fractional ppm of rare earths in zirconium are adsorbed on an H-form cation exchange resin from hydrofluoric acid solution, eluted with $6M$ hydrochloric acid, and then determined by emission spectrography.[151]

Second, the case is discussed where the matrix elements are adsorbed on an ion exchanger but the desired trace elements are not. The main consideration in this case is the breakthrough of the matrix elements from the column due to the saturation of the resin. Therefore, a relatively large amount of the ion exchanger is required compared with the case discussed previously. For example, impurities in uranium and plutonium are separated by this technique. Silicon in plutonium is separated by adsorption of plutonium on an H-form cation-exchange resin from $0.2N$ nitric acid–$0.05N$ hydroxylamine solutions, and determined by emission spectrography down to 1 ppm.[152] Parts per billion level boron in plutonium and uranium nitrate solutions are separated by cation exchange for emission spectrography.[153] A few ppm each of rare earths in uranium-233 dioxide are separated by adsorption of uranium on a Cl-form anion-exchange resin from $8M$ hydrochloric acid solution, and determined by emission spectrography.[154] Also about 1 ppm each of Mn, Pb, and Ni in uranium are separated by adsorption of uranium on a Cl-form anion-exchange resin from $9M$ hydrochloric acid solution for square-wave polarography.[155] Trace impurities in thorium–uranium and plutonium–thorium–uranium alloys are preconcentrated prior to emission spectrography by retention of the matrix elements on a column packed with an anion-exchange resin and a TBP-coated trifluorochloroethylene polymer in an $8M$ nitric acid medium.[156] Adsorption of uranium on cellulose phosphate from $1M$ hydrochloric acid is applied to the separation of a few ppm of Cu, Pb, Cd, and Zn in uranium for polarography.[157]

The most unfavorable is the last case, where both the desired trace elements and the matrix elements are adsorbed on the ion exchanger. Generally speaking, a relatively large column and careful manipulation are required to adsorb the desired elements quantitatively and separate them chromatographically from large tailed peaks of the matrix elements. However, separation can be carried out successfully by the proper choice of eluents. Various metallic impurities in uranium and plutonium are adsorbed on a Cl-form anion-exchange resin with the matrix elements from 12*M* hydrochloric acid solution, and eluted chromatographically with various concentrations of hydrochloric acid, nitric acid, and water. Appropriate fractions are then evaporated for emission spectrography.[158] Cadmium in uranium is adsorbed on a Cl-form anion-exchange resin from 3*M* hydrochloric acid solution with the matrix, and after elution of the matrix with 1*M* hydrochloric acid, the cadmium is eluted with 0.5*M* nitric acid.[159] By the use of square-wave polarography, the lower limit is 0.01 ppm.

A novel technique called precipitation ion exchange[160] is useful to the preconcentration of a number of trace elements from various kinds of matrix elements. After the adsorption of the trace and the matrix elements on a cation-exchange resin column, 12.2*M* hydrochloric acid is introduced into the column. Because of the insolubility of the matrix elements and the low distribution coefficient of the trace elements in 12.2*M* hydrochloric acid, the matrix elements are precipitated in the column and the trace elements are eluted before the breakthrough of the matrix elements. Many trace elements are concentrated in greater than 90% yields, matrix-free or with only small amounts of matrix. Traces of Al, Cd, Zn, Cu, Ni, Co, Ag, Mn, In, Mg, and Rb in sodium chloride, and Rb, Cd, Ag, Sn, Zn, Co, Na, and Sr in barium chloride are successfully concentrated. The use of organic–hydrochloric acid eluents extends the applicability of this technique.[161] As many as 40 trace elements in various matrices such as alkali and alkaline earth elements, Sc, Y, La, Ni, Cr, Mn, Pb, and Al are thus preconcentrated.

IX. MISCELLANEOUS METHODS

Various chromatographic methods other than ion exchange including columnar partition and adsorption chromatographies, paper chromatography, thin-layer chromatography, electrochromatography, and gas chromatography are sometimes useful in metals analysis. Some of them can be applied to the separation of trace elements from the matrix. For example, ppb amounts of In, Zn, Co, Cu, Cr, Mn, Fe, Ag, Cd, Sr, Sb, Cs, and Ba in uranium are separated by adsorption on a cellulose column from a 95:5 ether–nitric acid medium.[162] The matrix passes through the column, and the impurities adsorbed on the column are eluted with potassium thiocyanate and hydrochloric acid in organic solvents. This technique is also applied to the separation of trace amounts of rare earths from thorium. Various metals in uranium are adsorbed on a silica gel column treated with 6*M* nitric acid, and after selective elution of uranium with methyl isobutyl ketone, the metals are eluted with 6*M* nitric acid.[163] For the emission spectrographic determination of nanogram quantities of Be and rare earths in purified uranium compounds, an 8*M* hydrochloric acid solution of the sample is passed through a column of tri-*n*-octylamine supported on silica to remove the matrix.[164] Paper chromatography with an anion-exchange paper is successfully applied to the separation of low ppm of Co in nickel prior to spectrophotometry.[149]

Zone melting, which is currently the most useful technique in preparing ultrapure metals and compounds, offers promise for concentrating trace impurities in ultrapure materials prior to the determination.[165,166] Parts per billion level impurities in high purity metals, e.g., Ag, Cu, and Tl in refined bismuth, are preconcentrated by this technique for emission spectrography.[167,168] Fire or dry assay is widely used in the isolation of silver, gold, and the platinum metals in ores and various metallurgical products.[169]

References

1. I. M. Kolthoff and P. J. Elving, Eds., *Treatise on Analytical Chemistry*, Part I, Vols. 2 and 3, Interscience, New York, 1961.
2. *U.S. Atomic Energy Commission Reports, Nuclear Science Series*, NAS-NS, Department of Commerce, Washington, D. C.
3. J. G. Maltby, *Analyst*, **79,** 786 (1954).
4. R. E. Thiers, in *Methods of Biochemical Analysis*, Vol. 5, D. Glick, Ed., Interscience, New York, 1957, p. 273.
5. V. P. Guinn, S. Bellanca, J. D. Buchanan, J. C. Migliore, and R. R. Ruch, "Application of Neutron-Activation Analysis in Scientific Crime Detection," General Atomic GA-3491, Sept. 1962.
6. O. G. Koch and G. A. Koch-Dedic, *Handbuch der Spurenanalyse*, Springer, Berlin, 1964.
7. E. P. Laug, *Ind. Eng. Chem., Anal. Ed.*, **6,** 111 (1934).
8. R. J. Henry and E. C. Smith, *Science*, **104,** 426 (1946).
9. E. B. Butler and W. H. Johnston, *Science*, **120,** 543 (1954).
10. T. J. Chow and C. R. McKinney, *Anal. Chem.*, **30,** 1499 (1958).
11. R. E. Thiers, in *Trace Analysis*, J. H. Yoe and H. J. Kock, Jr., Eds., Wiley, New York, 1957, p. 637.
12. J. I. Hoffman and G. E. F. Lundell, *J. Res. Natl. Bur. Std.*, **22,** 465 (1939).
13. E. B. Sandell, *Colorimetric Determination of Traces of Metals*, 3rd ed., Interscience, New York, 1959.
14. H. F. Walton, in *Standard Methods of Chemical Analysis*, 6th ed., Vol. II A, F. J. Welcher, Ed., Van Nostrand, Princeton, N. J., 1963, Chap. 8, p. 201.
15. C. L. Luke, *Anal. Chem.*, **30,** 1405 (1958).
16. A. R. Eberle and M. W. Lerner, *Anal. Chem.*, **32,** 146 (1960).
17. M. Freegarde and J. Cartwright, *Analyst*, **87,** 214 (1962).
18. *The Determination of Gases in Metals*, Special Report No. 68, Iron Steel Inst., London, 1960.
19. B. D. Holt, *Anal. Chem.*, **32,** 124 (1960).

20. C. L. Luke, *Anal. Chem.*, **21,** 1369 (1949).
21. C. L. Luke, *Anal. Chem.*, **29,** 1227 (1957).
22. R. P. Larsen and N. M. Ingber, *Anal. Chem.*, **31,** 1084 (1959).
23. W. T. Rees, *Analyst*, **87,** 202 (1962).
24. F. A. Pohl and W. Bonsels, *Mikrochim. Acta*, **1960,** 641.
25. F. A. Pohl and W. Bonsels, *Mikrochim. Acta*, **1962,** 97.
26. R. E. Heffelfinger, E. R. Blosser, O. E. Perkins, and W. M. Henry, *Anal. Chem.*, **34,** 621 (1962).
27. C. L. Luke and M. E. Campbell, *Anal. Chem.*, **25,** 1588 (1953).
28. V. A. Brophy, L. W. Strock, and T. Peters, *Anal. Chem.*, **26,** 430 (1954).
29. C. L. Luke and M. E. Campbell, *Anal. Chem.*, **28,** 1340 (1956).
30. J. H. Oldfield and D. L. Mack, *Analyst.*, **87,** 778 (1962).
31. D. L. Mack, *Analyst*, **88,** 481 (1963).
32. R. H. Jones, *Analyst*, **71,** 60 (1946).
33. E. L. Bush, *Analyst*, **88,** 614 (1963).
34. N. Etten and J. Muschaweck, *Z. Anal. Chem.*, **206,** 17 (1964).
35. E. Schreiber, *Z. Anal. Chem.*, **210,** 93 (1965).
36. G. Bradshaw and J. Rands, *Analyst*, **85,** 76 (1960).
37. K. Nishimura, A. Tsuchibuchi, and T. Aoyama, *Japan Analyst*, **13,** 220 (1964).
38. J. H. Oldfield and E. P. Bridge, *Analyst*, **85,** 97 (1960).
39. W. G. Guildner, *Talanta*, **8,** 191 (1961).
40. M. W. Mallett, *Talanta*, **9,** 133 (1962).
41. P. D. Donovan, J. L. Evans, and G. H. Bush, *Analyst*, **88,** 771 (1963).
42. F. M. Evans and V. A. Fassel, *Anal. Chem.*, **35,** 1444 (1963).
43. R. K. Winge and V. A. Fassel, *Anal. Chem.*, **37,** 67 (1965).
44. H. Hartmann and G. Ströhl, *Z. Anal. Chem.*, **144,** 332 (1955).
45. A. K. Babko, A. I. Volkova, and O. F. Drako, *Tr.-Komis. Analit. Khim. Akad. Nauk SSSR*, **12,** 53 (1960); *Anal. Abstr.*, **8,** 3643 (1961).
46. L. Ducret and C. Cornet, *Anal. Chim. Acta*, **25,** 542 (1961).
47. S. Mandelstam, *Appl. Spectry.*, **11,** 157 (1957).
48. R. F. O'Connell, *Appl. Spectry.*, **18,** 179 (1964).
49. W. Geilmann, *Z. Anal. Chem.*, **160,** 410 (1958).
50. W. Geilmann and K. H. Neeb, *Z. Anal. Chem.*, **165,** 251 (1959).
51. W. Geilmann and A. A. Estebaranz, *Z. Anal. Chem.*, **190,** 60 (1962).
52. K. H. Neeb, *Z. Anal. Chem.*, **194,** 255 (1963).
53. W. Geilmann and H. Hepp, *Z. Anal. Chem.*, **200,** 241 (1964).
54. R. R. Marshall and D. C. Hess, *Anal. Chem.*, **32,** 960 (1960).
55. K. H. Neeb, *Z. Anal. Chem.*, **200,** 278 (1964).
56. K. H. Neeb, *Z. Anal. Chem.*, **211,** 334 (1965).
57. W. Koch, *Metallkundliche Analyse*, Verlag Stahleisen, Düsseldorf, 1965.
58. G. H. Morrison and H. Freiser, *Solvent Extraction in Analytical Chemistry*, Wiley, New York, 1957.
59. G. H. Morrison and H. Freiser, in *Comprehensive Analytical Chemistry*, Vol. IA, C. L. Wilson and D. W. Wilson, Eds., Elsevier, Amsterdam, 1959, p. 147.
60. H. Freiser and G. H. Morrison, in *Annual Review of Nuclear Science*, Vol. 9, E. Segrè and L. I. Schiff, Eds., Annual Reviews, Palo Alto, Calif., 1959, p. 221.
61. H. Irving and R. J. P. Williams, in *Treatise on Analytical Chemistry*, Part I, Vol. 3, I. M. Kolthoff and P. J. Elving, Eds., Interscience, New York, 1961, Chap. 31, p. 1309.

62. G. H. Morrison, H. Freiser, and J. F. Cosgrove, in *Handbook of Analytical Chemistry*, L. Meites, Ed., McGraw-Hill, New York, 1963, p. 10–5.
63. G. H. Morrison and H. Freiser, *Anal. Chem.*, **30,** 632 (1958); **32,** 37R (1960); **34,** 64R (1962); **36,** 93R (1964).
64. H. Freiser, *Anal. Chem.*, **38,** 131R (1966).
65. C. F. Coleman, C. A. Blake, Jr., and K. B. Brown, *Talanta,* **9,** 297 (1962).
66. H. Green, *Talanta,* **11,** 1561 (1964).
67. P. L. Kirk, *Quantitative Ultramicroanalysis*, Wiley, New York, 1950.
68. I. P. Alimarin and M. N. Petrikova (translated by M. G. Hell), *Inorganic Ultramicroanalysis*, Pergamon, Oxford, 1964.
69. L. C. Craig and D. Craig, in *Technique of Organic Chemistry*, Vol. III, Part I, 2nd ed., A. Weissberger, Ed., Interscience, New York, 1956, p. 149.
70. J. Růžička and J. Starý, *Talanta,* **8,** 228, 296, 535 (1961); **9,** 617 (1962).
71. J. Starý, J. Růžička, and M. Salamon, *Talanta,* **10,** 375 (1963).
72. J. Růžička and J. Starý, *Talanta,* **10,** 287 (1963); **11,** 697 (1964).
73. W. A. Lancaster and M. R. Everingham, *Anal. Chem.*, **36,** 246 (1964).
74. C. S. Haas, R. A. Pellin, and M. R. Everingham, *Anal. Chem.*, **36,** 245 (1964).
75. J. Rynasiewicz, M. P. Sleeper, and J. W. Ryan, *Anal. Chem.*, **26,** 935 (1954).
76. G. H. Morrison and R. L. Rupp, *Anal. Chem.*, **29,** 892 (1957).
77. D. C. Newton, J. Sanders, and A. C. Tyrrell, *Analyst,* **85,** 870 (1960).
78. C. L. Luke and S. S. Flaschen, *Anal. Chem.*, **30,** 1406 (1958).
79. C. A. Parker and W. J. Barnes, *Analyst,* **85,** 828 (1960).
80. M. L. Salutsky, in *Treatise on Analytical Chemistry*, Part I, Vol. 1, I. M. Kolthoff and P. J. Elving, Eds., Interscience, New York, 1959, Chap. 18, p. 733.
81. J. A. Hermann and J. F. Suttle, in *Treatise on Analytical Chemistry*, Part I, Vol. 3, I. M. Kolthoff and P. J. Elving, Eds., Interscience, New York, 1961, Chap. 32, p. 1367.
82. L. Gordon, M. L. Salutsky, and H. H. Willard, *Precipitation from Homogeneous Solution*, Wiley, New York, 1959.
83. D. T. Englis and B. B. Burnett, *Anal. Chim. Acta*, **13,** 574 (1955).
84. C. L. Luke, *Anal. Chem.*, **32,** 836 (1960).
85. C. L. Luke, *Anal. Chem.*, **27,** 1150 (1955).
86. R. P. Hair and E. J. Newman, *Analyst,* **89,** 42 (1964).
87. B. L. Clarke and H. W. Hermance, *Ind. Eng. Chem. Anal. Ed.*, **9,** 292 (1937), **10,** 591 (1938).
88. T. S. West, *Anal. Chim. Acta*, **25,** 405 (1961).
89. S. Kallmann, H. K. Oberthin, and J. O. Hibbits, *Anal. Chem.*, **32,** 1278 (1960).
90. C. L. Luke, *Anal. Chem.*, **31,** 572 (1959).
91. C. L. Luke, *Anal. Chem.*, **31,** 1680 (1959).
92. J. F. Duke, in *Ultrapurification of Semiconductor Materials,* M. S. Brooks and J. K. Kennedy, Eds., Macmillan, New York, 1962, p. 356.
93. T. J. Veleker and E. J. Mehalchick, *Anal. Chem.*, **33,** 767 (1961).
94. C. R. Veale and R. G. Wood, *Analyst,* **85,** 371 (1960).
95. J. J. Lingane, *Electroanalytical Chemistry*, 2nd ed., Interscience, New York, 1958.
96. N. Tanaka, in *Treatise on Analytical Chemistry*, Part I, Vol. 4, I. M. Kolthoff and P. J. Elving, Eds., Interscience, New York, 1963, p. 2417.
97. S. E. Q. Ashley, *Anal. Chem.*, **22,** 1379 (1950).
98. L. B. Rogers, *Anal. Chem.*, **22,** 1386 (1950).

99. I. Shain, in *Treatise on Analytical Chemistry*, Part I, Vol. 4, I. M. Kolthoff and P. J. Elving, Eds., Interscience, New York, 1963, p. 2533.
100. J. A. Maxwell and R. P. Graham, *Chem. Rev.*, **46,** 471 (1950).
101. J. A. Page, J. A. Maxwell, and R. P. Graham, *Analyst*, **87,** 245 (1962).
102. G. E. F. Lundell and J. I. Hoffman, *Outlines of Methods of Chemical Analysis*, Wiley, New York, 1938, p. 94.
103. J. F. Herringshaw and Z. M. Kassir, *Analyst*, **87,** 923 (1962).
104. S. Hirano, A. Mizuike, and M. Saeki, *Japan Analyst*, **8,** 827 (1959).
105. N. H. Furman, C. E. Bricker, and B. McDuffie, *J. Wash. Acad. Sci.*, **38,** 159 (1948).
106. T. Erdey-Gruz and A. Vazsonyi-Zilahy, *Z. Physik. Chem.*, **A177**, 292 (1936).
107. W. E. Schmidt and C. E. Bricker, *J. Electrochem. Soc.*, **102,** 623 (1955).
108. C. C. Casto, *Analytical Chemistry of the Manhattan Project*, C. J. Rodden, Ed., McGraw-Hill, New York, 1950, p. 511.
109. C. J. Sambucetti, E. Witt, and A. Gori, *Proc. Intern. Conf. Peaceful Uses At. Energy, Geneva, 1955*, Vol. 8, United Nations, New York, 1956, p. 266.
110. A. Mizuike and S. Hirano, *Japan Analyst*, **7,** 545 (1958).
111. A. Mizuike and S. Hirano, *Proc. 2nd Symp. At. Energy, Tokyo, 1958*, **3,** 6.
112. S. Hirano, A. Mizuike, and Y. Iida, *J. Chem. Soc. Japan, Ind. Chem. Sect.*, **62,** 1491 (1959).
113. J. K. Taylor and S. W. Smith, *J. Res. Natl. Bur. Std.*, **56,** 301 (1956).
114. S. Hirano and A. Mizuike, *Japan Analyst*, **8,** 746 (1959).
115. A. Mizuike, *Talanta*, **9,** 948 (1962).
116. E. J. Center, R. C. Overbeck, and D. L. Chase, *Anal. Chem.*, **23,** 1134 (1951).
117. E. W. Spitz, J. R. Simmler, B. D. Field, K. H. Roberts, and S. M. Tuthill, *Anal. Chem.*, **26**, 304 (1954).
118. R. C. Chirnside, H. J. Cluley, and P. M. C. Proffitt, *Analyst*, **82,** 18 (1957).
119. R. C. Rooney, *Analyst*, **83,** 546 (1958).
120. L. Meites, *Anal. Chem.*, **27,** 977 (1955).
121. H. B. Mark, Jr., and F. J. Berlandi, *Anal. Chem.*, **36,** 2062 (1964).
122. B. H. Vassos, F. J. Berlandi, T. E. Neal, and H. B. Mark, Jr., *Anal. Chem.*, **37,** 1653 (1965).
123. O. Samuelson, *Ion Exchange Separations in Analytical Chemistry*, Almqvist and Wiksell, Stockholm, Wiley, New York, 1963.
124. J. Inczédy, *Analytical Applications of Ion Exchangers*, Pergamon, Oxford, 1966.
125. K. A. Kraus, in *Trace Analysis*, J. H. Yoe and H. J. Koch, Jr., Eds., Wiley, New York, 1957, p. 34.
126. W. Rieman III and A. C. Breyer, in *Treatise on Analytical Chemistry*, Part I, Vol. 3, I. M. Kolthoff and P. J. Elving, Eds., Interscience, New York, 1961, Chap. 35, p. 1521.
127. H. F. Walton, in *Handbook of Analytical Chemistry*, L. Meites, Ed., McGraw-Hill, New York, 1963, p. 10–137.
128. R. Kunin, *Anal. Chem.*, **32,** 67R (1960); **38,** 176R (1966).
129. R. Kunin and F. X. McGarvey, *Anal. Chem.*, **34,** 48R, 101R (1962); **36,** 142R (1964).
130. H. F. Walton, *Anal. Chem.*, **36,** 51R (1964); **38,** 79R (1966).
131. G. Schmuckler, *Talanta*, **12,** 281 (1965).
132. K. A. Kraus, H. O. Phillips, T. A. Carlson, and J. S. Johnson, *Proc. 2nd Intern. Conf. Peaceful Uses At. Energy, Geneva*, United Nations, New York, 1958, Vol. 28, p. 3.
133. C. B. Amphlett, *Inorganic Ion Exchangers*, Elsevier, Amsterdam, 1964.

134. J. N. Van Niekerk, J. F. DeWet, and F. T. Wybenga, *Anal. Chem.*, **33,** 213 (1961).
135. W. T. Grubb and P. D. Zemany, *Nature*, **176,** 221 (1955).
136. C. L. Luke, *Anal. Chem.*, **36,** 318 (1964).
137. A. A. Smales and L. Salmon, *Analyst*, **80,** 37 (1955).
138. A. Mizuike, Y. Iida, K. Yamada, and S. Hirano, *Anal. Chim. Acta*, **32,** 428 (1965).
139. R. R. Brooks, *Analyst*, **85,** 745 (1960).
140. C. Feldman and T. C. Rains, *Anal. Chem.*, **36,** 405 (1964).
141. J. Starý and J. Růžička, *Talanta*, **8,** 775 (1961); **11,** 691 (1964).
142. K. A. Kraus and F. Nelson, *Proc. Intern. Conf. Peaceful Uses At. Energy, Geneva* 1955, United Nations, New York, 1956, Vol. 7, p. 113.
143. J. P. Faris, *Anal. Chem.*, **32,** 520 (1960).
144. J. P. Faris and R. F. Buchanan, *Anal. Chem.*, **36,** 1157 (1964).
145. E. A. Huff, *Anal. Chem.*, **36,** 1921 (1964).
146. R. R. Ruch, F. Tera, and G. H. Morrison, *Anal. Chem.*, **36,** 2311 (1964).
147. F. Nelson, T. Murase, and K. A. Kraus, *J. Chromatography*, **13,** 503 (1964).
148. A. Mizuike, Y. Iida, and S. Hirano, *Proc. 2nd Symp. At. Energy, Tokyo, 1958*, Vol. 3, p. 9.
149. A. Mizuike, Y. Iida, and S. Hirano, *J. Chem. Soc. Japan, Ind. Chem. Sect.*, **61,** 1459, (1958); **67,** 2042 (1964).
150. S. Hirano, A. Mizuike, and Y. Iida, *Japan Analyst*, **11,** 1127 (1962).
151. H. J. Hettel and V. A. Fassel, *Anal. Chem.*, **27,** 1311 (1955).
152. C. E. Pietri and A. W. Wenzel, *Anal. Chem.*, **35,** 209 (1963).
153. A. W. Wenzel and C. E. Pietri, *Anal. Chem.*, **36,** 2083 (1964).
154. F. T. Birks, G. J. Weldrick, and A. M. Thomas, *Analyst*, **89,** 36 (1964).
155. F. Nakashima, *Anal. Chim. Acta*, **30,** 167, 255 (1964).
156. E. A. Huff, *Anal. Chem.*, **37,** 533 (1965).
157. G. C. Goode and M. C. Campbell, *Anal. Chim. Acta*, **27,** 422 (1962).
158. J. K. Brody, J. P. Faris, and R. F. Buchanan, *Anal. Chem.*, **30,** 1909 (1958).
159. F. Nakashima, *Anal. Chim. Acta*, **28,** 54 (1963).
160. F. Tera, R. R. Ruch, and G. H. Morrison, *Anal. Chem.*, **37,** 358 (1965).
161. R. R. Ruch, F. Tera, and G. H. Morrison, *Anal. Chem.*, **37,** 1565 (1965).
162. R. A. A. Muzzarelli and L. C. Bate, *Talanta*, **12,** 823 (1965).
163. J. S. Fritz and D. H. Schmitt, *Talanta*, **13,** 123 (1966).
164. R. Krefeld, G. Rossi, and Z. Hainski, *Mikrochim. Acta*, **1965,** 133.
165. W. G. Pfann, *Zone Melting*, 2nd rev. ed., Wiley, New York, 1965.
166. W. G. Pfann and H. C. Theuerer, *Anal. Chem.*, **32,** 1574 (1960).
167. G. I. Abakumov and E. E. Konovalov, *Zavodsk. Lab.*, **29,** 1506 (1963); *Chem. Abstr.*, **60,** 6193 b (1964).
168. E. E. Konovalov, S. I. Peĭzulayev, G. P. Pinchuk, I. E. Larionova, and L. I. Kondrat'eva, *Zhur. Anal. Khim.*, **18,** 624 (1963); *Anal. Abstr.*, **11,** 2572 (1964).
169. N. H. Furman, Ed., *Standard Methods of Chemical Analysis*, 6th ed., Vol. I, Van Nostrand, Princeton, N. J., 1962, p. 475.

Chapter 3

VACUUM FUSION, VACUUM EXTRACTION, AND INERT GAS FUSION TECHNIQUES

MANLEY W. MALLETT, General Electric Company, St. Petersburg, Florida
and SILVE KALLMANN, Ledoux and Company, Inc., Teaneck, New Jersey

I. Introduction 70
II. The Vacuum Fusion Method 70
A. The Apparatus 71
1. Extraction Section 71
2. Power Supply 71
3. Furnace Assembly 72
4. Pumping System 73
5. Analytical Section 73
B. Method of Anaylsis 74
1. Sample Preparation 74
2. Furnace Degassing and the Blank 75
3. Gas Extraction and Analysis 75
4. Calculations 77
C. Recommended Techniques 77
1. Oxygen 78
2. Hydrogen 85
3. Nitrogen 87
III. Hot Vacuum Extraction 91
A. Hydrogen 91
B. Oxygen 95
C. Nitrogen 97
IV. The Inert Gas–Fusion Method 97
A. The Apparatus 98
1. Furnace and Reaction Tube 99
2. Oxidation of CO to CO_2 99
3. Analysis Section 99
B. Method of Analysis 103
1. Sample Preparation 103
2. Furnace Preparation 103

3. Analyzer Operation . . . 104
4. Determination of the Blank . . . 104
5. Instrument Calibration . . . 105
C. Sample Analysis . . . 107
1. Recommended Techniques . . . 107
2. Individual Element Analysis . . . 107
References . . . 112

I. INTRODUCTION

The determination of gases has become one of the most important analyses in the evaluation of metals. The need for accurate and precise analytical techniques for gases was intensified by the commercial development of high-purity reactive, refractory, and rare earth metals during the past quarter century. The refinement techniques often involve vacuum processing and zone refining designed to minimize the concentration of all impurities including that of the gases. Since the mechanical and physical properties of many of these metals are adversely affected by gas contents as low as 10 ppm, it is necessary to have adequate means for determining the common residual interstitial gases: oxygen, hydrogen, and notrogen.

Various methods including wet chemical and vacuum have been used for analysis for gases in metals. The wet chemical methods for determining oxygen and hydrogen are involved and unsatisfactory. The more favorable vacuum techniques may be traced rather directly back 100 years to the work of Thomas Graham, Master of the Mint in London. The work of Jordan and Eckman[1] of the U. S. Bureau of Standards in 1925 and of Oberhoffer and Hessenbruch[2] in 1927 combined vacuum melting with a carbon reduction reaction to produce the basic vacuum fusion method.

The vacuum fusion, vacuum extraction, and inert gas fusion methods are interrelated and are dealt with as a group in this chapter. Other modern determinations of gases in metals such as the spectrographic and neutron activation methods are described in other chapters.

II. THE VACUUM FUSION METHOD

Basically the vacuum fusion method involves the melting of a weighed metal sample in a graphite crucible under vacuum and collecting, measuring, and identifying the evolved gases. This method is used to determine gases in practically all metals and serves as a reference method for less direct analytical approaches. The techniques of the vacuum fusion method vary with the metal being analyzed. However, the fundamental of the method are exemplified by the following equations.

$$MO_x + C \rightarrow M \text{ (or MC)} + CO \tag{1}$$

$$[O] \text{ in } M + C \rightarrow M \text{ (or MC)} + CO \tag{1a}$$

$$MN_x \rightarrow M + \tfrac{1}{2}N_2 \tag{2}$$

$$[N] \text{ in } M \rightarrow M + \tfrac{1}{2}N_2 \tag{2a}$$

$$MH_x \rightarrow M + \tfrac{1}{2}H_2 \tag{3}$$

$$[H] \text{ in } M \rightarrow M + \tfrac{1}{2}H_2 \tag{3a}$$

These reactions describe only a fraction of those possible under operating conditions and show that oxygen-containing compounds such as oxides, silicates, aluminates, or oxygen in solution are reduced by carbon to produce CO which is evolved under vacuum. Nitrogen and hydrogen are released by thermal dissociation of their respective compounds or solutions. Other factors such as the heat of solution of the metal formed by the reactions or the heat of formation of MC may alter the thermodynamics of the reaction. However, the kinetics of degassing a sample cannot be rigidly defined from thermodynamic considerations. Factors such as thermal or electromagnetic stirring of the melt, free surface-to-mass ratio of the bath, sweeping action of evolved gas bubbles, sample volume to bath volume ratios, etc., have major effects on the extraction of gas from the sample. Optimum operating temperatures can be determined only by experiment.

A. The Apparatus

The vacuum fusion apparatus consists of (*1*) extraction and (*2*) analysis sections. A schematic diagram of a vacuum fusion apparatus is shown in Figure 1. Another sketch and detailed description of a typical apparatus appears in the ASTM *Book of Methods for Chemical Analysis of Metals.*[3]

1. EXTRACTION SECTION

The extraction section comprises the high temperature furnace in which the fusion is carried out, the power supply and heating element, the major portion of the pumping system, and a pyrometer for monitoring the temperature of the furnace. Vacuum gages on the furnace section and devices for loading samples during the operating cycle are optional.

2. POWER SUPPLY

Heating of the crucible is usually done by high-frequency induction, which made possible the first workable versions of the method. A number of early investigators used the graphite-spiral resistance-type heater which was comparatively inexpensive but characteristically yielded high furnace blanks. A graphite resistance furnace of modern design with a split cylinder

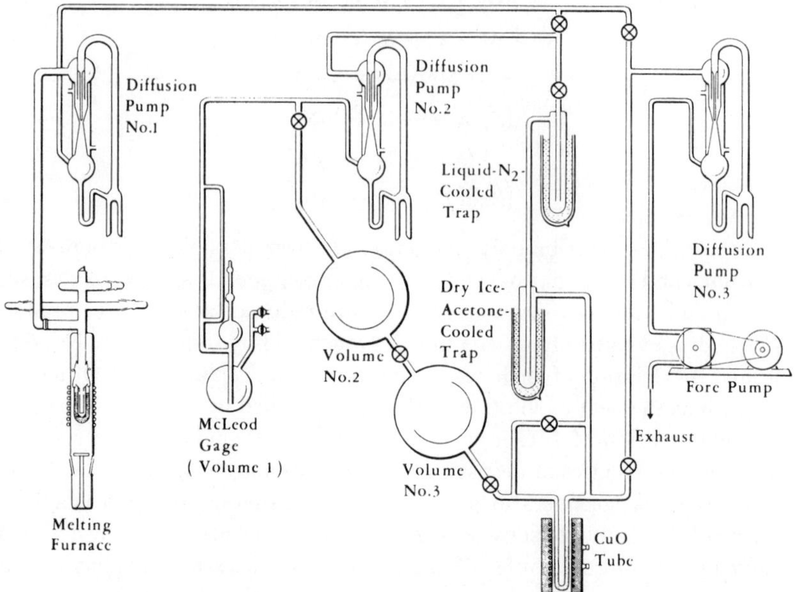

Fig. 1. Schematic diagram of a typical vacuum fusion apparatus.

resistance element proves useful for high-oxygen metals such as titanium but is limited to oxygen levels above about 50 ppm by its high furnace blank.

The high-frequency converter usually has a 6–8 KVA rating for the spark gap converter and 2–2½ KVA for the electronic converter. The former is relatively insensitive to work load changes such as those caused by variations in vacuum which cause the latter to cease oscillation.

3. FURNACE ASSEMBLY

A modified Guldner-Beach[4] furnace is rather universally used in the United States. The outer furnace wall is a Pyrex tube about 3 in. in diameter sealed at the bottom by a waxed, tapered ground-glass joint. The furnace wall is air-cooled by a blower. The graphite crucible is imbedded in graphite powder thermal insulation contained in a clear quartz thimble either suspended by platinum wire and glass hooks inside the furnace shell or by a quartz "egg-cup" pedestal resting on the bottom seal plug. The crucible is heated by a copper tubing induction coil surrounding the outer furnace shell and the graphite crucible positioned about midway on the length of the shell. A smaller Pyrex tube about 1 in. in diameter extends upward on the vertical axis of the shell. It contains an optically flat window at the top for use in temperature readings. Several side arms on the smaller tube act as loading and storage tubes for samples. One side arm contains a capsule-shaped sealed tube enclosing an iron slug. This tube is moved by a magnet and serves as a shutter to prevent fogging of the sight window by vapor from the crucible. Temperatures are read with an optical pyrometer. A mirror or prism may be placed at the window to facilitate readings.

4. PUMPING SYSTEM

A high-speed two-stage mercury diffusion pump is connected to the furnace by a short length of glass tubing of large enough diameter so as not to impede the removal of gases from the furnace. This serves to transfer gases to the evacuation line which contains a second diffusion pump and a mechanical vacuum pump. The furnace pump also transfers analytical gases directly to a three-stage diffusion pump in the analytical section.

5. ANALYTICAL SECTION

The analytical section contains a diffusion pump for packing gases into collection bulbs and for circulating gases through the analytical train. The pump has three stages to enable it to work against high back pressures of 25–40 torr. The extracted gases are compressed into a calibrated volume which includes a McLeod gauge for measuring low pressures. The gauge has multiple ranges to give precisions commensurate with the quantity of gas being measured. Typically the gauge has three scales with ranges 0–0.1 mm, 0–1 mm, and 0–10 mm.

All valving in the vacuum fusion apparatus is by mercury cutoffs. This eliminates the possibility of interference by gases being evolved or adsorbed by greases used for stopcock lubrication. Even where very low-pressure greases are used, solvents used for cleaning mating surfaces tend to be entrained in new applications of grease. The cut-offs are at the top of barometric height risers so that they are normally in closed position when the system is under vacuum.

The minimum collection volume consists of the McLeod gauge and its connecting tubing. The volume is about 400–500 cc, and may be increased to 1000 cc and 2000 cc by connecting one or two calibrated expansion volumes.

Following the collection volumes is a U-tube of Pyrex glass holding CuO. The U-tube is surrounded by a resistance-wound cylindrical furnace. The CuO oxidant[5] is activated by additions of rare-earth oxides. The CuO is maintained at temperature (325°C) continuously and may be regenerated daily by exposure to air for 30 min or may be used for as long as a year before regeneration.

Oxidized gases are most simply analyzed by passing into freezing traps. H_2O vapor is frozen out in a trap cooled by a Dry Ice–acetone mixture stirred by bubbling air through it. CO_2 is frozen out in a liquid nitrogen-cooled trap. Some operators use a solid desiccant, $Mg(Clo_4)_2$, to absorb H_2O vapor. Early equipment used P_2O_5 for H_2O and Ascarite to remove CO_2. The chemical absorbants permit operation in regions where refrigerants such as liquid nitrogen are unavailable. However, care must be taken to maintain their efficiency.

Many schemes have been used for analysis of the gas after it has been extracted and measured. These include older methods such as removal of the gases from the vacuum fusion apparatus for analysis on an Orsat-type apparatus, the standard low-pressure manometric analysis, and newer methods in which the gases are analyzed by mass spectrographic[6,6a,6b] or chromatographic[7] techniques. One must bear in mind that the advantages of the precision of the mass spectrographic method and the speed of the chromatographic method can improve the overall method only if complete extraction of the gas is obtained from the melt.

B. Method of Analysis

The details of procedure vary with apparatus and operator, therefore basic principles only are illustrated below.

1. SAMPLE PREPARATION

All metal samples are suspect as received. They must be representative to yield meaningful results. The advantages of the quartering system are generally not available to this analysis. Samples must be examined for atypical flaws, cracks, gaps, and visual inclusions. These are removed by a clean, sharp, flat file. The handling tools, forceps, hacksaw blades, files, vise grips, etc., should be degreased prior to use.

The rest of the sample surface also is abraded to remove surface contamination. Even samples that appear clean should receive this treatment. Further treatment of abraded samples is meaningless, however, some operators insist on a degreasing rinse with solvent. If used, the solvent should be residue free and should be warmed several degrees above room temperature so as not to chill the sample and cause moisture condensation.

The ideal sample is of one piece with a minimum surface-to-mass ratio. Of course, powders, crystals, flakes, and other finely divided materials often are analyzed. In this csse, the analysis represents both internal and surface gas and is legitimate since the materials are usually used in this form. Where surface oxidation films are to be evaluated, samples are analyzed in both the abraded and unabraded condition. Samples are weighed to 0.1% of total weight on an analytical balance and loaded into the storage arm of the apparatus. Loading is done shortly after cleaning and the apparatus is immediately evacuated. The initial pumping is done intermittently or slowly so that trapped gases may escape from the graphite powder without blowing the packing away from the crucible.

2. FURNACE DEGASSING AND THE BLANK

The furnace is degassed by gradually raising the crucible temperature to 2400°C. The slow initial heating avoids disturbance of the graphite packing. The furnace is held at maximum temperature until gas evolution decreases to a rate that indicates that a low evolution rate (blank rate) will be obtained at operating temperature. Normally this will require 1–4 hr. Some equipment may not reach the suggested degassing temperature and compensation for the lower temperature may be achieved by longer degassing times. However, maximum temperatures below 2150°C will generally prove unsatisfactory.

Following the degassing period the temperature is lowered to the operating temperature, e.g., 1950°C, and the blank gases at that temperature are collected and measured. The blank rate should be such that the blank correction for the sample extraction time does not exceed a few percent of the total. The blank gases are analyzed in the same manner as gases from the sample. Blanks should be reproducible both in size and analysis before and after a series of sample analyses. When metal baths such as platinum are used to dissolve high melting point samples, the blank rate for the empty crucible must first be satisfactory and then the bath metal is dropped and degassed. The working blank is taken on the degassed bath. Baths should not be degassed for extended times because they dissolve excessive carbon which precipitates graphite flakes that grow and thicken the bath.

3. GAS EXTRACTION AND ANALYSIS

Upon completion of the blank, with the forepumps valved off and the furnace diffusion pump pumping into the collection volume, the sample is dropped into the bath and melted. A normal extraction time—for example, 20 min—has been determined for each analysis. However, the extraction should be continued until the collection rate is equal to the blank rate. The pressure of the collected gas is recorded.

a. Manometry

The mercury cutoffs are set to allow the gas to circulate through the evacuated train including the hot CuO tube and the cold traps. The circulation is effected by the mercury-diffusion pump employed initially to pack the gas into the collection system.

Water vapor is removed first by the Dry Ice–acetone cooled trap. When the pressure of the circulating gases becomes constant, the cutoff isolating the collection volume is closed and the gases are collected until a constant pressure is obtained. The gases are then recirculated for a few minutes and recollected. When repeat pressures are found after recirculation, the loss in pressure in this stage is calculated to the volume of hydrogen. The residual gas is again recirculated through the hot CuO tube or through a bypass and CO_2 is removed by the second trap which is now cooled with liquid nitrogen. Pressure readings and recirculations are made as before. The pressure drop is calculated as CO and ultimately to oxygen content. The residual pressure is considered to be nitrogen.

These separations are summarized by the following equations.

(*1*) $P^*_{\text{total}} = P_{N2} + P_{CO} + P_{H2}$

(*2*) $P^*_{\text{total}} - (P_{N2} + P_{CO2})^* = P_{H2O} = P_{H2}$

(*3*) $(P_{N2} + P_{CO2})^* = P_{CO2} = P_{N2}{}^*$

(*4*) $(P_{N2} + P_{CO2})^* - P_{N2} = P_{CO2} = P_{CO}$

b. Mass Spectrography

In the mass spectrographic technique[6] a small quantity of gas in the collection volume is allowed to pass through a molecular leak into a continuously pumped mass spectrometer. The conductance of the molecular leak is small; therefore, the pressure of the collected gas often does not change significantly during the time required to take readings on the mass spectrometer. In cases where the pressure drop is significant, the calibration and test scans are run on the same time schedule to allow for loss of pressure of the sample. After calibration, the mass spectrometer can be used to measure pressures in the collection volume and thus monitor the quantity of gas and its rate of release during a vacuum fusion reaction.

One difficulty in the mass spectroscopic technique is that nitrogen and carbon monoxide both have mass-to-charge (m/e) ratios of 28. However, they may be distinguished from one another by their cracking patterns which are quite different. Thus, the peak for the m/e ratio of 12 is a measure of carbon monoxide concentration and the peak for an m/e ratio of 14 is a measure of nitrogen.

Carbon monoxide (CO^{++}) also gives a m/e 14 contribution. This is allowed for by calibrating the mass spectrometer with carbon monoxide and measuring both the 12 and 14 peaks; these are quite reproducible.

For a complete vacuum fusion analysis a full scan is made of masses 2 (H_2) through 44 (CO_2). From calibrations with pure gases and contributions of the individual gases to the recorded mass spectrometric pattern of the unknown sample, the partial pressure of each gas may be calculated by conventional mass spectrometric techniques. The percentage of each gas in the original sample may then be calculated by the ideal gas law.

c. Chromatography

The gas chromatograph also may be used to analyze the gases extracted by vacuum fusion. Unlike the mass spectrograph it cannot be used to monitor gas extraction rates. Rates and gas sample volumes are determined by manometric measurements in the vacuum fusion apparatus. An aliquot of the gas sample then is compressed to atmospheric pressure and passed into the chromatograph by standard techniques. The details of analysis of such a sample are given later in this article in section IV-A-3.

4. CALCULATIONS

Calculations are simplified by combining various constants into one multiplication factor for a given gas at a given room temperature. A set of such factors for one degree intervals covering the daily change in room temperature is computed for reference. For instance, at 25°C room temperature.

$$\text{ppm } H_2 = \frac{\text{cc-mm } H_2 \times 0.1084}{\text{sample wt (g)}}$$

$$\text{ppm } O_2 = \frac{\text{cc-mm CO} \times 0.86}{\text{sample wt (g)}}$$

$$\text{ppm } N_2 = \frac{\text{cc-mm } N_2 \times 1.51}{\text{sample wt (g)}}$$

The cc-mm unit is numerically the same as the micron-liter unit commonly used. The cc value corresponds to the calibrated volume of the collection system in use and the mm value is the pressure read directly from the gauge. Parts per million (ppm) units $\times$ 10^{-4} = percentage.

C. Recommended Techniques

Almost every metal requires a special analytical technique applicable only to itself and one or two closely related metals and their dilute alloys. These techniques are described by a listing of several critical factors, i. e., (*1*) sample size, (*2*) bath metal, (*3*) maximum sample-to-bath ratio, (*4*) operating temperature, (*5*) gas extraction time. Following are given lists of these factors for the individual metals and references in each case to one or more papers most closely approximating the recommended technique. Further details are given for the more basic techniques.

1. OXYGEN

a. Aluminum

Determine oxygen in aluminum by the copper bath method of Kopa.[8] Degas the crucible at more than 2000°C. Add 20 g copper (OFHC) for the bath at 1250°C and degas 15–30 min. Take a blank by lowering the temperature to 1050°C for 10 min and then progressively heat to 1550°C in 20 min to complete the total blank time of 30 min. Next drop the sample at 1050°C and lower a lid onto the crucible. Follow the heating schedule of the blank run. Do not exceed about 20% aluminum; above 30% aluminum distillation may occur. Four 1-g samples may be run in the same crucible. Aluminum forms a carbide if given time at the lower temperatures of the operating range. This lowers the partial vapor pressure of aluminum and suppresses distillation. Aluminum reacts (as silicon does) with the crucible, causing it to swell and crack. This interference is less with the copper than with the iron bath used by Sloman[9] and others. In fact the copper–aluminum residue is said to break clean from the crucible.

b. Beryllium

Determine oxygen in beryllium by the microvacuum fusion method of Booth and Parker.[10] Load up to six approximately 0.050-g samples of beryllium (filed clean), each with its accompanying 0.070 g of tin in the sample storage arm. Load several spare 0.070-g pieces of tin. Tin foil is unsatisfactory because of its large surface area. Also load 10 g of platinum rod or wire cut into suitable size for dropping into the crucible to form the bath. Degas the graphite crucible as usual, then lower the temperature to 1950°C; drop the platinum and degas for 10 min. Lower the temperature to a dark red heat and add 0.070 g of tin; heat to 1400 ± 50°C for 1 min and then to 1950°C for a minimum of 5 min (Booth and Parker state 2 min). Determine the blank by this heating schedule. If the low temperature schedule is not used a high blank will result. For the analysis, drop a sample plus tin into the cooled crucible and repeat the heating schedule. Successive samples are extracted in the same manner. Oxygen may be determined at the 100 ppm level with a relative standard deviation of 20%.

c. Bismuth

Use a modified vacuum fusion technique.[11] Replace the vacuum fusion furnace tube with a simple Vycor furnace tube having a side arm for sample storage. Heat with a resistance-wound cylindrical furnace which may be raised about the furnace tube. Place a graphite vacuum fusion crucible containing 0.2 g of graphite powder in the vertical section of the tube; evacuate and degas at 850°C until the blank rate is less than 0.01 ml/hr. Lower the furnace and drop 1–20 g of sample (depending on the oxygen level) into the crucible. Replace the furnace and heat to 750°C, collecting the gases in the vacuum fusion system for 30 min in the normal manner. Oxygen will be evolved as both CO and CO_2; therefore, measure and remove the CO_2 before oxidizing the rest of the gases. The experimental error of the analysis is estimated to be less than 5%.

d. Boron

Take 1-g samples and remove surface oxide by stirring in hot water for a few minutes. Rinse in lukewarm acetone and air dry. Wrap samples in platinum foil maintaining a minimum ratio of 4Pt:1B. Drop the sample and flux into the degassed graphite crucible and extract the gas at 1850°C for 15–20 min. The limit of detection of oxygen by this method[12] is about 5 ppm.

e. Chromium

Use chromium samples weighing 1–10 g depending on the expected amount of oxygen.[13] Samples may be wrapped in tin foil or dropped simultaneously with a piece of high purity tin weighing 0.3 g. Prepare a bath by degassing 30–50 g of iron at 1600°C. Start the blank determination by lowering the temperature to 1200°C; during the initial 3 min raise the temperature to 1600°C and complete the rest of the 20-min blank. Analyze the sample by following the same heating schedule. Make correction for the oxygen content of the tin. Analyses made entirely at 1600°C tend to be low.[14] The chromium concentration in the bath must be less than 20%.

f. Cobalt

Use the iron-bath technique similar to that described by Sloman.[15] (See method for iron and steels.) Degas a bath of 10–20 g of iron at 1600°C. Drop 1–5-g samples into the bath at 1600°C and extract the gas for 20 min. Keep the iron/cobalt ratio at greater than 3:1. Drop 1 g of tin between samples.

g. Copper

Melt samples weighing 1.0–10 g directly in a degassed graphite crucible at 1150[16,17] to 1250°C[18] for 15 min. Several samples may be analyzed consecutively in the same crucible. A mass spectrographic study[16] showed that the gas evolved from Chilean copper (approximately 250 ppm O_2) was pure carbon monoxide. However, in other studies[18] it was found that certain coppers at this high oxygen level especially those containing surface oxides or oxide inclusions sometimes evolved part of the oxygen as carbon dioxide.

Evolved gases should be analyzed for carbon dioxide as well as for carbon monoxide. The need for further study of this technique is indicated.

h. Germanium

Take samples weighing 1–10 g. Etch samples in a mixture of HNO_3, HF and CH_3COOH containing bromine.[19] Fuse the sample directly in a degassed graphite crucible at 1650°C for a 30-min extraction period. The reported sensitivity of the apparatus was equivalent to 2.6 ppm oxygen for a 1-g sample. The total blank was about 0.016 cc STP in 30 min.

i. Hafnium

Use the platinum-flux techniques[20] given in a later section for titanium except that the platinum/hafnium ratio may be as low as 5:1.

j. Iron and Steels

Melt 1–5 g of sample in the graphite crucible at 1650 ±50°C.[15,21] Extract the gas for 20 min. The first sample forms an iron bath for subsequent samples. High manganese contents may cause low oxygen results. This interference may be minimized by dissolving high manganese steels in an iron bath and by use of high-speed diffusion pumps and large diameter tubing for the pumping lines. Slow removal of oxygen also results when oxygen is combined as large Al_2O_3[22] or glassy SiO_2[23] inclusions. Extraction times of 30–40 min may be required.

If estimation of the distribution of the oxygen content among FeO, MnO, SiO_2, Al_2O_3, the common oxides in carbon, and low alloy steels is desired, gas fractions may be taken at 1070, 1170, 1320, and 1650°C[24,25] respectively. At best, the separation is inexact but in special cases such as the evaluation of weldments the technique is useful.

Nitrogen results by the iron bath technique have been questioned by many operators. A recently evaluated platinum bath technique[26] applied to iron yields quantitative nitrogen analyses and somewhat higher oxygen values than obtained with the iron bath. For the platinum bath technique, take 0.5–1-g samples. Prepare a degassed bath of 20–30 g of platinum. Condition the bath by addition of 3 wt% iron. When the blank rate at 1850°C returns to normal (in about 20 min) drop the first sample. Simultaneously add 0.5 g of platinum with each sample. This promotes quantitative extraction of oxygen. Correct for the oxygen content of the platinum addition. Maintain the cumulative platinum/iron ratio in the bath above 4:1.

A platinum-flux technique such as that described for titanium[27] also may be used. First wet the bottom of the crucible by melting and degassing 5 g of platinum. Wrap 0.5–1-g samples in platinum maintaining the platinum/iron ratio of the sample complex above 1:4. Extract the gas for 20 min at 1850°C.

k. Lanthanum

Use the nickel bath technique.[28] After degassing the crucible, lower the temperature to 1300–1400°C and drop in about 50 g of 99+ purity nickel. Slowly raise the temperature to 1900°C and hold for one to 1.5 hours until a steady blank of less than 01. ml STP/hr is obtained. Take a blank for the same temperature cycle to be used for the analysis. Up to six 1-g samples may be analyzed in this bath. To prevent loss of sample as hydrogen boils out lower the temperature to 1300°C before dropping the sample. Raise the temperature to 1900°C in 4 min and continue the extraction for 11 min more. Repeat the cycle for subsequent samples.

Consistent and accurate results could not be obtained at lower operating temperatures.[28] The technique showed 91% recovery from oxygen-doped samples with a probable error of 4.7%. The low recovery may be due to gettering by the vaporized nickel. This technique gives useful data within the limitations stated.

It appears that lanthanum would be amenable to analysis by the platinum-flux technique under the operating conditions given for titanium. Thus the interference from metal vapor would be avoided.

l. Lead

Analyze 2 g samples of lead in a tin bath.[12] Tin/lead ratio should be more than 2:1. Extract the gas for 20 min at 1200°C and analyze for both CO_2 and CO as recommended for the analysis of copper.

m. Manganese

Since the early days when the vacuum fusion method was applied only to iron and steels, manganese has been considered the worst getter of evolved gases. It maintains that reputation even now when the behavior of many more metals is known. Guldner concluded from his study of gettering[29] that it is not feasible to determine oxygen in manganese by vacuum fusion. However, a manganese steel containing 13% manganese apparently gave reliable results when introduced as the first sample to the crucible.[9,15] By limiting the analysis to only one sample in a crucible, success may be obtained with the following technique. Degas the crucible as usual; lower the temperature to 1550°C and add an iron bath large enough to keep the manganese content below 10%. Degas the bath and determine the blank rate. Introduce a 1–5-g sample and extract the gas for 10 min. Replace the crucible before proceeding with the next sample.

n. Molybdenum

The same operating conditions are also used for tungsten. Dissolve 2–10 g of sample in a degassed Pt–20% Sn bath at 1950 ±50°C. Extract gas for 20 min.[30] The final molybdenum or tungsten content of the bath shall not exceed 10%. Blanks of 4–10 μg of oxygen may be expected.

An alternative technique is the tentative method of Test for Oxygen in Molybdenum (ASTM Designation: E174-60T).[31] The method has been used with fair to good success by a number of operators. However, many have found the method inapplicable because of the tendency of the iron bath to thicken by formation of graphite flakes during prolonged degassing.[32] This is due to failure to maintain the same temperature during degassing and sample extraction. Briefly, dissolve the sample in an iron bath at 1650°C and collect the gas for 20 min. Drop 0.3 g tin between each sample and add 10 g of fresh iron after every second sample. The maximum molybdenum or tungsten content of the melt shall not exceed 30%.

o. Nickel

Use the same iron-bath conditions recommended for cobalt.[15]

p. Niobium

The same operating conditions are also used for tantalum. Dissolve 1–2 g of sample in a degassed platinum bath at 1950 ±50°C. The extraction time is 20 min. The niobium or tantalum content of the bath should not exceed 25 wt%. At the 100 ppm level, results with a relative standard deviation of less than 5% are readily obtained.[33]

q. Platinum

Drop a 5-g sample into a degassed graphite crucible. Extract the gas at 1900°C for 15 min. This should permit an accuracy of 1 ppm for the low oxygen contents normally found for platinum.

r. Plutonium

Analyze by the same technique as for uranium. However, the furnace section of the vacuum fusion apparatus should be located in a glove box filled at slightly below atmospheric pressure with nitrogen or inert gas.[18] A glass cloth filter should be placed in the vacuum line between the furnace and the first diffusion pump to remove all plutonium-bearing dust. The furnace parts and eventually the furnace proper must be treated as radioactive waste when removed from the glove box. The extracted gas is not radioactive.

s. Rhenium

Analyze by the platinum–tin bath technique described for molybdenum.

t. Silicon

Oxygen in silicon or high silicon ferroalloys may be extracted by fusion of the sample directly in the degassed graphite crucible. Solution of silicon in an iron bath causes precipitation of silicon carbide particles which thickens the bath and inhibits gas bubble evolution. The sequence of silicon-containing samples should be planned so as to decrease rather than increase the silicon content of the melt. Silicon tends to soak into graphite causing it to swell and eventually crack. The number of samples analyzed in a single crucible is limited by this reaction. Samples weighing 1–5 g are fused at 1500°C for a 10 min extraction period.[15]

Limited data[17] from fusion of 0.06-g samples in graphite capsules at 1800°C for 10 min show oxygen values somewhat higher than those by the platinum bath.

The platinum bath technique[17] for silicon calls for 01.–0.3-g samples to be used in a platinum bath at 1800°C for 10 min. Although not specified by the author, a 10-g bath should give satisfactory results.

u. Silver

Analyze by the technique given for copper. If the oxygen content is high, check evolved gases for carbon dioxide.

v. Thorium

In the degassed graphite crucible, prepare a bath of not less than 20 nor more than 40 g of a composite of 80% Pt : 20% Sn.[34] Pieces of 12-gauge platinum wire are convenient for this purpose. Add the bath metals intermittently at 1400°C to produce a quiet meltdown and degas at 1700°C until a satisfactory blank rate is obtained. Precondition the degassed bath by addition of 0.3–0.5 g of sample metal which is then degassed for the normal extraction time of 15–30 min. Carefully file clean several 0.1–0.5-g samples. Wrap each sample in 0.85 g of platinum wire or foil capsule. The platinum-flux blank should not exceed 8–10 μg oxygen and the furnace blank 2–4 μg. The thorium concentration of the bath should not exceed 10 wt%. This combination of Pt–Sn bath and platinum-flux technique yields optimal results. The same analytical conditions may be applied to yttrium.

w. Tin

Analyze by the technique given for copper. No carbon dioxide will be evolved.

x. Titanium

Use the platinum-flux technique[20,27,35] for the range 0.03–0.5 wt% oxygen. The sample normally weighs 0.25 g but not less than 0.1 g or more than 1.0 g. Clean in the usual manner and press into intimate contact with a clean piece of platinum weighing about 10 times the sample weight. Thin platinum may conveniently be in wire form wrapped about the sample. Melt the flux-sample unit at 1950°C and extract the gas for 20 min. Blanks for the platinum-flux as well as the usual furnace blank are subtracted from the total gas collection.

An alternative technique[36] which has been used for umpire purposes is the tin flux–vacuum fusion technique. It also is applicable to the range 0.03–0.5 wt% oxygen. The sample size is normally 0.25 g but not less than 0.1 or more than 1.0 g. Abrade the sample clean in the usual manner. Press together the sample and a piece of clean tin weighing about twice as much. Thus the sample and tin flux enter the crucible at the same time. The tin must have a known and reproducible oxygen content.

Add about 2 g of graphite chips or shavings (about 20 mesh) to the crucible and degas together. Provide a graphite cover for the crucible. Lower this into the neck of the crucible funnel for 10 min at the end of the degassing period.

Lower the temperature of the open crucible to 1200°C and determine the blank by collecting the gas evolved while heating the crucible as rapidly as possible to 1950°C which is maintained for the balance of the 30–45 min corresponding to the usual extraction time. A 0.5-g piece of tin may be dropped and briefly degassed to condition the crucible just prior to dropping the first sample. Additional 0.2-g pieces of tin may be provided for conditioning between samples.

Drop the sample and its tin flux and cover immediately. Collect the gas evolved while following the same heating schedule used for the blank. The cover must be moved occasionally to prevent sticking.

y. Tantalum

Use the method given for niobium.

z. Tungsten

Use the method given for molybdenum.

aa. Uranium

Prepare a 20-g iron bath at 1750°C; add 1 g of tin at 1600°C and degas about 10 min before analyzing the first sample.[37] Add the sample (1–2 g) to the graphite crucible at 1200°C. Raise the temperature to 1750°C over a period of 10 min and hold at temperature for 20–40 min until all gases are extracted. Add 1 g of tin between samples and degas for 10 min. Four analyses may be made in a crucible but the uranium content of the bath must not exceed 30 wt%.

bb. Vanadium

Wrap or press a 0.1–0.4-g sample into intimate contact with 0.5 g of platinum wire (about 12-gauge). Drop 20–60 g of platinum into the degassed crucible and degas at 1900°C until the furnace blank is less than 5 μg oxygen for the extraction period.[38] The blank for the platinum-flux should be less than 2 μg oxygen. Complete the extraction at 1900°C, for 15–30 min. Several samples may be run in the same crucible but the vanadium content of the bath should not exceed 15%.

cc. Yttrium

The same combination Pt–Sn bath and platinum-flux technique[34] given for thorium is best. The preconditioning of the bath will, of course, be done with yttrium metal.

dd. Zirconium

Use the same platinum-flux technique[20,27] as for titanium except that the platinum/zirconium ratio may be as low as 5:1.

2. HYDROGEN

The vacuum fusion method extracts hydrogen along with oxygen and nitrogen. Often the hydrogen is evolved even before the sample is molten. Hence, hydrogen extraction is usually complete even though only partial recovery of oxygen and nitrogen is obtained. The results are accurate but, in general, the hot vacuum extraction method is more convenient and rapid for routine analysis. In view of these facts, detailed techniques for the various metals are not presented here. However, two modified vacuum fusion techniques of considerable interest are given below.

a. Aluminum

The analysis for hydrogen in aluminum is complicated by the fact that fresh surfaces of the metal such as produced by machining or abrading immediately react with moisture in the air to produce a hydrated oxide. For practical purposes the hydrated layer reaches its maximum thickness in about 10 min. Hence, the surface hydrogen is constant for unit area of a given aluminum alloy. Both internal and external hydrogen may be determined by the tin bath technique[39] described below.

The apparatus consists of a Pyrex or Vycor furnace tube and crucible connected by a short length of 25 mm Pyrex tubing to a liquid nitrogen trap. This section replaces the usual furnace section of the vacuum-fusion apparatus. An alternative is to connect the furnace to a gas collection and analysis section such as that described by Griffith and Mallett.[39] The furnace tube is connected also to a side arm for sample storage and preheating of samples. About 200 g of 99.8% nominal purity tin is placed in the crucible. Two 6-g samples of the same material and of identical surface area are placed in the storage arm with a piece of degassed steel to serve as a magnetic pusher. The tin is degassed for about 1½ hr at 500–525°C while the samples in the side arm are heated for about 4 hr at 525°C to remove all the surface hydrogen and part of the internal hydrogen. All heating is done with wire-wound furnaces. After degassing, one of the samples is pushed into the tin bath along with one of the steel pieces. The bath is stirred occasionally by manipulating the steel with a magnet. Solution and degassing is completed in 1–1½ hr depending on the alloy. The hydrogen is determined in the analytical section of the apparatus.

The second partly degassed sample is removed from the side arm and freshly abraded along with an identical sample of the as-received sample. Both are then loaded into the storage arm and a new bath of tin is prepared in the crucible. These two samples are analyzed consecutively without preheat. Thus one obtains values for the total hydrogen in the as-received sample, the residual internal plus the external hydrogen of the preheated and reabraded sample and the residual internal hydrogen of the preheated sample. From these data may be calculated the internal hydrogen removed by preheat, the total internal hydrogen, and the external hydrogen.

This method is less involved than it may seem at first since the external hydrogen per unit surface area need be determined only once for each alloy. In subsequent analyses only the total hydrogen is determined and the known correction for surface hydrogen is applied. For some materials such as welding wire the total hydrogen from the wire in the condition just prior to welding may be the value of most meaning. In that case the wire is not abraded or otherwise prepared before analysis.

b. Iron and Steel

A tin fusion technique for iron and steels also has been reported.[40] Hydrogen is extremely fugitive and is lost from molten steel during sampling, solidification, and subsequent cooling, and even during storage at room temperature. Carney, Chipman, and Grant[41] found that some low alloy steels may be stored in Dry Ice for up to 12 hr; chilled austenitic steels may be stored up to at least 3 days without loss of hydrogen. These observations have led to the common practice of quenching samples to room temperature as rapidly as possible and immediately storing them in liquid nitrogen. Such refrigeration is essential to preservation of ferritic and low alloy steels but is not required for completely austenitic steels.

For hydrogen determination, use an inductively heated bath of 150 g of tin and 0.7 g of silicon in a fused silica crucible, at 1150°C. Two-gram samples of steel dropped into this bath will evolve all of their hudrogen in 5 min. The silicon suppresses the evolution of CO which may result from slow reduction of the crucible. The analysis may be carried out in a regular vacuum fusion apparatus with a modified crucible setup. The accuracy is about 0.1 ppm hydrogen. Ten or more samples may be analyzed in series.

3. NITROGEN

The vacuum fusion method has been directed largely to the analysis for oxygen. Chemical methods for oxygen are much more involved and are more subject to interference than chemical methods for nitrogen. Nitrogen values obtained incidental to oxygen vacuum fusion analysis have been reported in the literature although statements as to their accuracy often have been qualified. The only extensive comparisons of Kjeldahl wet chemical analysis and vacuum fusion nitrogen analysis have been in regard to plain carbon and alloyed steels.

a. Iron and Steel

A comprehensive study of chemical and vacuum fusion methods was made under the auspices of the British Iron and Steel Research Association.[42] The Kjeldahl analyses were performed by several laboratories and the vacuum fusion analyses by several other laboratories. Besides low and high carbon contents, the steels contained aluminum, boron, chromium, cobalt, copper, manganese, molybdenum, nickel, niobium, silicon, titanium, tungsten, vanadium, and zirconium. It appears that the vacuum fusion self-bath technique of melting in the otherwise empty graphite crucible was used. The conclusion was that the methods are equivalent. The exceptions were that the Kjeldahl method was not satisfactory for certain steels containing boron and silicon, and that vacuum fusion equipment employing resistance heating were unsatisfactory for steels containing tin. Goward[43] has pointed out that in those cases where the chemical method failed, evaluation of the vacuum fusion method was invalid. No comparisons were possible on zirconium steels because of segregation effects. The low nitrogen data (30–100 ppm) showed spreads of 10 ppm. This is to be expected since, in general, the Kjeldahl method shows a precision of no better than ±5 ppm. This points up the desirability of developing the accuracy of nitrogen vacuum fusion techniques which presently are capable of precisions of less than 1 ppm.

A cooperative study of analysis of 3% silicon steel has been reported.[44] It was concluded that for analysis of steels having certain heat treatments neither the Kjeldahl nor vacuum fusion results were satisfactory. A caustic fusion technique gave higher and more concordant analyses.

Masson and Pearce[45] reported a comparative study of vacuum fusion, isotope dilution, and Kjeldahl methods applied to high purity iron and low carbon manganese, aluminum killed, and high silicon steels. Most of the vacuum fusion analyses were made at 1650°C with no bath. It was concluded that the vacuum fusion method gave slightly lower results than the isotope dilution method. The recovery of nitrogen by vacuum fusion tended to increase with extraction temperature. Ihida[46] analyzed pure iron–carbon residues after vacuum fusion analysis at 1850°C and found 30–40 ppm nitrogen remaining. The total of these values and the vacuum fusion values equaled the Kjeldahl value.

A very significant study was that of Fassel, Evens, and Hill[26] who apparently have made the first concerted study of the platinum bath method applied to steels. With an extraction temperature of 1850°C this method is satisfactory for oxygen and nitrogen in all steels including those containing high silicon.

b. Refractory and reactive metals

Mallett[47] reported that the recovery of nitrogen from chromium by vacuum fusion is only about 80% complete. This analysis was made with an iron bath at 1650°C. Under the same conditions, Mallett and Griffith[48] obtained satisfactory vacuum fusion results on nitrogen-doped molybdenum samples only when the nitrogen was present in the surface layer. However, when the nitrogen was dispersed internally in the sample as nitride or solid solution the vacuum fusion results were low. Turovtseva and Kunin[49] reported satisfactory results for molybdenum samples doped with 400–600 ppm nitrogen. Analysis was with an iron bath at 1650°C for 15 min. The maximum molybdenum in the bath was 30%. In view of the findings of Mallett and Griffith[48] the method of doping may have led to fortuitous results.

Booth, Bryant, and Parker[50] tried a platinum bath at 1820°C to determine nitrogen in uranium. The results were low and scattered compared to those of the Kjeldahl method.

Sloman et al.[51] consider that nitrogen may be satisfactorily extracted from vanadium in an iron bath at 1560°C with less than 20% vanadium in the bath.

The analysis for nitrogen in niobium has been studied by both vacuum fusion and Kjeldahl method at the DuPont Experimental Station.[43] Vacuum fusion was by the platinum-flux technique at 1900°C with a 15-min extraction. Results at the 100 ppm level showed a standard deviation of 7.5% for vacuum fusion and 3.6% by Kjeldahl. Turovtseva and Kunin[49] analyzed niobium for nitrogen in the range 29–120 ppm. The vacuum fusion was by iron bath at 1650°C for 15 min. The niobium concentration was less than 20% in the bath.

No data are available for comparison of vacuum fusion and chemical analyses for nitrogen in tantalum or tungsten.

It is generally accepted that vacuum fusion analyses for nitrogen in titanium, zirconium, hafnium, and thorium are not reliable.[43] Certainly, in view of Ihida's work,[46] results by the iron bath must be questioned. However, Sloman et al.[51] have reported successful nitrogen analysis by the vacuum fusion iron-bath technique for titanium at 1900°C, zirconium at 1800°C, and thorium at 1750°C. Titanium and zirconium were limited to less than 20% and thorium to less than 30% in the bath.

Goward[43] reports satisfactory nitrogen values for a Zircaloy-2 chip sample. Analytical conditions were: Pt bath–Pt flux, 1900°C, less than 10% Zr in bath. Everett[52] also reports success in determining nitrogen in zirconium by platinum bath vacuum fusion at 1900°C.

c. Recommended methods

In view of the uncertainties in the determination of nitrogen in the metals specific conditions for the analysis of the individual metals will not be given. Iron bath techniques are questionable as are validation studies where nitrogen is not incorporated in the massive metal sample in a manner normal to that produced by usual melting and processing techniques. The platinum bath method as employed by Fassel, Evens, and Hill[26] is recommended for irons and steels with the qualification that the analysis of certain alloys should be validated before use in critical application. The conditions of analysis are: Pt bath, at 1850°C, for 20 min, less than 20% steel in the bath.

The platinum bath should also prove successful for most other metals and alloys including the reactive and refractory metals. Recommended analytical conditions are: 1950°C, Pt bath–Pt flux, less than 10% sample metal in the bath.

d. Thermodynamic considerations

The theoretical thermodynamic considerations for the vacuum fusion analysis for nitrogen in metals have been treated in considerable detail by Goward.[43] He also summarized the work of Sloman, Harvey and Kubaschewski[51] which constitutes the first definitive study of the thermodynamics and mechanisms of the vacuum fusion process. Of several reactions for the release of nitrogen, Sloman considered the thermal dissociation of the nitride and solution of the metal in liquid iron.

$$\text{MN solid} \rightarrow [\text{M}]\text{Fe} + \tfrac{1}{2}\text{N}_2$$

to be most important. Goward disagreed, pointing out that many of the metals dissolve relatively large amounts of nitrogen compared with nitrogen contents of samples normally being analyzed. He considered the dissociation pressure of the nitrogen solution in the vacuum fusion bath to be a more important factor in thermodynamics of the system. In other words, the problem may be reduced to estimating the feasibility of quantitatively extracting nitrogen from the vacuum fusion melt. Based on these considerations Goward described as an example the mechanism of the vacuum fusion process for nitrogen in niobium as follows:

$$[\text{N}]\text{Nbs} + (\text{Fe–C})\text{l} \rightarrow [\text{N}](\text{Fe–Nb–C})\text{l}$$

$$[\text{N}](\text{Fe–Nb–C})\text{l} \rightarrow \tfrac{1}{2}\text{N}_2$$

where [N]Nbs is nitrogen in solid solution in Nb, (Fe–C)l is the liquid bath, [N](Fe-Nb-C)(l) is the liquid Fe–Nb–C melt, and N_2 is nitrogen gas released from the bath.

A reasonable evaluation of the equation for the release of nitrogen may be made from the thermodynamic data of Chipman[53] and Pehlke and Elliott.[54] For such a multicomponent melt the system obeys Sieverts' law in the form $Kf_n = P_{N_2}{}^{1/2}/\%N$ for dilute solutions of nitrogen. Here f_n is the activity coefficient of nitrogen in the multicomponent melt. A figure presenting (for many metals) the logarithm of activity coefficient of nitrogen versus per cent metal in iron at 1600°C appears in Goward's paper.[43] A summary of some of the equilibrium nitrogen concentrations calculated from these data is given in Table 1. Assuming that an analysis is made with a 1 g metal sample in a 9 g iron bath, the unrecovered nitrogen based on the sample will be 10 times the value shown in the table, e.g., 23 ppm for titanium. Also if the ultimate vacuum (nitrogen pressure) of the system is assumed to be 100 times as high (10^{-4} mm Hg), then the values in Table 1 would according to Sieverts' law be increased by a factor of 10. The difficulty of nitrogen removal increases with increasing affinity of the solute metal for iron. The equilibrium data appear very favorable for most metals. One disturbing factor is that the pressure in the immediate vicinity of the melt surface is not known. Also the same interference from gettering observed for oxygen applies to nitrogen. A number of the metals show appreciable vaporization at 1600°C, the temperature of iron bath analysis. The result could be loss of nitrogen by gettering.

Table 1. Equilibrium nitrogen contents in carbon saturated (5.3% C) molten Fe–M solutions

System	Metal in bath (wt %)	Equilibrium N_2 content in bath at 10^{-6}mmHg pressure (ppm)
Fe	—	0.0017
Fe–Ni	10	0.0013
Fe–Al	10	0.0021
Fe–Cr	10	0.0044
Fe–Mo	10	0.0022
Fe–W	10	0.0018
Fe–U	10	0.015
Fe–Nb	10	0.0073
Fe–Ta	10	0.0037
Fe–Ti	2	0.032
Fe–Ti	5	2.3

The many thermodynamic data relating to iron alloys are available because of the very practical application of the knowledge to the steelmaking industry. Unfortunately, few such data are available for the platinum alloy systems and so thermodynamic evaluation of the promising platinum bath vacuum fusion technique is not possible at this time.

III. HOT VACUUM EXTRACTION

The hot vacuum extraction method has been used for the determination of hydrogen,[55] oxygen,[56] and nitrogen.[57] The method consists of heating the solid sample in vacuum and collecting and analyzing the gas that diffuses from it. Because of the prevalence of vacuum fusion apparatus this equipment is usually used for the extraction analysis.

A. Hydrogen

The extraction temperature for hydrogen ranges from about 600°C for steels to 1400°C for the reactive metals. The extraction crucibles usually are fused silica (or Vycor) and graphite (or molybdenum), respectively. The method has several distinct advantages. The low-temperature analyses require only a simple resistance furnace for heating. There is no interaction with the crucible so that it may be reused a number of times. In the case of hydrogen, the evolved gas may be pure and once this fact is verified the analysis is reduced to simple measurement of the total gas. Because of the low extraction temperatures for hydrogen the blanks are low or even negligible and gettering is largely eliminated.

If other gases are present, the analysis may be carried out by any of the techniques employed in vacuum fusion analysis. The simplest procedure is to separate the hydrogen by diffusion through a heated palladium alloy tube.

Since hot vacuum extraction is based on diffusion, the sample should be thin enough to provide short diffusion paths without producing an excessively large surface which could give measurable surface oxygen for some metals and surface hydrogen in the case of metals such as aluminum.

Even if the analysis can be reduced to mere measurement of the evolved gas, it is well to pump the gas from the furnace chamber to make the measurement. A partial pressure of gas over the sample prevents complete extraction. An exception to this rule is the equilibration method to be described later.

The samples are cut and abraded clean in the usual manner for vacuum fusion. Care is taken not to heat the sample during preparation. Some metals, particularly the steels, may lose hydrogen if heating occurs. Acid pickling of sample material must also be avoided since it may introduce hydrogen.

1. IRON AND STEELS

A comparison has been made of the determination of hydrogen in steel by vacuum fusion, by fusion in vacuum with tin, and by hot extraction.[58] It was concluded that it was preferable to remove the hydrogen by hot extraction from the solid sample rather than by the methods employing a molten sample. Martin et al.[58] studied analysis of the extracted gas by thermal conductivity, gas chromatography, pressure measurement before and after catalytic oxidation of hydrogen to water and removal of the water vapor, and pressure measurement before and after diffusion of hydrogen through a palladium membrane. The results showed that the palladium-membrane technique is best for routine use. The method[59] based on this principle follows.

Sample molten metal by the evacuated pin-tube or copper mold technique. Quench samples and immediately store in liquid nitrogen. In sampling large solid sections of ferritic steel keep the material cooled with Dry Ice during machining and filing and place samples under liquid nitrogen until analyzed.

The hydrogen in austenite is more stable and may not require refrigeration. Cut 2–8-g solid samples from the specimen. Avoid sheared edges or areas that have been heavily worked or heated. File the sample surfaces clean just prior to inserting in the furnace.

Use an apparatus containing the basic features of that presented in the ASTM Task Force Report, *Determination of Hydrogen in Steel*[59] and reproduced in Figure 2 through the courtesy of the Allied Research Laboratory, United States Steel, and ASTM. The apparatus includes BV, a ball valve for admitting samples, C_1–C_3, mercury cutoffs, DP_1, DP_2, mercury diffusion pumps, F, furnace tube of clear fused silica, G, McLeod gauge, Pd_1, Pd_2, palladium membranes, S_1–S_{10}, stopcocks, T_1, trap, V_1, V_2, vacuum pumps, and W, sample storage well. If one chooses catalytic oxidation and absorption to remove hydrogen, a hot CuO catalyst tube, a freezing trap, and an additional cutoff may be installed between cutoff C_3 and diffusion pump DP_1. and the palladium thimble, Pd_2, may be deleted from the connecting tubing,

Degas the furnace at 1100°C; determine the blank rate at 900°C from collection of gas passing through the palladium membrane for 30 min. The palladium thimble is heated at 400°C. The hydrogen pressure in the calibrated volume is measured with the McLeod gauge. After obtaining an acceptable blank rate, pass a weighed sample into the system through the ball valve. Hold the sample in the cool section of the system until the ultimate vacuum is reached, then isolate the calibrated volume and drop the sample into the 900°C furnace. Record the gauge pressure at 5-min intervals until the blank rate is reestablished. This indicates that the analysis of the first sample is concluded. Evacuate the entire apparatus and repeat the procedure for successive samples. Apply the method of calculation given for the vacuum fusion method.

2. TITANIUM

Hydrogen is quantitatively removed from a 1-g sample in 3–10 min at 1400°C.[60] The extraction is usually performed in a graphite crucible in a vacuum fusion apparatus. The extracted gas is entirely hydrogen, which is estimated by measuring its pressure in a known volume. The method is capable of high precision and the blank becomes negligible if the crucible assembly is degassed at 2200°–2400°C for about 2 hr. About five analyses per hour may be made following the degassing period. If some lower precision is tolerable, as in the case of high hydrogen contents, the sample size may be reduced to about 0.25 g. Then 20 or more samples may be run in a single loading. A similar method has been described for the analysis of zirconium.[61]

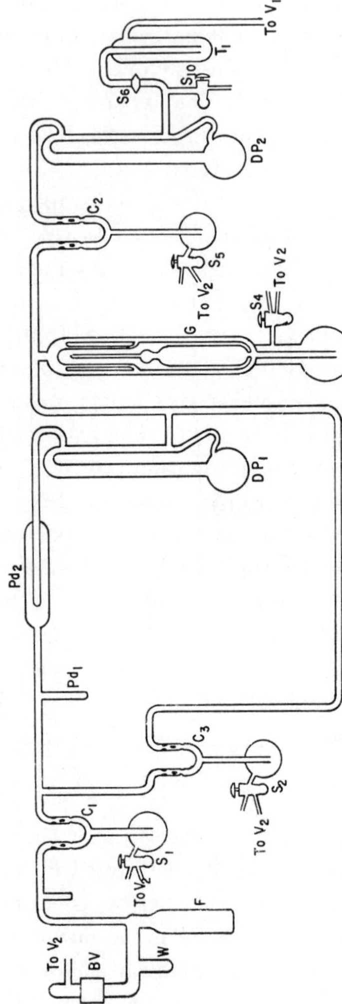

Fig. 2. Apparatus for determining hydrogen in steel by diffusion through palladium. (Courtesy ASTM, Committee E-3 and Applied Research Laboratory, U.S. Steel Corporation.)

3. REACTIVE AND REFRACTORY METALS

The reactive metals hafnium, vanadium, and zirconium; and the refractory metals niobium, tantalum, molybdenum, and tungsten, may all be analyzed for hydrogen by the hot vacuum extraction method given for titanium.

4. URANIUM

Waldron[62] has made a critical review of methods for the determination of hydrogen in uranium. His technique consisted of extraction of the gas at 800°C for 20 min from samples weighing 5–10 g. Analysis by palladium tube permeation and mass spectrographic techniques showed that only hydrogen was collected. Consequently the analysis requires only measurement of the evolved gas.

5. EQUILIBRIUM PRESSURE METHOD

A modification of the hot extraction method for determination of hydrogen in titanium is based on measurement of the equilibrium pressure of hydrogen over the metal in a closed system under predetermined conditions.[63] The typical apparatus contains the samples in individual silica tubes connected through a glass manifold to a multirange McLeod gauge and a vacuum system. The samples weigh 0.4–0.6 g and may vary in form, e.g., sheet, rod, sponge, or turnings. Evacuate the entire system to about 10^{-4} torr. Observe the pressure and isolate the pumps. If the pressure holds at less than 5×10^{-4} torr for 10 min pump out and close stopcocks to all tubes except tube *1*. Isolate the pumps and heat the first sample to 1000°C with a resistance heater in 6–7 min and hold to the end of 10 min. Observe the pressure then pump out and repeat the procedure for the next sample. Four samples may be completed in 1 hr.

To establish the hydrogen content of the sample, a predetermined curve of equilibrium pressure versus solubility at 1000°C must be consulted to obtain the concentration of hydrogen remaining in the sample. The hydrogen in the gas phase is calculated from the known pressure–volume relationship. Summation of these values yields the analysis of the sample. This method is satisfactory where many samples of known alloys are to be analyzed as in the control laboratory of a titanium production plant. However, this method is not applicable to samples of unknown alloy composition.

B. Oxygen

1. NIOBIUM

Oxygen may be determined by hot vacuum extraction in a conventional vacuum fusion apparatus[56] Use a 0.5-g sample. Degas the graphite crucible at 2300–2400°C; lower the temperature to 2000°C and determine if the blank rate is satisfactory. If so, lower the temperature to 1200°C and start collecting a blank. Increase the temperature to 2000°C over a 5-min period and complete the balance of a 30-min blank. Drop the sample at 1200°C and follow the same heating schedule as for the blank. Usually 30 min is long enough to remove all the gas. The oxygen is extracted as CO. The precision of the analysis is 0.001% for a 0.5-g sample. The precision would increase to 0.0001% for a 5-g sample. However, the sample should be thin so as to keep diffusion paths short.

2. MOLYBDENUM, TUNGSTEN, AND TANTALUM

Fagel, Witbeck, and Smith[64] report application of the hot vacuum extraction method to molybdenum, tungsten, and tantalum. In this method, use 0.2–2.0-g samples. Drop and extract the gas from each sample at 2000°C for a normal extraction time of 20–30 min for samples as thick as 5mm.

Friedrich[65] confirmed the applicability of the above method to molybdenum by use of weighed MoO_3 additions encapsulated in molybdenum. The oxygen content could escape only by diffusion through the molybdenum metal.

Table 2. calculated equilibrium pressures for 1 ppm nitrogen for various metals

Element	Temp., °C	Equilibrium N_2 pressure (mm)	$t_{0.5}$[a] (min)
δFe	1500	$4.8 + 10^{-2}$	1.6
liq. Fe	1600[b]	$4.8 + 10^{-3}$	1.4
Cr	1300	$9.9 + 10^{-6}$	—
liq. Cr	1600	$4.6 + 10^{-7}$	
Mo	2000	$6.9 + 10^{-2}$	7
W	2000	$6.3 + 10^{-2}$	31
Nb	2000	$1.2 + 10^{-9}$	1.9
	2420	$3.8 + 10^{-9}$	
Ta	2000	$2.7 + 10^{-9}$	2.8
	2960	$1.4 + 10^{-6}$	—

[a] 0.4 cm. diameter cylinder.
[b] Molten Fe, 1.6 cm. crucible diameter.

For comparison, the equilibrium pressure of 1 ppm of hydrogen in Nb is 1.5×10^{-2} mm at 1000°C and 4×10^{-1} mm at 2000°C

C. Nitrogen

Mallett and Griffith[48] reported the successful determination of nitrogen in molybdenum by hot vacuum extraction in a vacuum fusion apparatus. In this method 10 g of sample not exceeding 4 mm in thickness is used. The gas is extracted at 1800°C in a graphite crucible for 30–50 min. The precision of this analysis is ±1 ppm nitrogen at levels up to 40 ppm. Results by this method show agreement with Kjeldahl analyses for the range 20–900 ppm.

Fagel, Smith, and Witbeck[64] reported nitrogen obtained during hot extraction analysis for oxygen at 2000°C. Analyses of molybdenum, tungsten, and a Mo–0.5% Ti alloy appeared satisfactory. Nitrogen recovery was poor from tantalum, e.g., 10 ppm by extraction compared to 84 ppm by vacuum fusion. This behavior is now predictable from the study of basic principles of vacuum extraction made by Goward.[43] He presents the temperature–pressure–solid solubility data for a number of metals, the dissociation pressures of various nitrides and a table of equilibrium pressures for 1 ppm nitrogen for various metals. The latter is reproduced in Table 2 below through the courtesy of Goward and *Analytycal Chemistry.*

IV. THE INERT GAS FUSION METHOD

The inert gas fusion method is in principle related to the vacuum fusion method just discussed, inasmuch as it is based on the following reactions:

$$MO_x + C \rightarrow M \text{ (or MC)} + CO \tag{1}$$

$$[O] \text{ in } M + C \rightarrow M \text{ (or MC)} + CO \tag{1a}$$

The development of the inert gas fusion method is generally traced back to the work by Singer,[66] who in 1940 utilized a graphite crucible in which the sample was fused along with a tin flux. The carbon monoxide was swept by a carrier gas of nitrogen over heated copper oxide, after which the carbon dioxide was determined gravimetrically. Unfortunately, the method was limited in sensitivity. A breakthrough was achieved in 1955, when Smiley[67] devised a method that was sensitive to less than 1 μg of oxygen. In this method, the sample was dropped into a bath of molten platinum contained in a graphite crucible heated by an induction furnace. The carbon monoxide formed from the reduction of the oxygen in the sample was swept by a stream of argon at atmospheric pressure through a modified Schutze's reagent where it was oxidized to carbon dioxide. The carbon dioxide was condensed in a capillary trap and measured with a capillary manometer. In 1958, Peterson, Melnick, and Steers[68] utilized the method developed by Smiley to determine oxygen in steel. Shanahan and Cooke[69] used nitrogen as a carrier gas and measured the carbon dioxide gravimetrically. In 1959, Abresch and Lemm[70] determined oxygen in steel using argon as a carrier gas and measured the carbon dioxide by coulometry. In 1958, Laboratory Equipment Corp.[71] introduced a commercial version of an inert gas fusion oxygen analyzer. Argon purified by a standard train of sulfuric acid, Ascarite, and Anhydrone, followed by hot zirconium sponge, is passed over a graphite crucible insulated with carbon black and heated in a 3 kW induction furnace. Samples are introduced into the crucible through a loading head and the oxides are reduced to carbon monoxide. The CO is swept through I_2O_5, which oxidizes it to CO_2. The liberated iodine is absorbed by sodium thiosulfate. The CO_2 is measured conductometrically in a solution of barium hydroxide. A sample can be analyzed in about 5 min. The main disadvantage of the conductometric procedure is its comparatively poor sensitivity. Since the blank amounts to 15–30 μg of oxygen, the conductometric procedure is best suited for samples containing more than 50 ppm of oxygen. Exceptions are iron and copper where 10-g and even 20-g samples can be used.

Because of the need for determining 10 ppm or smaller amounts of oxygen in various materials, particularly the refractory metals, more sensitive equipment than the conductometric apparatus was investigated. Laboratory Equipment Corporation first intraduced a "Nitrox" analyzer.[72] In this device, the carbon monoxide formed by the reaction of the oxygen with graphite is carried by helium carrier gas into a trap containing a molecular sieve which is cooled to liquid nitrogen temperatures. Subsequently, the trap is heated and the evolved carbon monoxide is measured by gas chromatography. An improved version of this apparatus was introduced in 1965.[73] The method involves oxidation of CO to CO_2. The CO_2 is swept into a trap containing molecular sieve at room temperature. The trap is then heated and the CO_2 is measured by gas–solid chromatography. An improved version of this equipment involves solid state circuitry and provisions for the calibration for direct reading.[74] The Russian literature[75] mentions the introduction of an apparatus involving the inert gas fusion principle. In this method, the sample is introduced into small graphite capsules. The capsule is tightly closed and is subsequently heated for a few seconds at temperatures of about 3400°C by impulse heating. The CO is oxidized to CO_2 which is then measured by gas chromatography.

A. The Apparatus

The inert gas fusion apparatus consists of (*1*) a furnace in which the reaction takes place and (*2*) an analysis section in which the CO_2 is measured.

1. FURNACE AND REACTION TUBE

A typical furnace assembly is shown in Figure 3. It is somewhat different from that appearing in the ASTM book[76] which was used in connection with the conductometric measuring device. An induction furnace is used that utilizes a 220 V single-phase source with an imput of 3 kW and an output frequency of about 13 Mc/sec. When argon is used as the carrier gas (conductometric method), the power output is capable of heating the graphite crucible to 2700°C. With helium as the carrier gas (gas chromatographic method), the maximum temperature obtainable is about 2350°C. The induction coil is about 4 in. in height and 3 in. in diam containing 12 turns of ¼ in. copper tubing which is water-cooled. The graphite crucible is imbedded in carbon black insulation contained in a quartz thimble supported by a quartz pedestal. The thimble is contained in a reaction tube sealed at either end by silicone O-rings. The crucible is introduced by opening a bolt-action raising mechanism below the reaction tube. The temperature of the crucible is controlled by a phase shift network incorporated into the furnace. The upper end of the reaction tube is inserted into a sample-loading device that permits introduction of samples up to 0.5 in. in diameter without breaking the flow of carrier gas. A loading stopcock with a large bore capable of admitting samples was used in the model described in the ASTM book. The carrier gas (argon in the case of the conductometric finish, helium when CO_2 is measured gas chromatographically) is admitted to the apparatus by means of a two-stage regulator. It is passed through a purifying train containing copper oxide heated to 320°C, Ascarite, Anhydrone, and concentrated sulfuric acid. Removal of oxygen is accomplished by passing the carrier gas over titanium or zirconium sponge heated to 800°C.

2. OXIDATION OF CO TO CO_2

Carbon monoxide formed by the reaction of oxides of the sample with graphite is carried by the gas through rare earth copper oxide heated to 450°C. The resultant carbon dioxide is introduced into the analyzer and is retained in a stainless steel trap containing molecular sieve. In the earlier conductometric procedure, the CO was oxidized by iodine pentoxide heated at 140°C. Iodine released by the reaction $I_2O_5 + 5CO = 5CO_2 + I_2$ was absorbed by sodium thiosulfate, before the carrier gas containing the CO_2 was introduced into the conductometric cell (see Fig. 4).

3. ANALYSIS SECTION

For the gas chromatographic measurment of the CO_2, the analyzer is equipped with a thermal conductivity cell, oven temperature controls, com-

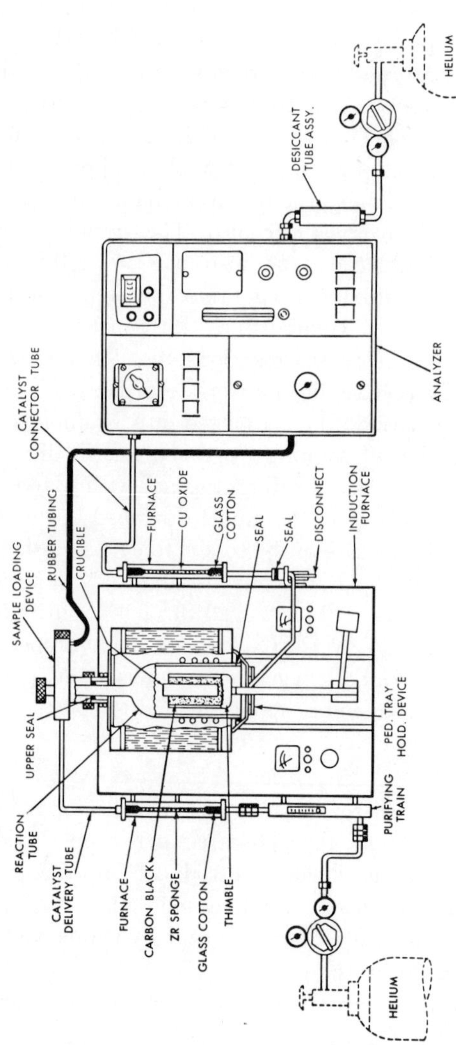

Fig. 3. Inert gas fusion.

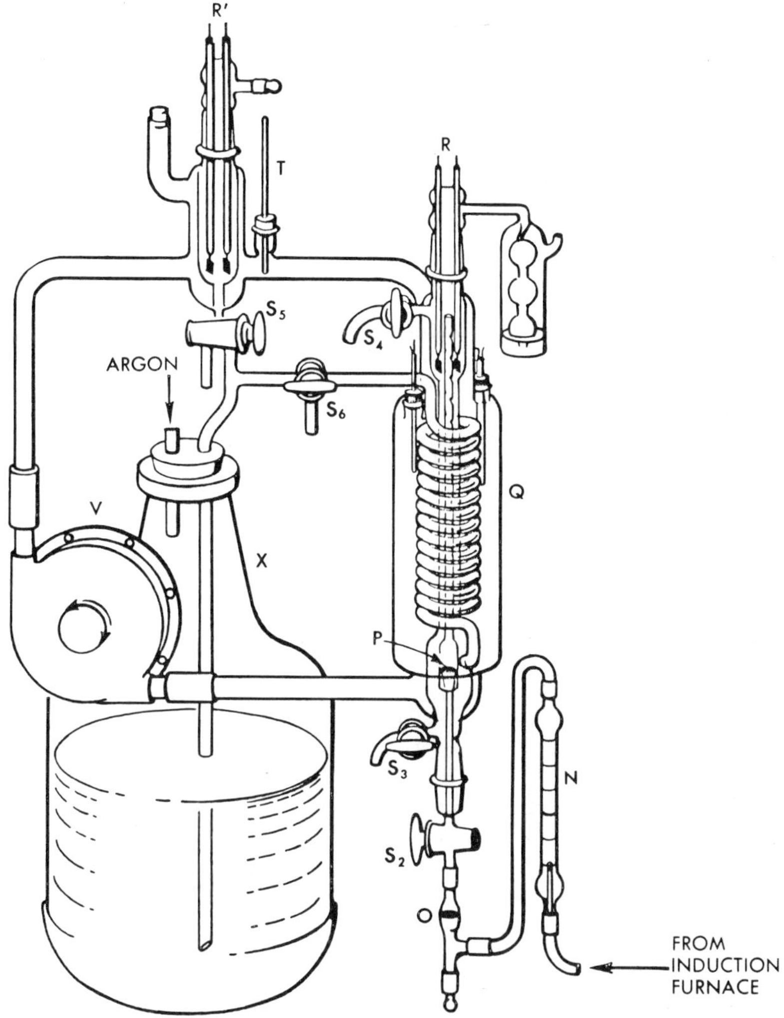

Fig. 4. Conductometric unit.

puter section, digital voltmeter, timer, molecular sieve trap, trap heater, chromatographic column, solenoid valves, and auxiliary power supplies. It essentially is a special-purpose gas chromatograph which utilizes its own helium supply. The computer section is programmed so that the digital voltmeter reads milligrams of oxygen. The instrument has a range of 0–1000 μg, and the least count is 0.0002 μg.

After all the carbon dioxide from the sample is collected in the trapped gas, the timers activate the solenoids to transfer the trap into the analyzer helium stream, and then the trap is electrically heated. The carbon dioxide leaves the molecular sieve trap, and enters the silica gel chromatographic column and finally the thermal conductivity cell, the output of which is a time–voltage curve. This output is processed by the computer to correct for blank and gain, and to integrate the area under the curve. This integral is read on the digital voltmeter. A flow schematic of the helium streams through the furnace and analyzer is shown in Figure 5.

For the conductometric measurement of the CO_2, the analyzer is equipped with a measuring cell containing about 85 ml and a reference cell containing approximately 40 ml of CO_2-free barium hydroxide (0.83 g/liter) or sodium hydroxide (0.22 g/liter). The standard cells are surrounded by water jackets and bath maintained at about 40°C. The cells contain platinum electrodes plated with platinum black. The CO_2 is absorbed by the barium hydroxide

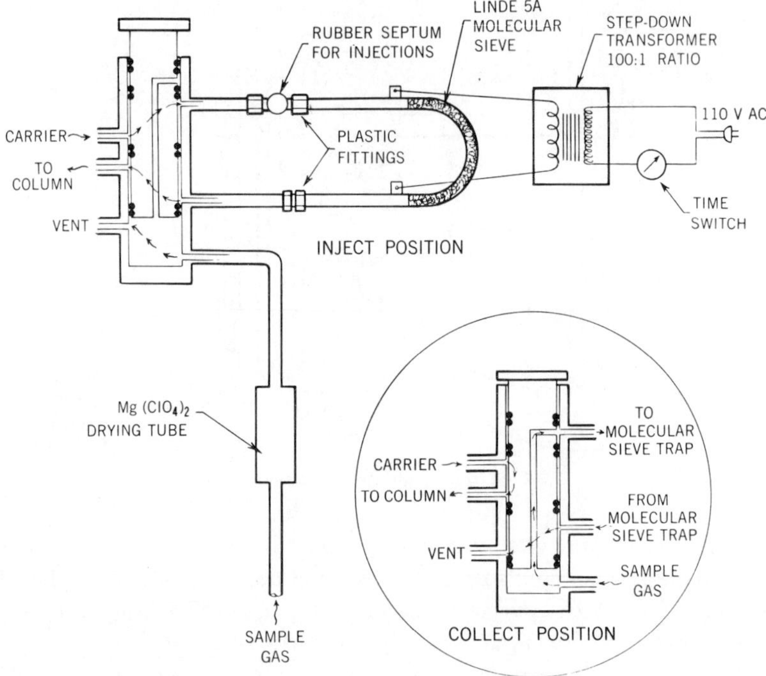

Fig. 5. CO_2 trapping system.

or sodium hydroxide solutions changing the resistance in the measuring cell. This is detected by a Wheatstone bridge arrangement with oscilloscope null detector.

B. Method of Analysis

1. SAMPLE PREPARATION

The techniques of sample preparation are the same as those used in vacuum fusion work. In addition, the samples should be cut to pass through the 0.5-in. diam hole in the plunger of the sample loading head. Larger samples can, of course, be accommodated by using alternative sample-loading devices.[77] All surface oxides should be removed by abrading with a flat file and the sample should not be subjected to excessive heat. Most analysts prefer to degrease abraded samples with a solvent. The necessity or even wisdom of the degreasing is questionable. Some analysts prefer to remove surface oxides by "pickling" with an appropriate acid mixture, particularly when the sample is of irregular shape. In Ledoux & Company's laboratory significantly lower oxygen results were obtained when beryllium samples were "pickled" with dilute sulfuric acid and "oxygen-free copper" (OFHC) samples with dilute hydrochloric acid.

Irrespective of whether abrading or a chemical surface treatment is employed, it is essential that the sample be introduced into the loading device without delay. Tests carried out in Ledoux & Company's laboratory[78] indicate that both platinum and copper sheets picked up a significant amount of apparent oxygen (probably moisture) in a very short time. It may therefore be advisable to expose abraded or pickled samples and flux to brief infrared heating, before the introduction into the furnace.

2. FURNACE PREPARATION

The following procedure is written for the new type furnace and the gas chromatographic equipment now produced by the Laboratory Equipment Corporation. A detailed description of a gas chromatographic recording technique is also provided. Details of the conductometric procedure are described in the 1965 ASTM book.[76]

The graphite crucible used for analysis is packed in a bed of carbon black in a quartz thimble. The carbon black functions as an insulator and is screened through a 20-mesh sieve prior to use. The crucible is floated in the carbon black firmly enough to prevent separation during heating, yet not so firm that heat losses through conduction will occur. The thimble is placed on the raising pan on the front of the induction furnace and raised into the reaction tube. Dallmann and Fassel[79] recently recommended a thimble machined from pyrolytic boron carbide. The boron carbide does not couple with the induction current and the relatively high blank caused by the reaction of hot carbon black with the quartz thimble ($SiO_2 + 2C = Si + 2CO$ is avoided). Prior to degassing, the furnace helium regulator is set to 10 lb and the flowmeter on the purifying train is set to a flow of 2 liters/min. The connection from the furnace to the analyzer is left open or broken, so that gases evolved during degassing will not needlessly pass through the catalyst. To "condition" the crucible, it is a common practice to add 1 g or more of platinum or copper during the degassing of the crucible. The furnace temperature is gradually increased by rotating the phase shift knob until 800 mA appears on the plate milliammeter of the furnace. The crucible is degassed for 20–30 min at 800 mA, after which the temperature is reduced to that required for sample analysis. The furnace is then connected to the analyzer. Temperatures may be read with an optical pyrometer and related to the plate current as read on the plate milliammeter.

3. ANALYZER OPERATION

The analyzer is operated with a helium gauge pressure of 20 lb. It is programmed to sequentially collect carbon dioxide in the molecular sieve trap, heat the trap, and elute the carbon dioxide for detection by the thermal conductivity cell. It then switches in a circuit to integrate the carbon dioxide time–voltage curve generated by the thermal conductivity cell, unclamps and reclamps the digital voltmeter so that the integral can be read in digital form, and activates a blower which cools the trap for the next analysis.

Contained on the instrument panel are controls for adjusting the timing sequence and for setting the blank and slope values for direct reading. Also contained on the instrument panel are switches for defeating the timer motor, clamping mechanism, and blank setting. A strip chart recorder can be attached to the instrument, as described below, which is useful not only in monitoring the output on the thermal conductivity cell, but also as an alternative to the digital voltmeter.

4. DETERMINATION OF THE BLANK

With the furnace at operating temperature, the timer is activated, placing the collection trap in series with the furnace. After a time interval which is set for 1 min or more the collection trap is transferred to the "analyze" mode in series with the thermal conductivity cell. Almost immediately, nitrogen will elute from the collection trap through the column and over the sensing leg of the thermal conductivity cell, as evidenced by peaks on the strip chart recorder, if one is employed. After these peaks appear, heat is automatically applied to the collection trap and carbon dioxide will elute from the trap into the column and over the sensing leg of the cell.

Immediately prior to elution of the carbon dioxide, the integrator circuit turns on, and immediately after the time–voltage peak is generated, the integrator turns off again. The digital voltmeter is automatically unclamped, the integral detected and the digital voltmeter reclamped. Upon reclamping of the digital voltmeter, the trap fan will turn on, cooling the trap. The reading that appears on the digital voltmeter is the blank obtained and can be negated by unclamping the digital voltmeter, setting the voltmeter to read zero, and reclamping the voltmeter.

5. INSTRUMENT CALIBRATION

Samples are analyzed by the same technique used for the blank determination. A standard sample containing close to 1 mg of oxygen is introduced into the plunger of the sample-loading head. The plunger is pushed in, but the sample is not dropped into the crucible until helium has flushed out the cavity. After 15–20 sec, the plunger is rotated, causing the sample to drop into the crucible. At the same time as the sample is dropped, the timer is activated and the instrument goes through the various steps described above under Determination of Blank. As soon as the reading appears on the digital voltmeter, the latter is unclamped manually and set to read the correct value in milligrams with the slope control. A sample is then analyzed in the 0.02-mg range, and the digital voltmeter is adjusted to read directly in milligrams with the blank control. The analyses of additional 1 mg and 0.02 mg samples with appropriate adjustments of the slope and blank controls will generally yield an adequate calibration.

Earlier models of the analyzer were equipped with an amplifier and voltage–time integrating clock. Operation of this instrument is essentially the same as that of the electronically integrating model now available with certain exceptions. Because of a limited range of response of the integrating clock it is necessary to attenuate the output signal of the thermal conductivity detector when more than about 0.01 mg of oxygen is to be determined. Attenuation factors of 2, 4, 8, 16, 32, and 64 are available. Each attenuation to be used must be calibrated by analyzing a standard sample. If a sample is run with insufficient attenuation, the integrating clock must be turned off to prevent burning it out. The sample, of course, must be rerun. Many users of this particular equipment have avoided the inconveniences of the integrating clock by resorting to potentiometric recording of the conductivity cell output. When using a 5 mV recorder, the attenuation factors provided by the analyzer have been shown to be accurate. Thus, standardization with 0.2 and 0.5 mg of oxygen provides a calibration factor valid for use at any attenuation. In addition to having a truly linear response, potentiometric recording permits one to "save" samples which are being run with insufficient attenuation. If the carbon dioxide peak is seen to be going off-scale, the attenuator can be switched to bring it back. As long as the peak itself is recorded, a peak height can be measured. A considerable gain in precision can be achieved by using an integral measurement of the peak area rather than simply using the peak height. The area can be estimated by use of an integrating recorder or by multiplying the peak height by the peak width at one-half the peak height.

Operation with a recorder is essentially the same as with the integrating circuit. The differences are:

(*a*) The blank is measured and expressed in area units,

(*b*) After performing the calibration runs, a factor is calculated.

$$F = \frac{\text{micrograms } O_2}{\text{peak area} \times (\text{attenuator setting} - \text{blank})}$$

(*c*) Sample results are calculated:

$$\text{ppm O} = \frac{[\text{peak area} \times (\text{attenuator setting} - \text{blank})]F}{\text{sample weight (g)}}$$

An advantage of using a recorder is that it provides a continuous monitor on the functioning of the analyzer. If the molecular sieve trap becomes poisoned or fails to heat properly, the peaks become broader. There is no critical timing requirement for starting and stopping of the integrator, since a complete time–voltage curve is recorded for evaluation. The calibration procedures described above are frequently carried out using tin capsules containing a known amount of potassium acid phthalate. These standards may be prepared as described in the ASTM book[76] or can be directly bought from Laboratory Equipment Corporation.

For high precision work, many analysts prefer to use standard samples of the same composition as the sample to be analyzed. U_3O_8 and PbO have also been recommended. Ag_2O, although widely used at one time, is less desirable because of its tendency to pick up H_2O which cannot be removed by drying.

The analyzer can also be calibrated by injecting a known amount of CO_2 at a convenient place between the furnace and the analyzer. This particular technique is especially recommended for "trouble shooting" purposes to determine the location of a leak or any other malfunction of the system. The furnace naturally must be cold if it is intended to inject the CO_2 into the furnace section because of the reaction of CO_2 with heated graphite.

C. Sample Analysis

Samples containing unknown amounts of oxygen are analyzed by the technique employed in the instrument calibration, except that, if the electronic integrating model is employed, the slope and blank controls are not adjusted. The results will be expressed in milligrams of oxygen. However, it is important that the instrument be calibrated using the same amount of flux or bath metal as would be introduced during sample analysis. For example, if 1 g of platinum is required for each sample of titanium, then the instrument must be calibrated with the introduction of 1-g of platinum with each standard. As in vacuum fusion work, various metals require variations from the basic techniques.

1. RECOMMENDED TECHNIQUES

Since the measurement of the CO_2 formed as a result of the inert gas fusion and the oxidation by copper oxide by gas chromatographic methods has been introduced only recently, there exists a lack of published methods. Therefore, the methods published here are largely based on those used in Ledoux & Company's laboratory. The conductometric measurement of CO_2 has been previously recommended for the determination of oxygen in zirconium and zircaloy[80] (platinum-flux technique), in beryllium[77] (nickel-copper flux technique), and in yttrium[81] (platinum-flux technique). Beck and Clark[82] employed graphite capsules to retain volatile decomposition products. Fluorides can be removed by the employment of traps.[83] Several ASTM task forces of Division I have also investigated the inert gas fusion-conductommetric approach to determine oxygen in niobium, tantalum, titanium, hafnium, copper, iron, and beryllium. Generally speaking, results have been satisfactory for oxygen contents greater than 50 ppm. Bath or flux requirements, however, limit the amount of sample which can be handled to the point where the blank of 15–25 μg makes results below 50 ppm uncertain. This is different in the case of such metals as iron and copper which require no flux and therefore permit the use of a sample large enough to disregard small variations in the blank.

2. INDIVIDUAL ELEMENT ANALYSIS

The following procedures are largely based on the vast experience of Ledoux & Company covering the determination of oxygen in a great variety of metallurgical products. Two basic procedures are employed, one involving a copper, and the other a platinum bath. Both metals can be obtained in sheet form containing only a few ppm of oxygen. The surface oxygen can be removed by pickling with HCl, washing with water and acetone and brief drying under an infrared lamp. Platinum can be ignited prior to analysis, to reduce the oxygen content to about 2 ppm.

Iron represents a less desirable bath because of the ready formation of viscous iron carbide. This is a much more serious defect in the inert gas fusion procedure than the vacuum fusion method because of significantly higher temperatures employed in the inert gas fusion procedure.

a. Aluminum

After degassing the crucible at about 2300°C, the temperature is lowered to about 1500°C. Twenty grams of OFHC copper, previously treated with hydrochloric acid, rinsed with water and acetone, and dried under an infrared lamp, is added. The crucible is heated to about 1800°C to remove any oxygen. A blank reading is taken, then the sample is introduced. Aluminum powder should be wrapped in previously cleaned OFHC copper. An extraction time of 3 min is sufficient. More copper should be added, if the Cu:Al ratio drops below 5:1. No gettering of the CO has been observed.

b. Beryllium

Previous work with the inert gas fusion method established nickel as the most desirable flux.[77] Additional work carried out by Ledoux & Company indicates that the greater sensitivity of the gas-chromatic approach allows the determination of 10 ppm of oxygen using a 0.1 g sample and 1.5 g of nickel as a flux. The furnace should be operated at about 2500°C. The nickel should be cleaned with HCl prior to its use.

c. Bismuth

The method is similar to that described earlier for the vacuum fusion technique. A combustion tube heated by resistance heating and a graphite boat containing the sample should be used. A little graphite powder should be added to the sample; heating to about 800°C is adequate.

d. Boron

Use a copper bath, as described for aluminum. A temperature of about 1900°C is adequate for complete recovery of the oxygen. Boron powder should be wrapped in copper foil.

e. Carbides, borides, and nitrides

It is recommended to use a platinum bath with a platinum to sample ratio of 20:1. The operating temperature should be at least 2200°C.

f. Chromium

Use a copper bath, as described for aluminum. A temperature of about 1900°C is adequate.

g. Cobalt

Melt samples in a copper bath at about 1900°C.

h. Copper

Melt samples at about 1500°C directly in a graphite crucible previously conditioned with copper.

i. Ferro-alloys

The copper bath technique is recommended. The operating temperature should be at least 2000°C.

j. Germanium

Melt samples in a copper bath at about 1900°C.

k. Gold

The sample should be dropped into a graphite crucible previously conditioned with platinum.

l. Hafnium

Use the platinum-flux technique employing a platinum/sample ratio of 10:1. Low-oxygen platinum should be used. The reaction of hafnium oxide with the graphite is sluggish and the furnace should be operated at a temperature of 2300–2350°C.

m. Iron and Steel

Heat a portion of the sample (at least 1 g) at about 1700°C for 4 min. The first portion acts as a bath for the next sample. Alternatively, a copper bath was found to provide a good medium for iron, since it stays fluid for a longer period of time than a pure iron bath. Recently, some laboratories have employed platinum. In Ledoux & Company's laboratory, the recovery of oxygen was not improved by the presence of platinum.

n. Lanthanum and Other Rare Earths

Both the platinum bath and platinum-flux techniques provide reliable oxygen results.

o. Lead

See Bismuth. If a tin bath is employed, the induction furnace can be used, as long as the temperature can be controlled to 1200 ±50°C.

p. Manganese

An iron bath and a furnace temperature of about 1600°C provides complete recovery of oxygen. If the iron/manganese ratio is kept above 5:1, no gettering is experienced.

q. Molybdenum

A platinum bath and a furnace temperature of about 2100°C are recommended. The concentration of the molybdenum in the bath should not exceed 10%. A platinum-flux technique works equally well. This requires, however, the use of low-oxygen platinum.

r. Nickel

A copper bath and a furnace temperature of about 1800°C are recommended.

s. Niobium

The platinum-flux technique is generally used. Niobium requires a platinum/sample ratio of only 3:1. The operating temperature should be about 2100°C. Low-oxygen platinum should be used.

t. Palladium

Use the method described for platinum.

u. Platinum

The sample should be dropped into a graphite crucible previously outgassed in the presence of some platinum. Operating temperature should be about 2000°C.

v. Rhenium

The platinum bath technique for molybdenum should be used.

w. Silicon

The copper bath technique described for aluminum is equally applicable to silicon.

x. Silver

A copper bath is recommended. Operating temperature should be about 1800°C.

y. Tellurium

Use the low-temperature method described for bismuth.

z. Thorium

Thorium oxide is very refractory. A combined platinum bath–platinum-flux technique is recommended.

aa. Tin

A copper bath is recommended. Operating temperature should be about 1800°C.

bb. Titanium

The platinum-flux technique is recommended. Platinum to sample ratio at least 8:1. Operating temperature should be about 2150°C.

cc. Tantalum

Use the method described for niobium, but use a platinum/sample ratio of at least 8:1.

dd. Tungsten

Use the method given for molybdenum.

ee. Uranium

Good results were obtained with the copper bath technique using an operating temperature of about 1900°C.

ff. Vanadium

The copper bath technique is recommended using an operating temperature of about 2000°C. The platinum-flux technique has also been applied successfully.

gg. Yttrium

Use the method described for lanthanum.

hh. Zirconium

Use the platinum-flux technique described for titanium.

References

1. L. Jordan and J. R. Eckman, Bureau of Standards Scientific Paper No. 514 (1925), *Ind. Eng. Chem.*, **18,** 279 (1926).
2. P. Oberhoffer and W. Hessenbruch, *Arch. Eisenhüettenw.*, **1,** 594 (1927).
3. *Book of Methods for Chemical Analysis of Metals*, American Society for Testing and Materials, Philadelphia, Pennsylvania, 1960, pp. 27–30.
4. W. G. Guldner and A. L. Beach, *Anal. Chem.*, **22,** 366 (1950).
5. D. I. Walter, *Anal. Chem.*, **22,** 297 (1950).
6. M. L. Aspinal, *Analyst,* **91,** 33 (1966).

6a. M. L. Aspinal and D. Hazelby, "Mass Spectrometric Method for Determining Trace Gases in Metals," *Transactions of the International Vacuum Metallurgy Conference,* American Vacuum Society, Boston, Mass., to be published.

6b. J. F. Martin, J. E. Friedline, L. M. Melnick, and G. E. Pellisser, *Trans. AIME,* **212,** 514 (1958).

7. L. L. Lewis and L. M. Melnick, *Anal. Chcm.*, **34,** 868 (1962).
8. Lubos Kopa, *Hutnicke Listy,* **14,** 322 (1959); UCRL-TRANS-774(L).
9. H. A. Sloman, *Metallurgia,* **32,** 223 (1945).
10. E. Booth and A. Parker, *Analyst,* **84,** 546 (1959).
11. C. B. Griffith and W. M. Mallett, *J. Am. Chem. Soc.*, **75,** 1832 (1953).
12. M. A. Van Camp, Battelle Memorial Institute, Columbus, Ohio, private communication.
13. J. M. Blocher, Jr., I. E. Campbell, D. J. Maykuth, R. I. Jaffee, and H. B. Goodwin, WADC Technical Report 53-470, C-2 (1954).
14. W. S. Horton and J. Brady, *Anal. Chem.*, **25,** 1891 (1953).
15. H. A. Sloman, *Engineering,* **160,** 385, 404 (1945).
16. W. M. Hickam, *Anal. Chem.*, **24,** 362 (1952).
17. Z. M. Turovtseva and N. F. Litvinova, *Proc. U. N. Intern. Conf. Peaceful Uses At. Energy 2nd Geneva,* P/2205 USSR, **28,** 593 (1958).
18. M. W. Mallett, Unpublished work at Battelle Memorial Institute, Columbus, Ohio.
19. C. D. Thurmond, W. G. Guldner, and A. L. Beach, *J. Electrochem. Soc.*, **103,** 603 (1956).
20. W. R. Hansen, M. W. Mallett, and M. J. Trzeciak, *Anal. Chem.*, **31,** 1237 (1959).
21. J. G. Thompson, H. C. Vacher, and H. A. Bright, *Trans. AIME,* **125,** 246 (1937).
22. J. G. Thompson and V. C. F. Holm, *J. Res. Natl. Bur. Std. (U.S.)*, **21,** 87 (1938).
23. M. W. Mallett, *Trans. Am. Soc. Metals,* **41,** 870 (1949).
24. L. Reeve, *Trans. AIME,* **113,** 82 (1934).
25. S. L. Hoyt and M. A. Scheil, *Trans. AIME,* **125,** 313 (1937).
26. V. A. Fassel, F. Monte Evens, and C. C. Hill, *Anal. Chem.*, **36,** 2115 (1964).
27. *Book of ASTM Standards,* American Society for Testing and Materials, Philadelphia, Pennsylvania, 1966, Part 32, p. 613.
28. D. T. Peterson and D. J. Beernsten, *Anal. Chem.*, **29,** 254 (1957).
29. A. L. Beach and W. G. Guldner, *Anal. Chem.*, **31,** 1722 (1959).
30. Navy-MAB Refractory Metals Round Robin No. 2 (1965).
31. *Book of ASTM Methods for Chemical Analysis of Metals,* American Society for Testing and Materials, Philadelphia, Pennsylvania, 1960, pp. 558–563.
32. W. H. Smith, *Anal. Chem.*, **27,** 1636 (1955).
33. M. W. Mallett, D. F. Kohler, R. B. Iden, and B. G. Koehl, Tech. Report WAL TR 823/5 (May, 1962).

34. V. A. Fassel, W. E. Dallmann, and C. C. Hill, *Anal. Chem.*, **38,** 421 (1966).
35. W. M. Albrecht and M. W. Mallett, *Anal. Chem.*, **24,** 401 (1954).
36. *Book of ASTM Standards*, American Society for Testing and Materials, Philadelphia, Pennsylvania, 1966, Part 32, p. 618.
37. C. B. Griffith, W. M. Albrecht, and M. W. Mallett, Battelle Memorial, Institute, Report BMI-1033, August 23, 1955, 8 pp.
38. V. A. Fassel, W. E. Dallmann, R. Skogerboe and V. M. Horrigan, *Anal. Chem.*, **34,** 1364 (1962).
39. C. B. Griffith and M. W. Mallett, *Anal. Chem.*, **25,** 1085 (1953).
40. D. J. Carney, J. Chipman, and N. J. Grant, *Trans. AIME*, **188,** 397 (1950).
41. D. J. Carney, J. Chipman, and N. J. Grant, *Trans. AIME*, **188,** 404 (1950).
42. British Iron and Steel Research Association, "Determination of Nitrogen in Steel," Special Report No. 62, Iron and Steel Institute, London, 1962.
43. G. W. Goward, *Anal. Chem.*, **37,** 117R (1965).
44. H. S. Karp, L. L. Lewis, and L. M. Melnick, *J. Iron and Steel Inst.*, **200,** 1032 (1962).
45. C. R. Masson and M. L. Pearce, *Trans. AIME*, **224,** 1134 (1962).
46. M. Ihida, *Bunseki Kagaku*, **8,** 786 (1959).
47. M. W. Mallett, *Talanta*, **9,** 133 (1962).
48. M. W. Mallett and C. B. Griffith, *Trans. Am. Soc. Metals*, **46,** 375 (1954).
49. Z. M. Turovtseva and L. L. Kunin, *Analysis of Gases in Metals*, Consultants Bureau, New York, 351 (1961).
50. E. Booth, F. S. Bryant, and A. Parker, *Analyst*, **82,** 50 (1957).
51. H. A. Sloman, C. A. Harvey, and O. Kubaschewski, *J. Inst. Met.*, **80,** 391 (1951–52).
52. M. R. Everett, *Analyst*, **83,** 321 (1958).
53. J. Chipman, *J. Iron Steel Inst.*, **180,** 97 (1955).
54. R. D. Pehlke and J. F. Elliott, *Trans. AIME*, **218,** 1088 (1960).
55. V. C. F. Holm and J. G. Thompson, *J. Res. Natl. Bur. Std.*, **26,** 245 (1941).
56. W. R. Hansen and M. W. Mallett, *Anal. Chem.*, **29,** 1868 (1957).
57. M. W. Mallett and W. R. Hansen, *The Metal Molybdenum*, J. Harwood, Ed., Amer. Soc. Metals, Cleveland, Ohio, 1958, Chap. 16, p. 391.
58. J. F. Martin, R. C. Takos, R. Rapp, and L. M. Melnick, *Trans. AIME*, **230,** 107 (1964).
59. ASTM Task Force Report, *Determination of Hydrogen in Steel*, Project No. 37.004-100(8), August 10, 1964, Applied Research Laboratory, United States Steel.
60. T. D. McKinley, *Trans. AIME*, **212,** 563 (1958).
61. *Book of ASTM Standards*, American Society for Testing and Materials, Philadelphia, Pennsylvania, 1966, Part 32, pp. 697–699.
62. H. F. Waldron, *Nuclear Sci. Eng.*, **13,** 366 (1962).
63. T. D. McKinley, *J. Electrochem. Soc.*, **102,** 117 (1955).
64. J. E. Fagel, F. F. Witbeck, and H. A. Smith, *Anal. Chem.*, **31,** 1115 (1959).
65. K. Friedrich, *Acta Chim. Hung.*, **28,** 187 (1961).
66. L. Singer, *Ind. Eng. Chem., Anal. Ed.*, **12,** 127 (1940).
67. W. G. Smiley, *Anal. Chem.*, **27,** 1098 (1955).
68. J. I. Peterson, F. A. Melnick, and J. E. Steers, *Anal. Chem.*, **30,** 1086 (1958).
69. C. E. A. Shanahan, F. Cooke; *J. Iron Steel Inst.*, **188,** 138 (1958).
70. K. Abresch, and H. Lemm, *Arch. Eisenhüettenw.*, **30,** 1 (1959).
71. LECO Conductometric Oxygen Analyzer, Laboratory Equipment Corp., St. Joseph, Michigan (1958).

72. LECO Nitrox Nitrogen–Oxygen Analyzer, Laboratory Equipment Corp., Form 1102 (1962).
73. Laboratory Equipment Corp., Form 184 (1965).
74. Laboratory Equipment Corp., 184B (1966).
75. A. M. Wasserman, Z. M. Turovtseva, and V. I. Vernadsky, *Inst. Geochem. Analy. Chem.*, USSR, Academy of Sciences, **20,** 1359 (1965).
76. *Book of ASTM Standards*, American Society for Testing and Materials, Philadelphia, Pennsylvania, 1965, Part 32, p. 690.
77. S. Kallmann and F. Collier; *Anal. Chem.*, **32,** 1616 (1960).
78. S. Kallman, R. Liu, and H. Oberthin; Air Force Report AFML TR-65-194 (1965).
79. W. E. Dallmann and V. A. Fassel; *Anal. Chem.*, **38,** 662 (1966).
80. P. Elbing and G. W. Goward; *Anal. Chem.*, **32,** 1610 (1960).
81. C. V. Banks, J. W. O'Laughlin, and G. J. Kamin; *Anal. Chem.*, **32,** 1613 (1960).
82. E. Beck and F. G. Clark, *Anal. Chem.*, **33,** 1767 (1961).
83. J. L. Potter, J. E. Murphy, and H. H. Heady, *Anal. Chem.*, **34,** 1635 (1962).

General References

Z. M. Turovtseva and L. L. Kunin, *Analysis of Gases in Metals*, Consultants Bureau, New York, 1961.

Special Report No. 68, *The Determination of Gases in Metals*, The Iron and Steel Institute, London, 1960.

Symposium on Determination of Gases in Metals, ASTM Spec. Tech. Publ. No. 222, 1957.

Chapter 4

COMBUSTION METHODS FOR CARBON AND SULFUR

WALTER G. BOYLE, JR., Lawrence Radiation Laboratory, University of California, Livermore, California

I. Introduction 116
II. General Principles 116
A. Chemistry of Carbon and Sulfur 117
1. Forms of Carbon 117
2. Forms of Sulfur 118
B. Release of Carbon 119
1. Condition of the Melt 120
2. Use of Fluxes 120
3. Types of Fluxes 120
C. Release of Sulfur 121
1. Adsorption of Sulfur Trioxide 121
2. Types of Fluxes 122
D. Errors and Standardization 122
1. Sources of Errors 122
2. Determination of the Blank 123
3. Standardization 123
4. Precision and Accuracy 124
III. Methodology and Instrumentation 125
A. Gas Purification 125
1. Purification of Oxygen 125
2. The Use of Manganese Dioxide 126
B. Construction Practices 126
C. Furnaces 128
1. Resistance Furnaces 128
2. Induction Heating 128
3. Containers 129
D. Detection Techniques for Carbon 129
1. Introduction 129
2. Conductometric Methods 129
3. Titrimetric Methods 131

4. Gas Chromatography 133
5. Other Methods 133
E. Detection Techniques for Sulfur 133
1. Chemical Methods 133
2. Physical Methods 134
IV. Selected Examples 135
A. Carbon Determinations 135
B. Sulfur Determinations 137
V. Concluding Remarks 139
References 139

I. INTRODUCTION

Sulfur and carbon can be determined in metallurgical samples by controlled oxidation of the sample and measurement of the resulting carbon dioxide and sulfur dioxide.[1] Under the impetus of modern metallurgy this method, originally used for the analysis of iron and steel, has been modified and expanded to include a wide variety of techniques. New alloys, many of them composed of the more refractory elements, have necessitated the use of higher combustion temperatures. Stringent purity considerations have required the use of extremely sensitive detection techniques. The wide variety of materials examined for carbon and sulfur has caused problems in developing methods which are sufficiently versatile and yet exhibit the required sensitivity and precision. It is the purpose of this chapter to describe some of these techniques and their applications.

II. GENERAL PRINCIPLES

A brief description of the combustion method is as follows. The sample is heated by a furnace and burned in a flowing stream of oxygen. The resulting carbon dioxide or sulfur dioxide is released into the flowing oxygen and determined quantitatively. Before discussing the details of equipment and detection techniques for this procedure, some of the more general problems associated with combustion methods are discussed. These are (*1*) the nature of carbon and sulfur compounds likely to be encountered, (*2*) the release of carbon and sulfur as the oxides, (*3*) sources of errors, and (*4*) standardization methods. The chapter concludes with some selected examples of procedures which illustrate the scope and applicability of the method.

A. Chemistry of Carbon and Sulfur

The problem in organic elemental carbon analysis has been to devise a combustion system which could be used on a large variety of organic compounds.[2] Because carbon and sulfur form many different inorganic compounds, it could be said that a similar problem exists in determining carbon and sulfur in metals and alloys. Therefore, it would be helpful to discuss briefly some of these different forms.[3]

1. FORMS OF CARBON

a. Graphite

Of the two allotropic forms of carbon, diamond and graphite, only graphite will be discussed. Graphite is composed of hexagonal layers of carbon spaced 3.35 Å apart with the C—C distances within a layer being 1.415 Å. The layers may be stacked in two different arrangements. In one arrangement, the stable form, the layers are stacked so that alternate layers are superposed, i.e., layers 1, 3, 5, etc. align. In the other arrangement every third layer is superposed. This can apparently be converted to the alternate layer form by heat.

The amorphous forms of carbon are in reality microcrystalline graphite. Thus charcoal, soot, and lampblack are all forms of graphite. Therefore it would be expected that so-called free carbon in metals and alloys would be graphitic in structure. The layerlike graphitic structure, however, lends itself to various types of chemical combinations.

b. Graphitic Compounds

When graphite is treated with strong oxidizing agents, e.g., fuming nitric acid, it forms an oxide. The graphitic oxide so formed does not conduct electricity and swells in water and alcohols. While the actual structure is uncertain, the oxygen to carbon ratios are about 1:2.

Graphitic fluoride, a somewhat similar compound, can be formed by the direct reaction of fluorine and graphite.

Besides these nonconducting compounds, there are a large number of electrically conducting compounds formed by the sandwiching of molecules or ions between the layers of the graphite itself. The heavier alkali metals, and many halides, oxides, and sulfides react spontaneously with graphite to form this type of compound.

c. Carbides

Carbides may be divided into three types, the saltlike carbides, the interstitial carbides, and the covalent carbides. The saltlike carbides can be considered as ionic compounds between a metal cation and negative C^{4-} or C_2^{2-} anion. They form crystals and are hydrolyzed by water or dilute acids. The C^{4-} anions give methane on hydrolysis and the C_2^{2-} anions give acetylene. Be_2C, Al_4C_3, BeC_2, and Ag_2C_2 are examples of these compounds. The saltlike carbides are too numerous for a detailed discussion of their properties here. It should be mentioned, however, that those carbides giving a variety of products on hydrolysis are usually metal carbides in unusually low oxidation states. Thus ThC_2 gives ethylene, methane, and hydrogen as well as acetylene on hydrolysis, possibly because of the further oxidation of Th(II) to Th(IV).

The interstitial carbides are formed when the carbon atoms occupy holes in the crystal lattice of the metal atom itself. The fundamental metallic properties of the atoms are not altered by this filling of interstitial holes. The crystal lattice, however, is in fact stabilized and this results in increased hardness and a higher melting point for the metal. It is estimated that a metal atom radius of $\backsim$1.3 Å or greater is required for this type of crystal stabilization. Some metals (Fe, Cr, Mn, Ni, Co) with radii smaller than this form a transitional type of compound with properties somewhat between ionic and interstitial carbides.

The extremely hard inert silicon carbide and boron carbide are examples of covalent carbides. There is also a certain degree of covalency in many of the ionic carbides thus bridging the gap between completely ionic and covalent compounds.

From this brief discussion, it can be seen that the problem of total carbon analysis becomes much more complex than merely arranging combustion of "buried" carbon. In fact, for precise trace work, procedures must allow for the release of some possibly very intractable forms of carbon under conditions which also allow the combustion and detection of hydrocarbons.

2. FORMS OF SULFUR

Sulfur and its compounds may also be present in a variety of forms. Some of these are briefly described.

a. Elemental Forms

The two crystalline forms of sulfur, monoclinic and rhombic, both contain S_8 cyclic molecules. Amorphous sulfur also exists. Liquid and vapor-phase sulfur have fairly complicated arrays of temperature-dependent species. Sulfur burns in air directly to give sulfur dioxide. It combines directly with the halogens and many metals and nonmetals.

b. Examples of Metallic Sulfur Compounds

Sulfur combines very easily with a large number of metals. It also forms compounds of great structural variety, the binary metallic sulfides probably being most often encountered in metals and alloys. The alkali and alkaline earth elements form sulfides which have properties indicating ionic-type structures. They dissolve in water, are extensively hydrolyzed by water, and crystallize in ionic-type lattices.

Most metallic binary sulfides, however, cannot be considered ionic. In fact, one cannot even use the simple approach used for discussing the metallic carbides. Many are nonstoichiometric or polymorphic and exhibit semi-metallic properties. The analogy between sulfur and oxygen breaks down when comparing oxides and sulfides. Some of the transition metal sulfides form structures in which each metal atom is approached fairly closely by two other metal atoms, thus imparting a number of metallic properties to the sulfide.

The disulfides contain S—S groups in the pyrite structure. FeS_2 and CoS_2 are examples of this type of compound.

c. Higher Oxides of Sulfur

The possibility of encountering oxidized forms of sulfur in metals and alloys also exists. Thus when using sulfur dioxide as the final form for measurement, it must be established that the conditions are suitable for the decomposition of sulfate and sulfite species to SO_2. Sulfur dioxide is the usual form of sulfur measured after combustion of a metallic sample. It undergoes further oxidation forming sulfur trioxide, SO_3, under the proper conditions. Small amounts of SO_3, are formed when sulfur is burned in air. Sulfur trioxide is formed chiefly by the reaction between sulfur dioxide and molecular oxygen in the presence of catalysts. Since with many detection methods SO_3 is either lost or not detected, it is important to realize that, although under the combustion conditions most often used, SO_2 is the expected form, SO_3 can possibly be formed. This will be discussed further in Section II-C.

B. Release of Carbon

It is fundamental that all of the carbon, both free and combined, in any form whatever must be combusted in oxygen to carbon dioxide. However, even if the carbon in the sample has been completely burned to CO_2, it does not necessarily follow that this CO_2 will be released.

1. CONDITION OF THE MELT

Green, Still, and Chirnside[4] have studied the release of carbon dioxide and described two conditions favorable for its formation and release from metallic oxide systems. One condition, and the most often employed, is that the oxides melt and remain molten during the entire combustion step. The other is that the metallic oxides form a finely divided powder. Induction furnaces employing fluxes and accelerators are usually used in the first case while resistance furnaces are sometimes more effectively employed in the second.

The surface temperature of iron burning in oxygen may be as high as 2300°C. In some cases the initial combustion temperature melts the sample. As the temperature of the melt drops to that of the furnace, the melt solidifies and some of the carbon is not released. When furnace temperatures can be controlled, the initial combustion temperature can also often be controlled to a certain extent. That is why it is sometimes more reasonable to use a resistance furnace. Some examples of this will be mentioned later.

2. USE OF FLUXES

The combustion temperatures of most oxides are too high for the powdery oxide condition to prevail. On the other hand, the melting temperatures of many of the oxides are too high for the molten state to exist long enough for carbon dioxide to be released quantitatively after the initial combustion is completed. Since most combustions involve a resolidifying of the melt, fluxes are employed to keep the melt molten.

Green, Still, and Chirnside[4] have also enumerated the factors which determine the choice of a flux. Besides the prime necessity of maintaining a melt, factors such as carbon content, crucible or container compatibility, and method of heating must be considered. If induction heating is used, then the oxide must couple with the induction furnace for the melt to remain molten. That is the oxide must be heated by induced current from the electric field of the furnace.

3. TYPES OF FLUXES

The most popular fluxes for induction furnace combustions are tin, copper, and iron. Tin is primarily an igniter. That is, the oxidation of the tin produces a temporary rise in temperature which acts as a "match" to ignite the sample and fluxes, if any. ACS reagent grade tin, 20 mesh, has been found to have a carbon content of between 2 and 4 ppm* and is suitable for this use.

* Parts per million by weight is abbreviated ppm and represents micrograms of impurity per gram of sample.

Iron is usually the flux of choice. Its oxide is a good coupler and it remains molten (about 1500°C) in most induction furnaces. Hydrogen reduced iron chips (G. Frederick Smith Chemical Co., Columbus, Ohio) have been found to contain about 8 ppm carbon and are excellent for fluxing. Copper is also a good coupler and remains fluid at induction furnace temperatures. All three of the above metals are often used in combination for induction furnace work.

Lead is often used as a flux with resistance furnace combustions. "Bedding compounds" such as Sinderite, a chromium oxide ore, are also sometimes recommended to maintain and distribute the heat to a sample in a resistance furnace.

It has been reported that the carbon content of lead and copper can be made negligible (< 1 ppm) by heating to bright redness immediately before use.[4]

C. Release of Sulfur

Most of the comments about the release of carbon as CO_2 can also be applied to the formation and release of SO_2. However, it has been found that in most cases only 90% or so of the sulfur can be recovered as SO_2.

1. ADSORPTION OF SULFUR TRIOXIDE

Fulton and Fryxel,[6] using a radioactive tracer technique and other methods, have investigated the combustion properties and recovery factor for sulfur in ferrous alloys. They found that combustion must occur at temperatures around 1550°C or appreciable amounts of sulfur are retained in the melt. They also found that if this minimum temperature was maintained, then most of the sulfur loss could be traced to adsorption on glassware. A detection procedure involving the absorption and oxidation of SO_2 by hydrogen peroxide and its subsequent titration as sulfuric acid with sodium hydroxide was used. By washing the walls of the glassware, then titrating with base, it was ascertained that sulfur trioxide was the substance being adsorbed. With the path length of the glass apparatus made as short as possible and oxygen flow rates of 3–5 liters/min employed, recoveries were considerably improved. Rinsings of the glassware still recovered some sulfur, however. Rooney and Scott have conducted a similar investigation also using tracer techniques.[7] Their conclusions were in general agreement with those of Fulton and Fryxell.

Rice-Jones[8] has mentioned some of the factors involved in the reaction of oxygen with SO_2 to produce SO_3. He has calculated some theoretical equilibrium constants and predicted that at 1427°C SO_2 should be produced at 98% yields. Another factor mentioned was the rate of sample combustion.

If a sample burns rapidly the temporary drop in oxygen will theoretically favor the formation of SO_2.

The problem of the formation of sulfur trioxide and its adsorption on glass surfaces is not as serious as it sounds. As Fulton and Fryxell have pointed out, the adsorption seems to be more dependent on flow rate and possibly the temperature of the glass than anything else. The correction factor is remarkably constant for any given piece of apparatus ($\pm 1\%$) and is around 90%.[9] That is, recovery is around 90%. The precision in trace determinations would therefore be sufficiently high for practically all cases. For precise determinations of macro amounts of sulfur, wet chemical methods are probably preferable.

2. TYPES OF FLUXES

The fluxes and accelerators mentioned in the section on release of carbon are useful for sulfur analysis. Some granulated tin shows considerable variation in trace sulfur (from 2 to 15 ppm). A low-sulfur iron powder furnished by the Laboratory Equipment Corporation (Leco), St. Joseph, Michigan, is quite satisfactory. Copper rings, also provided by Leco, contain only a few micrograms of sulfur. However, for precise low level trace work using the iodate–iodate system copper should probably be avoided.[9]

D. Errors and Standardization

Some of the sources of errors and methods of standardization encountered in combustion methods and flowing gas systems will be described.

1. SOURCES OF ERRORS

The problem of incomplete combustion and carbon or sulfur "lock up" has already been discussed. One of the easiest ways to recognize this phenomenon is to apply a detection system which indicates continuously the amount of carbon dioxide or sulfur dioxide in the gas stream. A combustion which after about 10 min still indicates a level or very slowly decaying reading above the blank will usually indicate slow or incomplete burning and melting.

Inhomogeneity of the melt can sometimes be seen by visual examination of the solidified melt after running the sample. An uneven appearance or obvious presence of separate phases probably means poor combustion has taken place. On the other hand, the absence of these signs does not necessarily indicate a complete combustion.[10]

Assuming that the combustion of the sample, either by the addition of flux or raising the temperature or both, is satisfactory, the other sources of error in the system may be considered. Leaks in the gas train and combustion system can lead to erratic results. These can be checked by pressurizing the system, closing it off, and applying a soap solution to the joints, or by using a mercury column and checking for a pressure drop with time. An important consideration can be contamination of the sample or container especially for trace carbon work. Dust and atmospheric CO_2 must be carefully guarded against. Samples must be carefully degreased and crucibles prefired. Despite these precautions a certain amount of the apparent CO_2 or SO_2 during each determination will be attributable to a "blank." The reproducibility of this blank and to a lesser degree the size of the blank determines the precision of the method.

2. DETERMINATION OF THE BLANK

A blank determined with the furnace cold can be useful for checking the sweep time necessary to remove atmospheric contamination from an apparatus after sample loading. This should be a very constant, low reading, and a system should approach this value within a few minutes of sweeping after being opened to the atmosphere.

However, this blank is not the one "seen" by the detector when the furnace is in operation. A means must be devised which enables the blank in a hot furnace to be determined. This blank is the apparent content of CO_2 or SO_2 in the gas stream at the temperature reached by the furnace for a particular sample. The blank may be estimated by allowing the furnace to remain on after the containers and fluxes have apparently released their SO_2 or CO_2. The system will then have reached a steady constant value representing a certain blank value per unit time for the hot furnace. For resistance furnaces this is relatively simple. The sample holders and fluxes are run without the sample at the predetermined temperature and the appropriate measurement taken. However, for induction furnaces the matching of the temperature of the blank and sample is more difficult, and must be found by trial and error. For the most accurate trace work, the ideal situation is to have the capability to add the sample directly to such a prefired crucible and flux without further handling. The flushing of the apparatus can then easily be followed by observing the cold blank and the hot blank per unit time employed to correct the results, the total time the sample spent in the furnace having been measured.

For routine cases and where traces are not so important, the crucibles are prefired in a muffle furnace and the results corrected for the residual carbon or sulfur content.

3. STANDARDIZATION

The direct standardization of the detection system is a relatively straightforward matter. A solution of sodium bicarbonate may be added from a buret to a sulfuric acid solution and the resulting CO_2 swept by oxygen from the gas train into the detection system. This system bypasses most of the furnace components and allows an independent calibration of the detector.

In another method, which employs the entire combustion train, a dilute solution of an appropriate salt may be evaporated on tinfoil. The sample can be combusted in the furnace and the value found taken as the CO_2 measure. This is the procedure most often used both for CO_2 and SO_2. Such standards are commercially available.* Care should be taken in the choice of compounds to be used as standards. The criteria for a standard substance of this kind are that they should be pure, easily dissolved, and most important, readily and smoothly combustible. Some organic compounds have been used. Solutions of potassium acid phthalate, for instance, can be aliquoted onto porous crucible covers. These are then dried and combusted.[10] Na_2CO_3 and K_2CO_3 can be more easily combusted if powdered silica is present.[12] For induction furnace work, a quartz enclosed graphite crucible is sometimes handy for combusting such samples.

The direct addition of CO_2 by means of gastight syringes has been investigated[13] using gas chromatography. This technique was compared to the combustion of primary standard potassium acid phthalate. A septum was provided for the injection of carbon dioxide into the incoming side of the induction furnace. Thus the CO_2 followed the same path and was exposed to the same conditions as CO_2 from combustion. The syringe was calibrated with mercury at 30 μl setting. Multiple injections at this setting were used for larger additions. The potassium acid phthalate was combusted by means of platinum heated in an induction furnace. The results showed good agreement at the levels investigated which were 15–45 μg. Some evidence suggested that the simplicity of the injection technique gave it more precision at the lower carbon level. Certainly this is a convenient way to check out the components in a combustion train. By injecting CO_2 at various points in the system a rather thorough test of its functioning can be obtained.

4. PRECISION AND ACCURACY

Although the calibration of the detection system and the gas train and furnace are relatively easily accomplished, the precision and accuracy of these procedures do not indicate whether all of the carbon in a given sample has been indeed combusted. If there has been no noticeable sign of incomplete combustion, an attempt must be made to provide a standard reasonably similar to the sample. A standard sample cannot be conveniently "mocked up" as is the case in solution work. Moreover, the complexity of

* General Catalogue, Laboratory Equipment Corporation, St. Joseph, Michigan.

the forms of carbon and sulfur and their interaction with other materials in the sample often precludes quantitative information from the addition of some standard form of carbon to the sample.

Ferrous-type alloys that have been analyzed for carbon and sulfur by different analysts using several independent methods are sometimes available as standards. For example, NBS standard sample 166 B (0.019% C) is a low-carbon stainless steel alloy especially used for standardizing and checking carbon procedures.

The problem is not so easily solved with other alloys, especially for trace determinations. Comparison of independent methods of analysis can be helpful but many of these methods themselves, especially in trace work, require some kind of standardization also.

One method which is useful for detecting matrix effects is varying the sample size and plotting the apparent carbon or sulfur content of the sample against the sample weight. A departure from a straight line beyond experimental error, especially at the extremes of the sample weight range, usually indicates problems in the procedure.

III. METHODOLOGY AND INSTRUMENTATION

A combustion train consists of a purification system for the oxidizing gas, a furnace for combusting the sample, a means of purifying and separating, if necessary, the gases formed in the combustion process, and finally a detection and measuring system.

A. Gas Purification

1. PURIFICATION OF OXYGEN

Incoming oxygen is most often purified by combusting carbon impurities over wire-form copper oxide which is heated to about 800°C. Completeness of combustion depends on the size of the purification system relative to the flow rate of oxygen required. A simple method of checking the purity of the oxygen is to pass it through the detection system after combustion; for a given flow rate the amount of CO_2 per unit time should level off as the temperature (or size) of the copper oxide furnace is increased.

After combustion in the copper oxide furnace, the oxygen plus any carbon dioxide is passed through a tube containing a solid absorbent for carbon dioxide (Ascarite, Caroxite), and a desiccant (Anhydrone, Dehydrite). The absorbent system should be such that no CO_2 is found by the detection system.

2. THE USE OF MANGANESE DIOXIDE

The combustion products from the sample furnace may also have to be purified depending on the detection system employed. Most detection systems can allow certain types of impurities to be present in the main gas stream of oxygen and carbon dioxide. One of the principal absorbents employed for CO_2 systems is a specially prepared form of manganese dioxide. Manganese dioxide removes oxides of nitrogen and sulfur from flowing gas streams. This external absorbing medium has been developed for microorganic chemists faced with the problem of determining carbon and hydrogen accurately in milligram (or smaller) amounts of sample.

Manganese dioxide must be specially prepared to function as an efficient absorber of oxides of nitrogen and sulfur. Commercial crystalline grades do not work. Generally 10–14 mesh granules are recommended and they must be acid precipitated.[14,15] Manganese dioxide absorbs only nitrogen dioxide. Other oxides of nitrogen will pass through this absorbent and can either cause trouble in detection systems themselves or gradually combine with oxygen to form more nitrogen dioxide.

Manganese dioxide is usually used in the oxygen purification system for sulfur determination and in both the incoming and outgoing oxygen flow in the carbon procedures. However, sulfamic acid has been used to destroy nitrogen dioxide when using a colorimetric finish for sulfur dioxide.[16]

It should be pointed out that the gas purification steps are especially important in trace work. Always dependent on the detection system, the presence of a trace component other than CO_2 or SO_2 could easily be expected to overbalance a sensitive system and this would not necessarily be indicated in any standardization or blanking procedure.

B. Construction Practices

In order to indicate some of the techniques employed in constructing a combustion train, it is appropriate at this point to present an example of such a system.

Figure 1 shows a complete schematic[17] for the determination of CO_2. This particular system employs titrimetric detection with a second absorption cell shown connected in series with the first. Most of the connections on the incoming oxygen side of the small copper oxide furnace were constructed of $\frac{1}{8}$ in. stainless steel tubing which was silver-soldered to copper, and copper–glass housekeeper seals with glass spherical joints. The spherical joints were sealed with Apiezon W black hard wax. The oxygen pressure was controlled

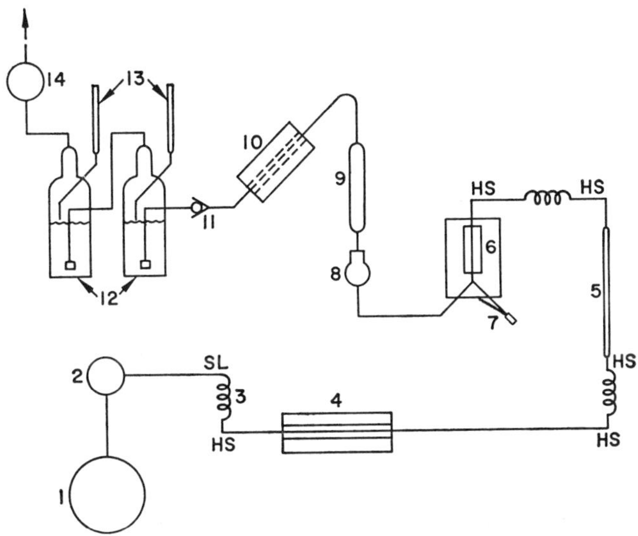

Fig. 1. Schematic of apparatus.[17] (1) Oxygen cylinder. (2) Pressure regulators. (3) ⅛-in. SS tubing. (4) Large furnace over quartz tube containing CuO. (5) Pyrex tube containing Ascarite and Anhydrone. (6) Quartz combustion tube and induction furnace. (7) Spring-loaded handle. (8) Dust trap. (9) Trap containing MnO_2. (10) Small furnace over quartz tube containing CuO. (11) One-way check valve. (12) Absorption cells in series. (13) Burets. (14) Flowmeter. SL, Swagelok. HS, Housekeeper seals.

with a low-pressure pancake-type regulator after the usual oxygen pressure regulator. The oxygen of 99.5% purity was led by means of a coil of ⅛ in. stainless steel tubing to a 1¼ × 24 in. quartz tube filled with wire-form copper oxide and heated to 800–900°C with a resistance furnace which was effective over 12 in. of the quartz tube. The oxygen plus any carbon dioxide impurity was then run through a 1 × 18 in. Pyrex tube filled with Ascarite and Anhydrone, and from there entered the quartz combustion tube in a Leco induction furnace equipped with a Variac control. The coils of stainless steel tubing imparted a good deal of flexibility to an otherwise rigid system. A piece of flexible tubing connected the outlet of the induction furnace to a dust trap consisting of a small flask filled with glass wool, a MnO_2 trap for absorbing oxides of sulfur and nitrogen, and finally to a smaller copper oxide furnace heated to 800–900°. The oxygen then passed through a one-way check valve which was connected to the absorption cell by a short length of Tygon tubing. Plastic quick-connectors were convenient for adding a cell or repositioning two cells.

C. Furnaces

There are two primary methods of oxygen combustion in use: the induction furnace technique and the resistance furnace technique. The induction furnace allows rapid heating of the sample, and rather high temperatures, but sacrifices some degree of control over the rate of combustion. In addition some materials must be indirectly heated, that is, heated by the addition of an appropriate conducting material. The resistance furnace technique gains in control of the heating rate but is quite slow and difficult to operate in the high temperature region.

1. RESISTANCE FURNACES

These are wire-wound furnaces shaped to contain a combustion tube. The furnace can be made of a variety of materials depending on the operating temperature desired. Heating elements are usually nickel-chromium alloy (nichrome) or platinum. They are usually wound on alundum blanks.

For most purposes commercially available nichrome wire furnaces operative up to around 1100 or 1200°C are sufficient. Materials which oxidize at higher temperature are usually best heated by an induction process. As will be seen, those materials easily oxidized are probably best done in a resistance furnace. However, special furnaces employing tungsten mesh heating elements and using thorium or zirconium oxide tubes and crucibles are commercially available.[18] Temperatures of about 2500°C can be attained, and the possibility of using this furnace for carbon determinations in flowing oxygen is being investigated.[19]

2. INDUCTION HEATING

Induction heating is resistance heating caused by electric currents induced in a conductor by an alternating current. The sample or object which is heated can be considered as the secondary of a transformer. The primary of this transformer is a copper coil. There is no physical contact between the sample and the heating element. The heating effect takes place primarily in the surface layers of the heated sample.

The depth of the heating is dependent on a number of parameters, the most fundamental being the frequency of the alternating field. The higher the frequency the less penetrating is the heating effect. Frequencies of 2–10 kHz give fairly deep penetration while frequencies above 500 kHz are primarily effective only for surface heating. Frequencies of the order of 50

MHz produce heating by a dielectric effect. The Leco induction furnace, used for ordinary oxygen combustion type work, for example, is a high frequency electronic oscillator type. It furnishes 1.5 kW of power. This power can be controlled to a certain extent by the use of a variable transformer in the primary side of the high voltage transformer.

3. CONTAINERS

Combustion tubes are often constructed of fused silica (1400°C maximum operating temperature) or mullite ($3A\ {}_2O_3 \cdot 2SiO_2$, 1800°C). Sample containers are usually porcelain or zirconia. Crucibles and containers are commercially available for a variety of special applications.

Of course, the induction furnace also requires a material for containing the oxygen and enclosing the crucibles. Since the heat is originated in the sample itself, quartz can be used as the enclosing material even though temperatures of the sample might be well above 1500°C.

D. Detection Techniques for Carbon

1. INTRODUCTION

The collection of carbon dioxide on a suitable absorbent and its subsequent weighing is a useful technique that is still often employed, especially for high-carbon alloys. The low pressure combustion method is also used. The samples are burned in oxygen, the released carbon dioxide is frozen out by liquid nitrogen and is then determined by measuring the pressure at constant volume. It is employed mostly for samples containing small amounts of carbon.[20] Another technique employs a liquid nitrogen trap and capillary manometer for a highly accurate procedure.[21] However, these procedures can sometimes be cumbersome and time-consuming. For this reason, many other detection systems have been devised for the determination of carbon dioxide. Among these are the change in conductivity of an alkaline solution as CO_2 is absorbed, acid–base titrations, gas chromatography, and the absorption of infrared radiation. Since these techniques form an integral part of any combustion system, some of them will now be discussed.

2. CONDUCTOMETRIC METHODS

Barium hydroxide or sodium hydroxide solution absorbs CO_2, converting the hydroxide anion to water and increasing the resistance of the solution. This increase in resistance is more sensitive the more dilute the solution, but the effect can be assumed linear over a well-defined range. Thus by measuring the change in resistance using known amounts of CO_2 a calibration or analytical working curve may be prepared.

Still et al. have described such a method for the conductometric determination of carbon in metals.[22] The carbon dioxide envolved from the combustion is absorbed in a solution of sodium hydroxide or barium hydroxide and the change in the electrical conductivity of this solution is measured. The detection system consists of an absorbing unit for insuring the complete reaction of carbon dioxide with the base, and a means of measuring the conductance of this solution. The change in conductance is measured by making the absorbing solution one arm of an ac Wheatstone bridge. The electrodes are made of platinized platinum. The cell system (electrodes and solution) is thermostated at 25°C.

Still et al.[22] regarded 1 mg as being the maximum amount of carbon that could satisfactorily be determined with barium hydroxide in their cell. Sodium hydroxide was used for samples containing more than 1 mg of carbon and was also preferred by them for work with smaller amounts of carbon, even though barium hydroxide is somewhat more sensitive. A determination took 25–40 min not counting the time required to bring the cell to temperature. The same absorbing solution could be used for more than one analysis. Results on steels and cast iron using this system indicate a range of applicability from 300 ppm to 3% and a detection limit of about 5 μg. Flow rates were 50–100 ml of oxygen per minute.

A commercial conductometric apparatus has become quite popular in recent years and has been used for carbon analysis in a wide variety of materials and especially for carbon at the 100 ppm level and less.[23] The apparatus, manufactured by the Laboratory Equipment Corporation, St. Joseph, Michigan, employs a number of innovations which improve the speed of operation.

One arm of the Wheatstone bridge contains a cell and electrodes and is used as a compensating cell. The hydroxide solution in this cell is only changed when the main stock solution is changed. A second arm of the bridge contains the measuring cell as usual. The difference in resistance between the two cells is amplified and displayed on a small cathode ray tube as a separation of a pair of lines. Balancing the bridge is accomplished by making the lines coincide while changing the resistance in the compensating cell arm. Thus the carbon dioxide evolution can be followed by observing the separation of the lines. Barium hydroxide solution is stored in a large reservoir and a portion of the oxygen from the gas train is used to transfer the solution into the measuring cell.

Violante[24] has suggested saturating barium hydroxide with carbonate before using it in a conductometric determination. This reduces the effect of carbonate ions to decrease the resistance and eliminates the possibility of the absorbing solution becoming saturated with $BaCO_3$ during a determination which would change the calibration curve. Violante also points out that sensitivity cannot be increased (by dilution of the absorbing solution) without limit because the change in resistance then becomes extremely dependent on other factors such as temperature. A solution of 0.6 g of $Ba(OH)_2 \cdot 8H_2O$ per liter was suggested by Violante for use with the Leco instrument. The sensitivity is estimated to be $\pm$ 1 μg of carbon.

Bruins[25] has mentioned the difficulties associated with the electrodes when using conductometric methods. The platinum electrodes require frequent cleaning and platinizing. They can become covered with precipitated $BaCO_3$ which will change the apparent conductivity measurements. For this reason he has designed a high frequency oscillometric method which effectively measures the conductivity of the $Ba(OH)_2$ solution by use of external electrodes.

Engelsman et al.[26] has described a very sensitive conductometric system. Two identical thermostated absorption cells, based on the design of Still et al., are used for temperature compensation. By equalizing the evaporation of water from the cells and using a highly sensitive measuring bridge with a stabilized voltage source, a detection limit of 0.1 μg of carbon was obtained.

3. TITRIMETRIC METHODS

Carbon dioxide may be determined by a standard acid–base titration method. Barium hydroxide is used to absorb carbon dioxide and the remaining barium hydroxide is titrated with standard acid. This method is relatively insensitive, difficult to follow continuously, and is time consuming.

A number of nonaqueous acid–base titration methods for the determination of carbon dioxide in flowing gas systems have been proposed. Patchornik and Shalitin[27] have described a titration of carbon dioxide in benzylamine and applied it to biological samples. Grant, Hunter, and Massie[28] used dimethyl formamide as a solvent and used a titrimetric procedure for carbon in limestone, dolomite, organic compounds, and a high-carbon steel. Jones et al.[29] have described a ttiration using dimethyl formamide as a solvent and applied it to carbon in iron and steel samples.

Blom and Edelhausen[30] have shown that the continuous titration of carbon dioxide in acetone with a methyl alcohol–pyridine solution of sodium methoxide using thymol blue indicator is a sensitive and accurate procedure for the determination of carbon dioxide in flowing systems. They used it for determining carbon in organic compounds. White[12] showed that the endpoint drift in this system was due to the presence of water and excess methanol in the solvents and successfully applied the titration to microgram quantities of carbon derived from organic compounds and carbonates.

Boyle et al.[31] modified the method of White and have demonstrated its usefulness for a combustion–titrimetric method. The titration cell is a gas washing bottle of about 250 ml capacity. It contains a stainless steel hypodermic needle pushed through a septum for use as a buret tip. A Teflon gas dispersion tube[32] is used as the gas bubbling system. The cell is filled with about 150 ml of dry acetone. The titrant for trace work is 0.005*M* sodium hydroxide in pyridine and a minimum amount of methanol. The system is used to continuously titrate CO_2 as it comes over from the oxygen stream and can handle flow rates of 1 liter/min. Blanks can be determined in the same way, and since the method operates continuously, a drift rate or blank per unit time may be obtained which can be related back to the time required to run the actual sample.

Two cells may be used: one for the generation of a known amount of CO_2 by the addition of a known amount of $NaHCO_3$ to sulfuric acid, and the second to titrate this carbon dioxide as it is swept over into the oxygen stream into the second cell. This standardizes the sodium methoxide under flowing conditions independent of furnace parameters. Two cells may also be connected in series—the second cell to check the completeness of the CO_2 absorption in the first. As long as the titration is able to keep up with the incoming carbon dioxide, very little CO_2 appears in the second cell.

The effectiveness of this method for trace determinations of carbon depends on the magnitude and reproducibility of the drift at the endpoint during the titration. That is, a blank exists which not only depends on the amount of CO_2 from extraneous sources, but is also inherent in the detection system. White[12] has shown that water and excess methanol causes a slow change from blue to yellow in this system.

Using a glass frit and with a hot furnace, a blank or drift rate of about 0.026 ml of 0.005*M* sodium methoxide/min is obtained. One ml of 0.005*M* sodium methoxide is equivalent to 50 μg of carbon, so the magnitude of the blank, allowing 10 min to run a sample, would be about 13 μg. However, the maximum spread is only 0.1 ml, and if during a 10 min run this drift changes the maximum amount, an error of 5 μg total would be expected.

A Teflon gas dispersion tube[32] shows considerable improvement. It is constructed by plugging an 8-mm i.d. Teflon tube with a piece of Teflon rod and punching two rows of 15 holes around the end of the tube with a fine needle. When this dispersion tube is used, the drift becomes quite low and reproducible. The drift rate is about 0.012 ml/min, corresponding to a total of 10 min blank of 6 μg of carbon with a reproducibility of plus or minus 0.002 ml/min maximum spread or 0.02 ml equivalent to $\pm$ 1 μg.

4. GAS CHROMATOGRAPHY

Two approaches can be made in the application of gas chromatography to the determination of CO_2 in flowing oxygen. In one approach, the carbon dioxide is removed by means of liquid nitrogen or oxygen traps subsequent to its determination by injection into a sample column. In another approach the CO_2 plus oxygen is swept through a column which retains CO_2, and the CO_2 is later measured by raising the temperature and sweeping with helium through the detectors. In both cases, the parameters for proper combustion must be well worked out since the release of CO_2 cannot be followed during combustion.

The second method based on the work of Walker and Kuo (33) utilizes a 4-ft Molecular Sieve 5A column. The CO_2 is trapped in the column at 100°C. After combustion is complete the column is swept with helium to purge oxygen, and finally the temperature is raised to 275°C, the CO_2 removed from the column, and the peak integrated. This technique was used for ferrous metals and found to be quite sensitive down to about 2 or 3 μg of carbon. It has abroad detection range from 0.0005% to the order of 20%.

Other similar chromatographic methods have been described especially for ferrous-type alloys.[34–36]

Some analyzers, especially for control purposes, are based on the thermal conductivity of O_2 compared to O_2–CO_2. Hence the gas purification step can be avoided altogether, but at the expense of assuming that the CO_2 is the only contribution to the detection signal and that other variables have been taken care of in the standardization.

It is interesting to note that many of these approaches parallel the development of automated C, H, and N analyzers used in organic microanalyses.[37,38]

5. OTHER METHODS

An automatic apparatus has been described by Tipler[39] which uses the absorption of infrared radiation by CO_2 for its determination in an oxygen stream after induction furnace combustion. It is specific for CO_2. An accuracy of $\pm$ 0.0005% absolute at the 0.01% level was reported. A manual form is also discussed. Some of the other ways of detecting trace carbon, not necessarily applied to combustion systems, are isotope dilution,[40] activation analysis,[41] and spark-source mass spectrometry.[42]

E. Detection Techniques for Sulfur

1. CHEMICAL METHODS

Sulfur dioxide has been determined by a wide variety of chemical methods,[43,44] and the chemical reactivity of sulfur dioxide lends some emphasis to this approach. One of the standard methods for determining sulfur in metal and alloys is the hydrogen sulfide evolution method.[1,45] The sample is dissolved in a nonoxidizing acid and the hydrogen sulfide evolved is determined. The method allows the determination of sulfur as sulfides only and does not involve combustion of the sample.

When the sample is combusted SO_2 is formed. This must be trapped and determined in a stream of oxygen at flow rates from 20 ml/min up to 2 liters/min. One of the most popular methods of accomplishing this is by iodine oxidation. Sulfur dioxide is continuously titrated with iodine, usually generated by the acid–iodide–iodate reaction, in a combined gas absorption, titration cell as it comes over in the gas stream. The reactions are

$$6H^+ + 5I^- + IO_3^- \rightleftharpoons 3I_2 + 3H_2O$$

$$SO_2 + I_2 + 2H_2O \rightleftharpoons H_2SO_4 + 2HI$$

The titration is followed by the blue starch–iodine color. Commercial instruments are available which perform the titration automatically, or manually using a photocell to follow the color change and to keep it at some preselected color intensity. This titration can be made quite sensitive[9]: a few micrograms of sulfur can be detected.

A sensitive colorimetric method developed by West and Gaeke[46] for sulfur dioxide in the atmosphere shows promise as a technique for sulfur in metals.[47] The sulfur dioxide is absorbed in sodium tetrachloromercurate solution as the stable nonvolatile disulfitomercurate(II). This SO_2 is then measured by means of the color developed with pararosaniline and formaldehyde. The sensitivity is about 2 μg.

Sulfur dioxide may be absorbed in hydrogen peroxide and the resulting sulfate determined in a number of ways. For instance, the barium sulfate precipitate may be measured either gravimetrically or turbidimetrically, or by adding excess barium and titrating the excess with EDTA. These are all fairly standard methods. There are many others and the interested reader is referred to the *Treatise on Analytical Chemistry*[44] for these as well as numerous other techniques for sulfur dioxide.

2. PHYSICAL METHODS

Although chemical methods seem most popular for sulfur dioxide measurements, a number of physical methods have been described. Gas chromatography has been used for sulfur determinations, usually in conjunction with carbon determinations. Stuckey and Walker[36] have described a system for determining sulfur as well as carbon in ferrous metals after combustion using a silica gel column and thermal conductivity detectors. The products are eluted from the column by temperature programming after using helium to purge oxygen. Galwey[35] discusses the possibilities of determining sulfur by means of a chromatographic method employing argon as a sweep gas for liquid O_2 trap which was used for carbon determinations. Malissa and Schmidts[48] have described a conductometric procedure for determining carbon, hydrogen, oxygen, and sulfur from the combustion of organic compounds. Presumably this could be applied to combusted products from metal analysis.

Other physical methods including activation analysis, mass spectrometry, and emission spectroscopy have been applied to the detection of sulfur, although not necessarily in combustion processes.

IV. SELECTED EXAMPLES

In order to illustrate some of the ideas presented here and to demonstrate the different approaches which one may use in combustion methods, a few selected examples of specific determinations will be presented. This is not a complete bibliography but a representative sample of the type of analyses being done by these methods.

When sensitivity is mentioned, it will usually be an estimate of the lowest amount of carbon in weight percent or parts per million that can be detected successfully. This depends on the allowable sample size as well as the inherent sensitivity of the particular detection technique employed. For instance, a method for carbon in uranium is often more sensitive than a method for carbon in beryllium, other things being equal, simply because uranium is more dense and thus allows more sample by weight to be taken for each analysis. However, if one were to consider atomic percent, which is probably a more realistic measure of trace element effect, then the differences in sensitivity are much less significant.

A. Carbon Determinations

Unlike sulfur, carbon analyses in metals and alloys are more often accomplished by oxidative combustion and measurement of CO_2 than by other methods.

Pepkowitz and Moak[49] have determined carbon in a number of iron and steel samples as well as manganese, molybdenum, copper, lead, and tin without flux. They used a high frequency induction furnace, a gas buret (to regulate the flow rate during combustion), and a differential freeze-out technique for isolating carbon dioxide. They measured the carbon dioxide by means of a McLeod gauge. They indicated a precision of $\pm 2\ \mu g$ and used sample weights of about 1 g. Besides the elements mentioned above, attempts were made to combust zirconium and titanium, but without the addition of flux these elements burned too fast for proper CO_2 trapping, and the authors recommended the use of manganese as a bedding material which extended the combustion time. This method is time limited for a given oxygen flow. The volume of the gas buret determines the length of time of the combustion, which is about 5 min.

The determination of carbon in molybdenum without a flux by this method is noteworthy since molybdenum sublimes and tends to interfere with most procedures somewhere in the analytical system. Lead has been used[4] as a flux for molybdenum to correct this condition.

Green et al.[4] have described some specific conditions for the determination of carbon in molybdenum, tungsten, silicon, stainless steels, nickel and some of its alloys, zirconium, uranium, titanium, beryllium, and several other materials. They used either a high frequency induction furnace or a resistance furnace depending on the material being analyzed and employed a conductometric finish. Some of their recommendations are as follows. Carbon in stainless steels, nickel and its alloys, zirconium, and uranium can all be determined relatively easily using an induction furnace and an iron or iron–copper flux. As mentioned previously, lead should be used when combusting molybdenum, and they have found that if lead is used when combusting silicon, it makes the combustion possible even with a resistance furnace. They use a copper–lead flux for titanium. Tungsten is burned in a tube furnace with no flux. Beryllium is combusted with an iron copper flux but this is not entirely satisfactory. They indicated the precision of their method was about ± 10 μg and sample weights ranged from 200 mg to 1 g.

Massey[50] has described a conductometric method for carbon in beryllium similar to the one above. Sample weights were 300 mg and the precision was ± 15 μg of C. Aronin et al.[51] have also developed a method for carbon in "high purity" beryllium employing a combustion–conductometric technique. By using large samples, they were able to gain an increase in sensitivity. They could detect 25 ppm C in beryllium. They also prepared several batches of beryllium to be used as standards in the 25, 100, and 120 ppm C range. The determination of nitrogen was also investigated.

A method for carbon in boron has been developed by Kuo et al.[52] A powdered boron sample was burned with tin accelerator using an induction furnace, and the carbon dioxide was determined conductometrically. The crucibles were prefired with NBS steel and tin to control the blank. Boron could be cracked in a "diamond mortar" and selected particles taken by sieving. 100 mesh and smaller boron samples gave excellent reproducibility but larger samples showed evidence of incomplete combustion. There was some evidence that steel inhibited the combustion of boron. Flow rates of 50–200 ml/min were investigated and found to have no effect on the determination. A precision of about ±5 to ±25 ppm was indicated for the range 350–1000 ppm carbon. Later Walker et al.[53] applied these combustion conditions using a gas chromatographic finish having a maximum sensitivity of around 5 ppm carbon.

Gordon et al.[10] developed a method for determining less than 10 ppm carbon in tungsten. They question the necessity of adding fluxes to tunsten and "other refractories" and present some evidence for incomplete combustion of tungsten in the presence of iron chips. They used 2–8 g samples and an induction furnace–conductometric approach, and showed that the complete oxidation of a number of metals including tungsten could be done by direct induction heating, providing the particle sizes of the samples were between 16 and 80 mesh. Larger tungsten particles did not heat effectively because of the surface heating phenomena of high frequency induction heating. Small particles provided such a discontinuous path that they also were heated ineffectively. However, if the smaller particles were raised to the ignition temperature, the reaction became exothermic and was then self-sustaining. Platinum disks placed in the bottom of the crucible were heated by induction and were used for this purpose. Some experiments were done with tantalum, niobium, zirconium, titanium, chromium, and copper, indicating similar results could be expected with these metals.

In concluding this section on carbon determinations, the method of Solet[54] should be mentioned. He employed a temperature of 600°C and a conductometric finish to oxidize and determine surface carbon on nickel strips. By this method the diffusion of bulk carbon to the surface could be observed.

B. Sulfur Determinations

There are a number of methods available for sulfur in metals and alloys other than the conbustion techniques described here; an appropriate review should be consulted for more details.[44] The most popular alternative method is the H_2S evolution method discussed in the ASTM Methods for Chemical Analysis of Metals.[1] A few specific applications of the combustion-in-oxygen method will be described to indicate its scope.

The combustion of the sample in oxygen and the subsequent determination of sulfur dioxide by the iodate–iodide method is described for both resistance and induction heating in the ASTM Methods of Chemical Analysis.[1] Originally proposed for metals containing 50–500 ppm, its sensitivity can be extended by photometric reproduction of the blue iodate–iodide color before and after absorption of sulfur dioxide. This method is very popular and a number of commercial instruments are available which employ this procedure. It has been applied to a large variety of materials by the Laboratory Equipment Corporation which has published a collection of methods.*

Boyle et al.[9] have applied the iodate–iodide combustion method to the determination of sulfur in beryllium oxide and beryllium oxide–uranium oxide fuel elements. They were able to extend the range of the standard method below 10 ppm by some simple modifications of a Leco iodimetric titrator. A flux consisting of 2.2 g of iron powder and 2.2 g of granulated tin was used in this application. The same technique has been employed for the determination of trace sulfur in a number of other metals[56]. One-gram samples were generally used.

Recently a number of investigators have employed the pararosaniline and formaldehyde colorimetric endpoint after combustion of the sample. Burke and Davis[47] have employed this system for the determination of sulfur in nickel and some of its alloys as well as a variety of iron and steel samples. They used a flux consisting of 1 g of iron powder, 1.5 g of tin and one copper ring. A 30 μg range at about $\pm$ 1 μg was covered.

Barabas and Kaminski[57] made a rather complete study of this method as applied to sulfur in blister and refined copper. They used a resistance furnace at a temperature of 1150 $\pm$ 10°C. This temperature is about 70° higher than the melting temperature of copper. No fluxes were necessary and the copper samples melted promptly. Some interference studies were made with selenium, tellurium, arsenic, and antimony. No interferences were found at concentrations up to 10 times that of sulfur. Sensitivity was about 1 μg.

Acs and Barabas[58] applied the method to the determination of sulfur in selenium samples. Here the principal problem was to devise a trap which would remove selenium from the oxygen stream during combustion and absorption of the SO_2. Using sample sizes of 0.5–0.7 g, a range of 5–600 ppm sulfur was covered.

Kahin[59] was able to determine sulfur in solid lubricants containing molybdenum by a combustion flow method employing wet oxygen. He was able to separate and determine sulfur contained on the surface of alloys, ceramics, and quartz using oxidizing temperatures rising to 550°C. A 10–50 mg sample of MoS_2 gave satisfactory results.

Cějchan and Vorliček[60] have studied the thermal stability and decomposition rates of some sulfides and sulfates and from this have estimated optimum conditions for combustion procedures. They have also attempted to define precisely the term "combustible sulfur."

* These are available on request from the Laboratory Equipment Corporation, St. Joseph, Michigan.

V. CONCLUDING REMARKS

It is important when developing a combustion procedure for a new material, or when employing such a procedure in the analysis of a wide variety of methods, that a detection system be used which enables one to follow the combustion process. Those techniques which involve trapping combustion products subsequent to final measurement can lead to erroneous results. Once conditions for a specific material have been established, then of course these detection systems can be used.

Methods of combustion other than those performed in flowing oxygen systems have not been discussed. These include techniques such as oxide bed combustions and wet chemical oxidations. The latter procedure offers some advantage for standardizing methods and determining accuracy at trace levels since it involves total sample dissolution. The former procedure[61] gives more control over oxidizing conditions.

Besides the ever-increasing demands for lower levels of detection, future trends will probably include more emphasis on accuracy and an increasing interest in the chemical forms of carbon and sulfur which may be contributing to the properties of the sample.

References

1. *ASTM Methods for Chemical Analysis of Metals*, American Society for Testing and Materials, Philadelphia, 1960.
2. G. Ingram, in *Treatise on Analytical Chemistry*, Part II, Vol. 11, I. M. Kolthoff and P. J. Elving, Eds., Interscience, New York, 1965, Section B–1, p. 297.
3. F. A. Cotton and G. Wilkinson, *Advanced Inorganic Chemistry*, Interscience, New York, 1962, Chaps. 11 and 21.
4. I. R. Green, J. E. Still, and R. C. Chirnside, *Analyst*, **87**, 530 (1962).
6. J. W. Fulton and R. E. Fryxell, *Anal. Chem.*, **31**, 401 (1959).
7. R. C. Rooney and F. Scott, *J. Iron Steel*, (*London*), **195**, 417 (1960).
8. W. G. Rice-Jones, *Anal. Chem.*, **25**, 1383 (1953).
9. W. G. Boyle, Jr., L. J. Gregory, and W. Sunderland, Report No. UCRL–7204, Lawrence Radiation Laboratory, University of California, Livermore, California, 1963.
10. W. A. Gordon, J. W. Grant, and Z. T. Tumney, *Anal. Cheml*, **36**, 1396 (1964).
12. D. C. White, *Talanta*, **10**, 727 (1963).
13. E. J. Merkle and J. W. Graab, *Anal. Chem.*, **38**, 159 (1966).
14. A. M. G. MacDonald, *Ind. Chemist*, **35**, 193 (1959).
15. T. T. White, V. A. Campanile, E. J. Agazzi, L. D. TeSelle, P. C. Tait, F. R. Brooks, and E. D. Peters, *Anal. Chem.*, **30**, 409 (1958).
16. P. West and F. Ordoveza, *Anal. Chem.*, **34**, 1324 (1962).
17. W. G. Boyle, Jr., F. B. Stephens, and W. Sunderland, Report No. UCRL–12201, Lawrence Radiation Laboratory, University of California, Livermore, California, 1964.

18. Sylvania Electric Products, Inc., Towanda, Pennsylvania.
19. C. J. Morris and J. W. Frazer, Lawrence Radiation Laboratory, University of California, Livermore, California, private communication, 1966.
20. R. M. Fowler, W. G. Guldner, T. C. Bryson, J. L. Hague, and H. J. Schmitt, *Anal. Chem.*, **22**, 486 (1950).
21. J. W. Frazer and R. T. Holzman, Report No. UCRL–6020, Lawrence Radiation Laboratory, University of California, Livermore, California, 1960.
22. J. E. Still, L. A. Dauncey, and R. C. Chirnside, *Analyst*, **79**, 4 (1954).
23. *Instruction Manual* for the 515–000 Conductometric Analyzer, Laboratory Equipment Corporation, St. Joseph, Michigan.
24. E. J. Violante, *Anal. Chem.*, **36**, 856 (1964).
25. E. H. Bruins, *Anal. Chem.*, **35**, 934 (1963).
26. J. J. Engelsman, A. Meyer, and J. Visser, *Talanta*, **13**, 409 (1966).
27. A. Patchornik and Y. Shalitin, *Anal. Chem.*, **33**, 1887 (1961).
28. J. A. Grant, J. A. Hunter, and W. H. S. Massie, *Analyst*, **88**, 134 (1963).
29. R. F. Jones, P. Gale, P. Hopkins, and L. N. Powell, *Analyst*, **90**, 623 (1965).
30. L. Blom and L. Edelhausen, *Anal. Chim. Acta*, **13**, 120 (1955).
31. W. G. Boyle, Jr., F. Stephens, and W. Sunderland, *Anal. Chem.*, **37**, 933 (1965).
32. W. G. Boyle, Jr. and W. Sunderland, in Report No. UCRL–14767, Analytical Chemistry Section Progress Report, Jan. 1-Dec. 3, J. Harrar, Ed., Lawrence Radiation Laboratory, University of California, Livermore, California, 1965, p. 22.
33. J. M. Walker and C. W. Kuo, *Anal. Chem.*, **35**, 2017 (1963).
34. L. L. Lewis and M. J. Nardozzi, *Anal. Chem.*, **36**, 1329 (1964).
35. A. K. Galwey, *Talanta*, **10**, 310 (1963).
36. W. K. Stuckey and J. M. Walker, *Anal. Chem.*, **35**, 2015 (1963).
37. W. Simon, P. F. Sommer, and G. H. Lyssy, *Microchem. J.*, **6**, 239 (1962).
38. C. F. Nightingale and J. M. Walker, *Anal. Chem.*, **34**, 1435 (1962).
39. G. A. Tipler, *Analyst*, **88**, 272 (1963).
40. K. Y. Eng, R. A. Meyer, and C. D. Bingham, *Anal. Chem.*, **36**, 1832 (1964).
41. E. Ricci and R. L. Hahn, *Anal. Chem.*, **37**, 742 (1965).
42. W. L. Harrington, R. K. Skogerboe, and G. H. Morrison, *Anal. Chem.*, **38**, 821 (1966).
43. D. F. Boltz, *Colorimetric Determination of Nonmetals*, Interscience, New York, 1958, pp. 261–308.
44. B. J. Heinrich, M. D. Grimes, and J. E. Puckett, in *Treatise on Analytical Chemistry*, Part II, Vol. 7, I. M. Kolthoff and P. J. Elving, Eds., Interscience, New York, 1961, pp. 1–135.
45. M. Codell, G. Norwitz, and C. Clemency, *Anal. Chem.*, **29**, 1496 (1957).
46. P. W. West and G. C. Gaeke, *Anal. Chem.*, **28**, 1816 (1956).
47. K. E. Burke and C. M. Davis, *Anal. Chem.*, **34**, 1747 (1962).
48. H. Malissa and W. Schmidts, *Microchem. J.*, **8**, 180 (1964).
49. L. P. Pepkowitz and W. D. Moak, *Anal. Chem.*, **26**, 1022 (1954).
50. E. M. Massey, Report Y–1321, Union Carbide Nuclear Company, Y–12 Plant, Oak Ridge, Tennessee, 1960.
51. L. R. Aronin, F. A. Bauman, and A. K. Wolff, Report AFML–TR–65–195, Nuclear Metals, West Concord, Mass., 1965.
52. C. W. Kuo, G. T. Bender, and J. M. Walker, *Anal. Chem.*, **35**, 1505 (1963).
53. J. M. Walker, J. Spigarelli, and G. T. Bender, *Anal. Chem.*, **37**, 299 (1965).
54. I. S. Solet, *Anal. Chem.*, **38**, 504 (1966).
56. W. G. Boyle, Jr. and W. Sunderland, unpublished work.

57. S. Barabas and J. Kaminski, *Anal. Chem.*, **35**, 1702 (1963).
58. L. Acs and S. Barabas, *Anal. Chem.*, **36**, 1825 (1964).
59. I. L. Kalnin, *Anal. Chem.*, **36**, 886 (1964).
60. O. Cějchan and J. Vorlíček, *Sb. Praci Vuzdhe*, **5**, 155 (1964); *Anal. Abstr.*, 6448 **12**, (Dec. 1965).
61. J. G. Sen Gupta, *Anal. Chem.*, **35**, 1971 (1963).

Chapter 5
ELECTROANALYTICAL TECHNIQUES

JACKSON E. HARRAR, Lawrence Radiation Laboratory, University of California, Livermore, California

I. Introduction 144
II. Potentiometry and Potentiometric Titrations 144
A. Direct Potentiometry 146
B. Potentiometric Titrations 146
III. Voltammetry, Polarography, and Related Techniques 151
A. Introductory Considerations 151
1. Mass Transfer Processes and Classification of Methods 151
2. Working Electrode Materials 152
3. Analytical Sensitivity and the Charging of the Double Layer 154
B. Chronoamperometry 154
1. Current–Time Relationship 156
2. Current–Potential Relationships 157
C. Classical Polarography 159
D. Potential-Sweep Chronoamperometry 164
E. Stripping Analysis 166
F. Derivative Polarography 167
G. Alternating Current Polarography 168
H. Square-Wave, Pulse, and Tast Polarography 168
I. Cyclic Techniques 169
J. Chronopotentiometry 169
K. Voltammetry with Forced Convection 170
IV. Amperometric Titrations 172
V. Coulometry 173
A. Constant-Current Coulometry 174
B. Controlled-Potential Coulometry 175
VI. Electrogravimetry and Electroseparations 179
VII. Electrography and Electrospot Testing 179
VIII. Conductometry and Conductometric Titrations 180
General References 181
References 182

I. INTRODUCTION

As a branch of electrochemistry, electroanalytical chemistry is based on the utilization of electrochemical phenomena for chemical analysis, and its techniques deal with the measurement of the fundamental quantities of voltage, current, resistance, and time. Many different electroanalytical techniques have been developed for relating the measured value of one of the electrical parameters to the identity or quantity of substance determined. These techniques are characterized by a diversity of application, both in the types of material that are analyzed and in the quantities or concentrations of substances that are determined. Electroanalytical methods may be employed for the examination of virtually any material that can be brought into solution, and the field ranges from procedures suitable for the analysis of trace contaminants in highly pure specimens, to the techniques of accurate assay of major constituents. In addition to providing quantitative and qualitative information on material composition, several electroanalytical procedures are also extensively used as separation or preconcentration techniques in conjunction with other analytical methods. Closely allied with the analytical applications, electrochemical methods have been developed concurrently as techniques for the study of the kinetics and thermodynamics of electrode reactions.

II. POTENTIOMETRY AND POTENTIOMETRIC TITRATIONS

Potentiometric analytical methods are based on the fact that the electromotive force or potential (emf, E) of a galvanic cell depends on the activities, or concentrations, of the species present in the cell solution. Measurements are made of the emf existing between an electrode of known or constant potential, a *reference electrode*, and another *indicator electrode* whose potential is a function of the concentration of the analyzed constituent.

Consider, for example, a cell represented by the following:

$$\text{(I)} \qquad \mathrm{Hg, Hg_2Cl_2} \mid \mathrm{KCl\ (sat'd)} \parallel \mathrm{HCl, Fe(II), Fe(III)} \mid \mathrm{Pt}$$

In this notation the single vertical line indicates a boundary between solid and liquid solution phases, and the double line symbolizes the boundary between two solutions or a *liquid junction*. The total voltage of this cell, measured at terminals connected to the mercury and platinum, will be the sum of (*1*) the potential of the left-hand half-cell, the saturated calomel electrode, SCE; (*2*) the liquid junction potential, which arises because of the differences in concentrations and mobilities of the ions of two solutions; and (*3*) the potential of the right-hand cell, consisting of a platinum electrode in contact with a solution of ferrous and ferric ions.

Because the potential of the platinum indicator electrode is established by the electrochemically reversible couple Fe(II)/Fe(III), the electrode potential can be measured with respect to the SCE and related to changes in the ratio of the activities of the iron species. The manner in which the potential, E, of the platinum electrode varies is given by a form of the Nernst equation

$$E = E^0 + (RT/nF) \ln [a_{Fe(III)}/a_{Fe(II)}] \tag{1}$$

where E^0 is the *standard potential* of the system, R is the gas constant, T is the absolute temperature, F is the value of the faraday, 96,487 C/g-equiv, and n is the number of electrons involved in the reaction. The prefix of the logarithmic term has a value of $0.05915/n$ V for log to base 10 at 25°C. To be analytically useful in either direct potentiometry or potentiometric titrations, the potential-determining reaction must reach equilibrium rapidly with the indicator electrode.

Ideally, electrode potentials would be determined with the passage of zero current through the cell so that the equilibrium at the indicating electrode is not disturbed, and reaction kinetics or mass transfer effects do not influence the measured values. Such *polarization* effects are avoided in practice by the use of null-balancing potentiometers or electronic indicating instruments with high impedance input stages.

Half-cell electrode emf's and junction potentials must always be measured with respect to some reference electrode; these numerical values are based on a scale with a zero established arbitrarily as the potential of the normal hydrogen electrode (NHE). The hydrogen electrode consists of a platinized platinum electrode in contact with hydrogen gas at 1 atm partial pressure and immersed in a solution containing hydrogen ions at unit activity. For example, the emf of the cell

$$\text{(II)} \qquad \overbrace{\text{Pt} \mid \text{H}_2\text{(1 atm)}, \text{H}^+(a = 1)}^{\text{NHE}} \parallel \text{KCl (sat'd)} \mid \text{Hg}_2\text{Cl}_2, \text{Hg}$$

establishes the numerical value of the saturated calomel electrode potential, which at 25°C is 0.2444 V, including the liquid junction potential.[1] Because it is stable, reproducible, and convenient to prepare, the SCE is the most commonly used reference electrode in electroanalytical chemistry. Another widely used reference electrode consists of a silver wire coated with AgCl in contact with a solution of chloride ion at a fixed concentration.

From the measured cell potentials and a knowledge of the standard potentials, an approach could be made to calculate concentrations directly from equation 1. This absolute method would be disadvantageous, however, because of the need for activity coefficient information and the difficulty in evaluating the effect of liquid junction potential potentials. Thus the applications of direct potentiometry are limited to techniques in which the electrolytic cell-measurement system is calibrated or standardized with known concentrations of the analyzed substance. Still better accuracy and precision are obtained when the potentiometric system is made the basis of a titration method.

A. Direct Potentiometry

The measurement of pH by means of the glass electrode is the most familiar example of a direct potentiometric method; this technique involves the type of cell

$$\text{(III)} \qquad \text{Hg, Hg}_2\text{Cl}_2 \,|\, \text{KCl} \,||\, \overbrace{\text{H}^+ \ (a = ?) \,\vdots\, }^{\text{Glass membrane}} \underbrace{\text{HCl} \,|\, \text{AgCl, Ag}}_{\text{Glass electrode}}$$

The special glass membrane develops a potential which is sensitive to changes in the hydrogen ion activity in the same manner as the hydrogen ion–gas electrode, i.e., its potential follows the relation

$$E = E^0 + 0.05915 \log a_{\text{H}^+} = E^0 - 0.05915 \text{ pH} \qquad (2)$$

Thus the pH of an unknown solution can be found after standardization of the cell–instrument system with a solution of known pH.

Another practical application of direct potentiometry is in the determination of various anions with silver electrodes coated with the insoluble silver salts of the anions. For example, the potential of the silver–silver chloride electrode is a Nernstian logarithmic function of the activity of chloride ion. Similar responses can be obtained with silver bromide, iodide, and sulfide electrodes.

Recently a number of electrodes have been developed and are commercially available which are sensitive to specific ions and groups of ions in specific valence states. These electrodes, like the glass pH electrodes, are composed of a reference electrode together with a special cationic or anionic sensitive membrane. The electrodes are used with conventional potentiometric instrumentation such as pH meters. A summary of the characteristics of some of the available specific ion electrodes, including the pH and Ag,AgCl electrodes, is given in Table 1.[2]

B. Potentiometric Titrations

Titration procedures involving potentiometric detection of the equivalence point can be applied to a variety of analytical problems. A very large number

Table 1. Specific Ion Electrodes

Type of electrode	Ion	Concentration range	Manufacturer
pH glass membrane	H^+	$0.5–10^{-14}M$	Several
Silver–silver chloride	Cl^-	$0.5–10^{-5}M$	Several
Glass membrane	Na^+	$1.0–5 \times 10^{-7}M$	Beckman Instruments, Inc.[a]
Glass membrane	Monovalent cations, e.g., Ag^+, K^+, Rb^+, Cs^+, Li^+	$1.0–5 \times 10^{-7}M$	Beckman Instruments, Inc.[a]
Liquid ion-exchange membrane	Ca^{2+}	ca. $1–10^{-5}M$	Corning Glass Works[b] Orion Research, Inc.[c]
Liquid ion-exchange membrane	Divalent cations, e.g., Ca^{2+}, Mg^{2+}	$2.0–10^{-5}M$	Orion Research, Inc.[c]
Rubber membrane	SO_4^{2-}, PO_4^{3-}, Ni^{2+}	$0.1–10^{-5}M$	National Instruments Laboratories, Inc.[d]
Rare earth-doped crystal	F^-	$1.0–10^{-6}M$	Orion Research, Inc.[c]

[a] Fullerton, Calif.
[b] Medfield, Mass.
[c] Cambridge, Mass.
[d] Rockville, Md.

of indicator electrode systems have been devised for the monitoring of various types of reactions: acid–base, precipitation, oxidation–reduction, and complex formation. These titration methods, indeed electrometric titration methods in general, frequently possess advantages in selectivity compared with visual indicator techniques, and may be applicable to highly colored solutions when visual equivalence-point detection is difficult. As with classical volumetric procedures, a routine accuracy and precision of 0.1% of the amount of the constituent analyzed is usually possible, provided 50–100 mg of the constituent is available for analysis.

Cell I is an example of an actual cell that is used for the potentiometric titration of ferrous iron. In this determination the Fe(II) is titrated and oxidized to Fe(III) by means of a strong oxidizing agent such as Ce(IV). The total cell potential is followed during the titration and is found to vary as shown in Figure 1. Because a constant potential reference half-cell is used,

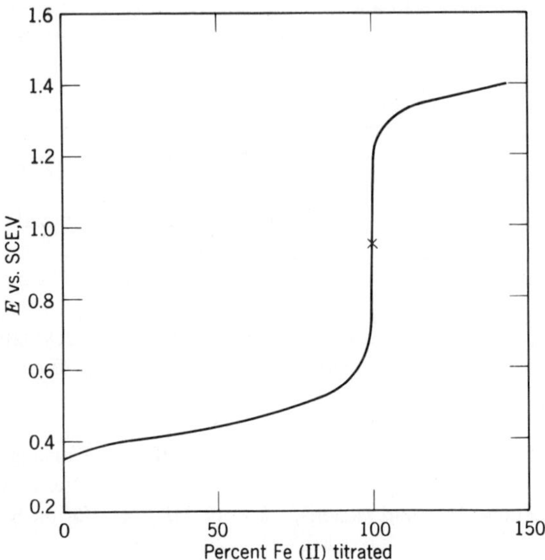

Fig. 1. Potentiometric titration curve for the determination of ferrous ion with ceric ion. (X) Equivalence point.

changes in the total cell potential reflect changes only at the platinum indicator electrode. These concentration changes cause the electrode potential to vary in accordance with equation 1 and the corresponding equation for Ce(IV)/Ce(III). As the concentration of Fe(II) approaches zero in the titration, a large change in cell potential occurs; this equivalence point denotes that a stoichiometric amount of Ce(IV) has been added. The actual potential of the equivalence point in this case is the mean value of the standard potentials of the two couples.

A requirement for a successful potentiometric titration is that there be sufficient potential change at the equivalence point; this depends on the completeness of the titration reaction, as indicated by its equilibrium constant. In addition, equilibrium should be established rapidly between the reactants and between the electrode and the potential-determining species. Variations of the basic potentiometric method, in which the indicator electrode potential is measured at virtually zero current, include the use of a small polarizing current between the electrodes and the use of two metallic indicator electrodes with polarizing current. The polarizing current has been found to speed the establishment of equilibrium in titrations involving irreversible couples such as thiosulfate/tetrathionate. The most commonly used indicator electrodes for oxidation–reduction titrations are the noble metals such as platinum or gold.

Potentiometric titrations are important in the analysis of many substances with acidic or basic groups, with the glass pH electrode usually functioning as the indicator electrode. The magnitude of the potential break near the equivalence point depends on the concentrations and ionization constants of the acidic and basic species. For practical work in 0.1N solutions, acid or base ionization constants of at least 10^{-8} are required for titration with, respectively, strong bases and strong acids.[3]

Precipitation reactions are the basis of a number of potentiometric titrations; most of the available methods involve insoluble salts of silver or mercury and indicator electrodes of these metals. The silver electrode, for example, may be used for the determination of silver as well as for the titrations of many of the anions which form insoluble silver salts. In this type of titration the smaller the solubility product of the salt, the larger the potential change at the equivalence point. A useful method for the determination of zinc is based on its precipitation with ferrocyanide; however, in this case a platinum indicator electrode is used, with the electrode potential being established by the $Fe(CN)_6^{3-}/Fe(CN)_6^{4-}$ couple.

The advent of chelating agents such as ethylenediaminetetraacetic acid (EDTA) as standard titrating agents in recent years has been accompanied by the development of a valuable potentiometric method for detecting the equivalence points in the titrations. The indicator electrode is a small mercury pool or mercury-coated gold or platinum electrode.[4–6] The mercury indicator electrode for EDTA titrations devised by Reilley and co-workers[5,6] consists of the following half-cell:

$$\text{(IV)} \qquad Hg \,|\, Hg(EDTA)^{2-}, \quad M(EDTA)^{2-}, \quad M^{2+}$$

where M^{2+} is the metal ion which is titrated. The $Hg(EDTA)^{2-}$ is added to the solution in a small amount and an equilibrium is established which causes the electrode to function as a pM electrode in a manner exactly analogous to a pH electrode. By appropriate choice of titrating conditions, many different metal ions can be titrated with this technique and a few mixtures can be titrated with 1–2% accuracy for each component. Some metals that can be determined by this procedure are listed in Table 2. This table is a brief compilation of some of the most popular zero current potentiometric titration methods.

Several instruments have been designed for carrying out the operations of a potentiometric titration automatically. Equipment is commercially available both for automatically recording the change in potential as a function of titrant volume, and for automatically terminating the titration when the equivalence point is reached.[7] The automatic titrators are based either on titration to a preset potential, or titration until the maximum rate-of-change of potential is detected.

Table 2. Potentiometric Titrations

Substance determined	Indicator electrode	Titrant	Ref.
	Oxidation–Reduction		
Mn(II)	Pt	$KMnO_4$	a
Fe(II)	Pt	Ce(IV), $K_2Cr_2O_7$	b
Cr(VI)	Pt	Fe(II)	c
V(IV)	Pt	Ce(IV)	d
U(IV)	Pt	$K_2Cr_2O_7$	e
Pu(III)	Pt	Ce(IV)	f
	Acid–Base		
B as H_3BO_3	Glass(pH)	NaOH	g
	Precipitation		
Ag	Ag	KCl	h
Zn	Pt	$Fe(CN)_6^{4-}$	i
	Complexometric		
CN^-, forming $Ag(CN)_2^-$	Ag	$AgNO_3$	j
Ca, Sr, Ba, Mg, Zn, Cd, Hg, Cu, Pb, Mn, Co, Ni, V, Al, Ga, In, Tl, Cr, Sc, Y, rare earths, Zr, Hf, Th, Bi	Hg(pM)	EDTA	k

References:

[a] J. J. Lingane and R. Karplus, *Ind. Eng. Chem., Anal. Ed.*, **18**, 191 (1946).

[b] L. M. Melnick, in *Treatise on Analytical Chemistry*, Part II, Vol. 2, I. M. Kolthoff and P. J. Elving, Eds., Interscience, New York, 1962, p. 277; J. J. Lingane, *Electroanalytical Chemistry*, 2nd ed., Interscience, New York, 1958, p. 135.

[c] W. H. Hartford, in *Treatise on Analytical Chemistry*, Part II, Vol. 8, I. M. Kolthoff and P. J. Elving, Eds., Interscience, New York, 1963, pp. 327–328.

[d] N. H. Furman, *J. Am. Chem. Soc.*, **50**, 1675 (1928).

[e] C. J. Rodden, *Analytical Chemistry of Essential Nuclear Reactor Materials*, U.S. Government Printing Office, Washington, D. C., 1964, pp. 92–96.

[f] C. F. Metz and G. R. Waterbury, in *Treatise on Analytical Chemistry*, Part II, Vol. 9, I. M. Kolthoff and P. J. Elving, Eds., Interscience, New York, 1962, pp. 352–358.

[g] R. D. Strahm, in *Treatise on Analytical Chemistry*, Part II, Vol. 12, I. M. Kolthoff and P. J. Elving, Eds., Interscience, New York, 1965, pp. 203–205.

III. VOLTAMMETRY, POLAROGRAPHY, AND RELATED TECHNIQUES

A. Introductory Considerations

1. MASS TRANSFER PROCESSES AND CLASSIFICATION OF METHODS

Voltammetry, used as both a generic and specific term, refers to electroanalytical techniques in which the electrolysis current at a *working electrode* is determined as a function of the electrode potential. In contrast to potentiometry, which is concerned theoretically only with the thermodynamics of electrode processes, techniques involving appreciable electrolysis currents are influenced by the kinetics of the electrode processes and the nature of the flow of matter and electricity within the electrolytic cells. Three modes of mass transfer can be operative in an electrolytic cell: *diffusion* of molecular or ionic species due to concentration gradients, *convection* or bulk movement of the solution due to stirring or density gradients, and *migration* of charged species due to electric field gradients. Most voltammetric work is conducted with solutions in which the electroactive species is present at low concentrations with a large excess of an inert *supporting electrolyte* species so that migration effects are negligible. The supporting electrolyte also serves to lower the resistance of the medium and reduce the effects of ohmic iR losses in the solution. Either diffusion or convection is thus the dominant mode of mass transfer and voltammetric techniques are differentiated accordingly.

In the case of voltammetric methods which do not involve deliberate stirring of the solution or motion of the electrode during the electrolysis, diffusion is the sole mode of mass transfer. Natural convection is avoided by carrying out the electrolysis on a short time scale (< 60 sec). Since the diffusion gradient, which is established by the electrolysis, changes continuously, the current or potential varies as a function of time. This group includes *chronoamperometry*, in which the current is measured as a function of time at constant or controlled potential; and *chronopotentiometry*, in which the controlled variable is current and the measured variable is the electrode potential. Classical *polarography* with the dropping mercury electrode (DME) also involves the measurement of transient currents, but the periodic

---

[h] E. P. Przybylowicz and C. W. Zuehlke, in *Treatise on Analytical Chemistry*, Part II, Vol. 4, I. M. Kolthoff and P. J. Elving, Eds., Interscience, New York, 1966, p. 59.

[i] I. M. Kolthoff and N. H. Furman, *Potentiometric Titrations*, 1st ed., Wiley, New York, 1926, pp. 268–271.

[j] *Ibid.*, pp. 170–173.

[k] Reference 6.

renewal of the electrode creates a pseudo steady state so that current-potential relationships are readily obtained without special timing circuits.

The contrasting feature of voltammetry with forced convection is that, at constant potential, a steady state current is obtained. Because the rate of transfer of the electroactive species to the electrode is greater in convection than in diffusion, convection methods in principle have greater analytical sensitivity; however, these techniques are somewhat difficult to implement for small mercury electrodes and are less subject to rigorous theoretical treatment. Thus the principal applications of these methods are in analysis with solid electrodes and in electrometric titrations.

2. WORKING ELECTRODE MATERIALS

Of fundamental importance in all electroanalytical methods is the nature of the working electrode material. Because of the differences in the solvent reaction and electrode decomposition potentials, various electrode materials have different ranges of application. This is illustrated in Figure 2 for platinum and mercury. In the case of platinum the useful potential range is established by the decomposition of the solvent: the evolution of oxygen at the anodic limit and the evolution of hydrogen at the cathodic limit. The evolution of hydrogen occurs on platinum at nearly the thermodynamic reversible potential; oxygen evolution on the other hand requires an overvoltage of about 0.5 V. In the case of mercury a high overvoltage for the evolution of hydrogen extends the practical cathodic range of this metal by 0.8 V. The anodic range of mercury, however, is more restricted than that of platinum because of the greater ease of oxidation of mercury. The anodic range of mercury is further limited when anions such as chloride, which form insoluble mercury salts, are present.

The behavior of platinum as an electrode material is also profoundly influenced by the fact that, as shown in Figure 2, the metal forms oxides above a certain potential. Although the processes are not completely understood, these oxides form as a surface film on the electrodes[8]; they also exchange electrons in the electrode reactions as would the pure metal and in some cases take part in the mechanism of the electrode reactions. Platinum electrodes thus do not function in an ideally inert manner; moreover, irreversibility in the oxide film formation may cause the characteristics of a given electrolysis to depend on the past history of the electrode. Similar effects are exhibited by platinum electrodes in chloride media as a result of platinum chloride films.[9] Thus careful experimental technique directed toward reproducing the surface condition of the electrode is frequently required in order to obtain meaningful measurements on platinum.

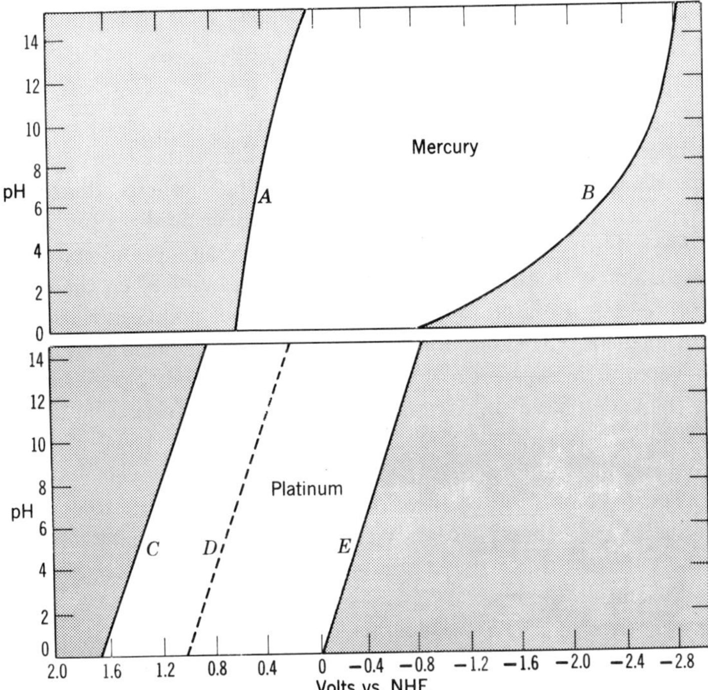

Fig. 2. Practical potential–pH range for mercury and platinum electrodes in aqueous solutions. *A*, $Hg + H_2O \rightarrow Hg_2^{++}$, HgO, $HHgO_2^-$; *B*, $2H_2O + 2e^- \rightarrow H_2 + 2OH^-$; $2H^+ + 2e^- \rightarrow H_2$; *C*, $2H_2O \rightarrow O_2 + 4H^+ + 4e^-$; *D*, $Pt + H_2O \rightarrow PtO$, PtO_2; *E*, $2H_2O + 2e^- \rightarrow H_2 + 2OH^-$; $2H^+ + 2e^- \rightarrow H_2$.

Gold forms oxide layers[10] at slightly more anodic potentials than platinum and has proved to be a better electrode material in some applications; however, gold has the disadvantage of dissolving anodically in halide media. The search for solid electrode substances with better characteristics than the noble metals has led to the use of such materials as pure carbon[11], wax-impregnated carbon[12], boron carbide[13], pyrolytic graphite[14], and glassy carbon[15], but none has replaced platinum for general utility.

Mercury is a nearly ideal electrode material. The liquid form allows mercury electrodes to be easily renewed so that a reproducible electrode state is achieved: this advantage is embodied in the polarographic DME and in mercury-pool electrodes. The perfectly smooth surface eliminates surface roughness as a factor in electrolytic measurements. The large cathodic potential range encompasses the regions of electrode reaction of the majority of the chemical substances of interest.

3. ANALYTICAL SENSITIVITY AND THE CHARGING OF THE DOUBLE LAYER

The direct potential and current measuring procedures, such as classical polarography, typically are applicable to solutions of electroactive species in the concentration range of 10^{-2} to 10^{-5} mole/liter with a routine accuracy and precision of the order of a few percent. The methods, therefore, are usually not employed in assay or major constituent analysis but are quite useful for impurity analyses. However, greater accuracy and reproducibility are possible with careful differential techniques and when the measurement procedures are used to determine equivalence points in titrations.

Much of the recent effort in devising new electrochemical techniques has been devoted to increasing the sensitivities of the methods. Special apparatus and methodology have in a few cases extended the detection limit down to about 10^{-10} mole/liter. In general there is no difficulty in the measurement of electrical signals at the level, e.g., 0.001 μA, generated by very low concentration constituents. The problem is in separating or extracting the signal due to the electrode reaction of the electroactive species, the *faradaic current,* from the current due to the charging of the electrical *double layer* at the electrode–solution interface.

The existence of a charge separation or double layer at an electrode–solution interface, which behaves like a capacitor, is one of the important phenomena of electrochemistry.[16] Values of electrode capacity are of the order of 20 μF/cm^2, and this represents a rather large capacitance that must be charged whenever a potential is applied to an electrode. Typical capacity or *charging currents* in polarography are of the order of 0.1 μA, which is a significant correction for measurements of faradaic currents even in the millimolar range. A difficulty in correction or compensation for the effects of the charging current is that the value of the electrode capacity is itself a complex function of electrode potential. This charging current, together with any other extraneous faradaic currents, constitutes the *residual* or *background* current which tends to complicate measurements of constituents at trace concentration levels. The practical sensitivity of the low level electrochemical method is determined primarily by the reproducibility of this residual current.[17]

B. Chronoamperometry

The most simple electroanalytical technique in principle is chronoamperometry, in which the electrolysis current at a working electrode is measured as a function of time at constant electrode potential, under the experimental conditions of purely diffusive mass transfer. The technique itself is relatively unimportant in analysis; however, a discussion of its theory and methodology serves to introduce some of the fundamental ideas of electroanalytical chemistry.

Schematic diagrams of two types of experimental arrangements that are used in chronoamperometry are shown in Figure 3. In the two-electrode system of Figure 3*a*, the potential of a working electrode, *W*, is established with respect to that of a reference electrode, *R*, by means of a potential source, *E*, and the resulting current is measured by means of a suitable device, *I*, such as a chart recorder or galvanometer. The reference electrode, although it conducts the cell current, *i*, is constructed so as to remain unpolarized and at constant potential. The cell current–potential characteristics are therefore determined by the properties of only one electrode in the cell, the working electrode. If the cell current and the resistances of the solution and current measuring device are sufficiently low so that ohmic *iR* losses are negligible, the potential of the working electrode will be equal to that of the potential source, *E*. The two-electrode system is the basic configuration used for most analytical constant potential methods such as polarography.

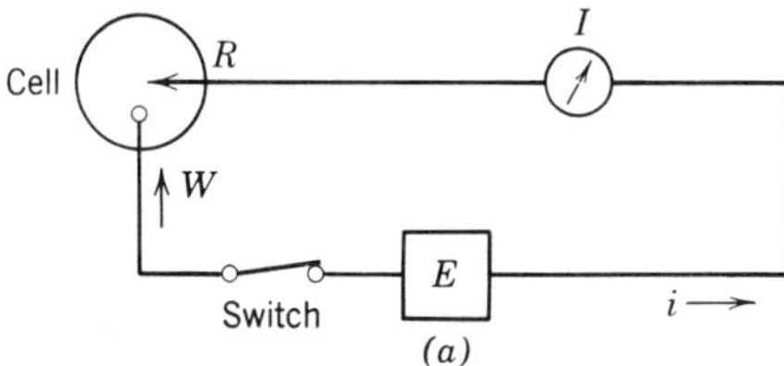

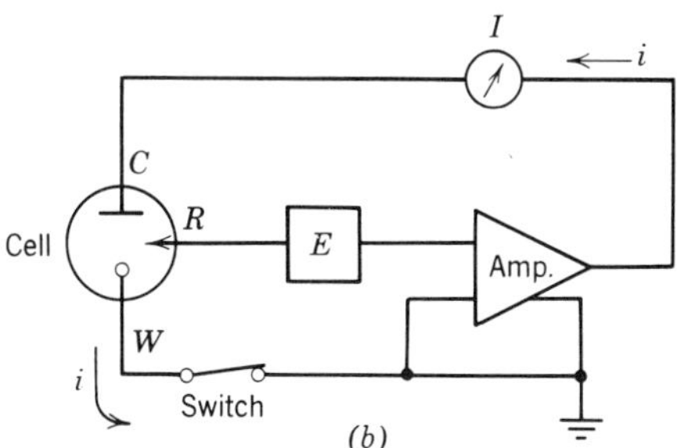

Fig. 3. Schematic diagrams of apparatus for study of current–potential characteristics at constant or controlled potential. (*a*) Two-electrode system; (*b*) controlled-potential three-electrode system. *R*, reference electrode; *W*, working electrode; *C*, counter electrode; *I*, current-measuring device; *E*, controlled-potential source; *Amp*, dc amplifier; *i*, cell current.

The three-electrode arrangement of Figure 3*b*, which utilizes a feedback controller known as a *potentiostat*, is useful for automatically compensating for most of the effects of iR losses in the circuit. Included in the cell is a counter electrode, C, which, together with the working electrode, constitute the principal current path through the cell. The heart of the potentiostat is a high-gain, high-impedance-input, dc amplifier (Amp.), connected in a negative feedback arrangement. The potentiostat maintains the working electrode vs. reference electrode potential at the selected value, E, by delivering the required cell current via the counter electrode. The sensing of the electrode potential is essentially potentiometric so that physically small reference electrodes can be used without danger of polarization. The current–potential behavior of the counter electrode is not critical, and again the measured properties of the cell are only those of the working electrode. The three-electrode potentiostatic system is coming into increasing use for analytical work, especially in high-resistance nonaqueous solutions, and the technique is always employed in electrode kinetics investigations.

1. CURRENT–TIME RELATIONSHIP

The variation of the working electrode current as a function of time at constant potential in a chronoamperometric experiment can be treated mathematically in terms of the laws of diffusion. Considering a generalized electrode reaction:

$$O + ne^- \longrightarrow \mathrm{R} \tag{3}$$

where O denotes the substance being reduced and R is the reaction product, the simplest case is that in which the electrode potential is sufficiently cathodic so that virtually all of the molecules of substance O which reach the electrode are reduced. A diffusion gradient is established which supplies substance O to the electrode, where the concentration of O becomes virtually zero. Under these conditions the magnitude of the current is determined only by the rate of diffusion and is called the *diffusion current*, i_d; and the electrode is said to be completely *concentration polarized*. Further increase of the electrode potential in the cathodic direction does not increase the current. The equation for the diffusion current as a function of time is obtained by Laplace transform methods applied to the partial differential equations appropriate to the particular electrode geometry.[18] For a plane electrode and linear diffusion, the solution is the Cottrell equation[19]:

$$i_d = nFAC^0 D_O^{1/2}(1/\pi^{1/2}t^{1/2}) \tag{4}$$

The new symbols and units of equation 4 are as follows: i_d in μA; A, the area of the electrode, in cm^2; C^0, the bulk concentration of species O, in mmoles/

liter; D_O, the diffusion coefficient of species O, in $cm^2\ sec^{-1}$; and t, the elapsed time since the beginning of the electrolysis, in sec. The principal analytical significance of this equation is that the current is directly proportional to the concentration of substance O. The current is seen to decrease continuously with time without reaching a steady state. The equation has been found to agree with experiment very closely at times in the range of about 0.05 sec to less than about 1 min. Convection processes due to density gradients, vibrations, etc. are difficult to exclude at the longer times.

2. CURRENT–POTENTIAL RELATIONSHIPS

When the electrode is polarized at a potential less cathodic than the value required to maintain the diffusion current, the concentration of O at the electrode surface is finite and the current becomes a function of potential as well as time. Two cases are distinguished according to whether equilibrium among the reacting species is established at the electrode surface. If the electrode reaction reaches equilibrium rapidly compared with the rate of mass transfer, and the Nernst equation is applicable, the electrode process is said to be *reversible*. If the characteristics of the process are governed by the rate and kinetics of the reaction so that electrochemical equilibrium is not attained, the process is said to be *irreversible*.

When substance O in equation 3 undergoes reversible reduction, the concentrations of the species O and R *at the electrode surface* ($x = 0$), are related to the electrode potential according to the Nernst equation (making the simplifying assumption that the ratio of the activity coefficients are close to unity):

$$E = E^0 + (RT/nF)\ \ln\ (C_O/C_R)_{x=0} \tag{5}$$

where E^0 is the standard potential of the electrode reaction. The concentrations C_O and C_R may refer to concentrations in the aqueous phase or in the electrode as is the case with a mercury amalgam. Using equation 5 as a boundary condition in a solution of the linear diffusion equation results in an expression for the current i:

$$i = i_d \frac{1}{1 + [D_O/D_R]^{1/2} \exp\,[(nF/RT)(E - E^0)]} \tag{6}$$

This equation reduces to equation 4, $i = i_d$, when E becomes increasingly more negative than E^0. A plot of equation 6 for i, *at a particular value of the time* t, as a function of $E - E^0$, yields an S-shaped curve of the form shown in Figure 4. This current–potential *wave* has two important features: the wave height, equal to i_d, from which the concentration of the electroactive species is determined; and the *half-wave potential*, $E_{1/2}$, which is the potential

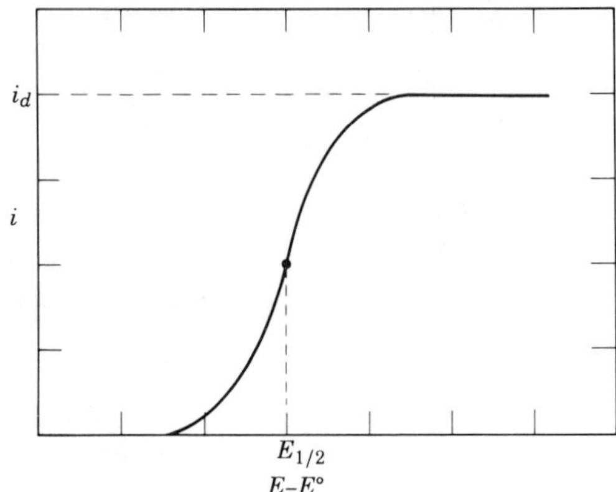

Fig. 4. Theoretical current–potential curve for a reversible process.

at which $i = 1/2\ i_d$. Rearrangement of equation 6 in terms of the half-wave potential gives:

$$E_{1/2} = E^0 + (RT/nF) \ln [D_R/D_O]_{1/2} \tag{7}$$

The half-wave potential is characteristic of the particular reaction and serves to define the position of the wave on the scale of potentials. Since the ratio of the diffusion coefficients is usually not greatly different from unity, the half-wave potential is approximately equal to the standard potential of the couple. Standard potential data thus are useful in predicting the electrode reaction potentials of various species; conversely, unknown substances can often be identified from their half-wave potentials.

Irreversible electrode processes exhibit current–potential waves which tend to rise less steeply than reversible waves, and which are displaced (cathodically for reduction processes, anodically for oxidations) from what would be the reversible half-wave potential, as shown in Figure 5. Irreversibility, however, does not detract from the analytical usefulness of the electrode process, provided a region of diffusion current is present within the limits of the electrode potential range.

Rigorous theoretical treatments have been developed for a variety of more complex systems than the foregoing linear diffusion, reversible reaction case. These include reversible systems involving complex ions, hydrogen ions, and insoluble substances; irreversible systems involving homogeneous chemical reactions and catalytic processes; and other electrode geometries such as the cylinder, the stationary sphere, and the expanding sphere.

In the construction of current-potential curves or measurement of diffusion currents in chronoamperometry, the time of the electrolysis at which

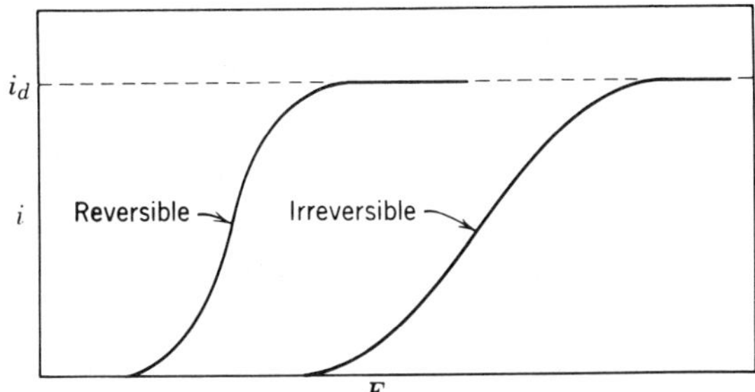

Fig. 5. Comparison of reversible and irreversible current–potential waves.

the current is measured must be carefully reproduced. In the classical technique of polarography, this electrolysis time is automatically reproduced by the periodic renewal of the mercury electrode.

C. Classical Polarography

Polarographic experiments were first carried out by Heyrovský at the Charles University in Prague, and were reported in 1922.[20] The unique advantages and relative simplicity of the DME technique have made it an extremely popular electroanalytical method.

The experimental arrangements shown in Figure 3 for chronoamperometry are also used for polarography, and many different instruments for manually or automatically recording polarograms are available commercially.[21] A widely used form of DME assembly and H-cell[22] are shown in Figure 6. The capillary through which the mercury flows is marine barometer tubing with a bore diameter of about 0.05 mm, selected so that a new drop of mercury forms approximately every 5 sec. Both a sintered glass disk and an agar gel plug are necessary in this cell to prevent significant cross contamination between the two cell compartments; the typical reference half-cell is an SCE. Since oxygen is reduced at the DME over most of the applicable potential range, it is necessary to deaerate polarographic solutions with an inert gas such as nitrogen. Because the temperature coefficient of diffusion currents is about 1%/°C, polarographic cells are usually thermostatted.

The equation for the diffusion current at the DME was first derived by Ilkovič[23]:

$$i_d = 708nC^0 D_O^{1/2} m^{2/3} t^{1/6} \tag{8}$$

and is interesting to compare with the linear diffusion equation 4. The coefficient 708 includes the value of the faraday and several constants arising in the derivation. The new variable in equation 8 is m, the rate of flow of mercury, in mg sec^{-1}. Here t is the time elapsed since the beginning of drop life,

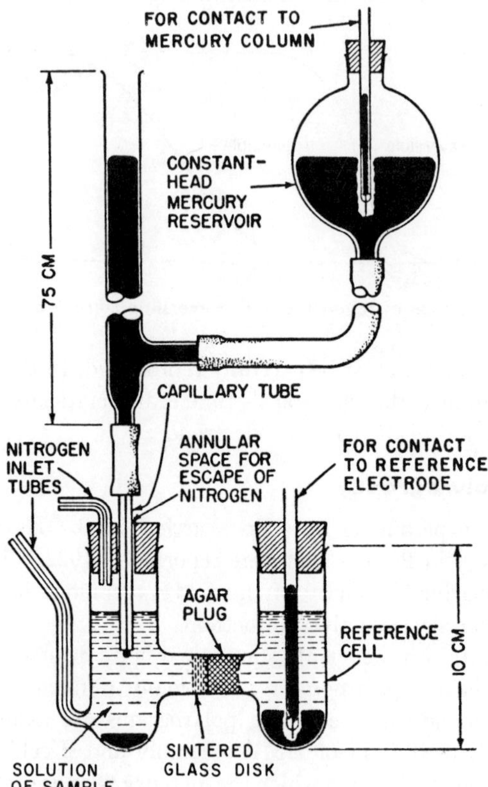

Fig. 6. A polarographic electrode and cell assembly (see reference 22).

in seconds. Because the derivation of this expression involves a number of assumptions that are not quite experimentally true (for example, that the drop is perfectly spherical and that each drop emerges into a perfectly fresh solution), the original Ilkovič equation is not exact and has been the subject of investigation and refinement.[24] However, the equation is adequate for predicting the salient features of the polarographic currents. In contrast to the diffusion current at the plane electrode, which decreases with the square root of time, the current at the DME increases as $t^{1/6}$. The reason is that the increase in area as the drop grows, which varies as $t^{2/3}$, more than compensates for the $t^{-1/2}$ depletion effect.

The variation of current during drop life is shown in Figure 7 for two successive drops. In the measurement of the DME current, common practice is to record with a highly damped galvanometer or recorder so that the average

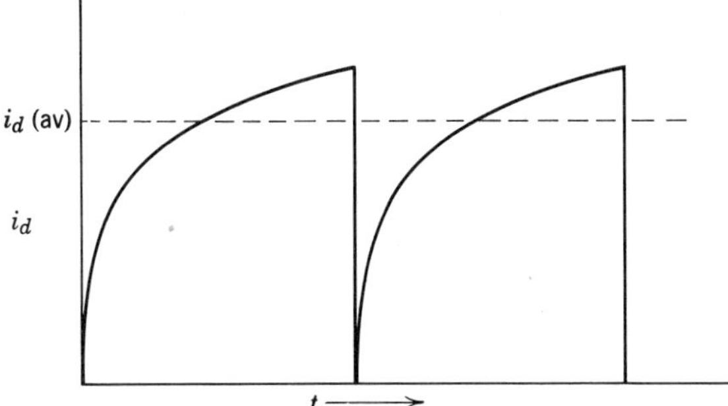

Fig. 7. Current–time characteristics of the dropping mercury electrode.

current, i_d (av), can be determined. The average current is theoretically equal to 6/7 of the current at the end of drop life, and the Ilkovič equation is then used in the average current form. The peak current can also be accurately measured and used for analytical determinations. In the recording of a polarogram, the applied potential is scanned linearly at a slow rate of about 100 mV/min so that the polarographic wave is formed from the measured currents of a number of DME drops. Because the potential of an individual drop is essentially constant, the half-wave potential and diffusion–current theory outlined above is applicable.

An example of an actual polarogram of a solution of a mixture of species is shown in Figure 8. For the Ag(I) wave, the baseline is the residual current; for succeeding waves, the diffusion current can be determined by subtracting the diffusion current for the previous wave. Various techniques have been developed for measuring the heights of waves in mixtures, when the waves are ill-defined, and for applying corrections for residual currents.[25] The concentration of the desired constituent can be determined from wave height data by several methods,[26] including direct calculation of C^0 from the parameters of the Ilkovič equation, the use of a standard curve, and the standard addition procedure. The fourth wave of Figure 8, which is due to the reduction of Ni(II) to Ni, exhibits the drawn-out shape caused by an irreversible electrode reaction.

The polarographic method has been used to determine a great many substances—gases, inorganic and organic compounds, metals and nonmetals—and a vast literature of the polarographic characteristics of these materials has accumulated. Many substances which are not electroactive, or are difficult to analyze by direct electrode reaction, have also been determined by polarography after reaction with an electroactive species. For example, sulfate can be determined by reaction with excess lead and determination of the excess. The complexation of aluminum with a reducible dye Pontachrome

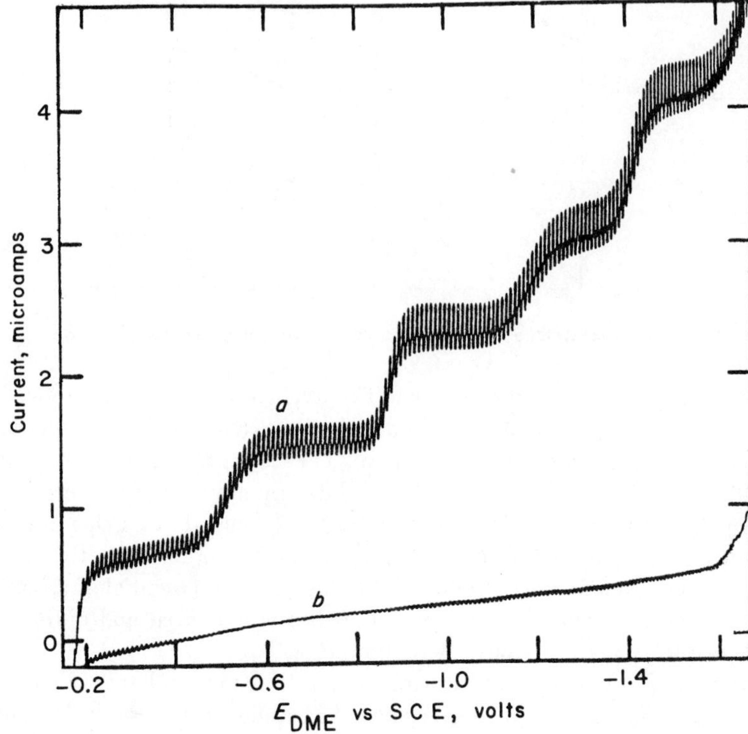

Fig. 8. Polarograms (see reference 25) of (*a*) approximately 0.1 m*M* each of Ag(I), Tl(I), Cd(II), Ni(II), and Zn(II), listed in the order in which their waves appear, in 1*M* NH_4OH–1*M* NH_4Cl; (*b*) the supporting electrolyte alone.

Violet RS, is the basis of a polarographic method for the determination of this element.[27]

Tabulations of polarographic half-wave potential data constitute useful summaries of the characteristics of electroactive species in various supporting electrolytes. These data indicate the feasibility of separation and determination of substances not only by polarography, but by other electroanalytical methods as well. In Table 3 are listed a selection of half-wave potentials of some well-defined reactions which occur at the DME. Most of the heavy metal and transition metal ions give analytically useful waves. Because the alkali and alkaline earth metal ions are reducible only at extremely negative potentials, salts of these metals are normally used as supporting electrolytes. However, polarographic determinations of these metals have been developed which involve materials such as the tetraalkylammonium salts as supporting electrolytes in aqueous and nonaqueous media. With the exception of cerium

Table 3. Half-wave potentials of some inorganic substances at the DME[a]

Reaction	Supporting electrolyte	Half-wave potential, V vs. SCE
Ba(II) → Ba	0.1*M* $(CH_3)_4NCl$	−1.92
Bi(III) → Bi	1.0*M* HCl	−0.09
Cd(II) → Cd	0.1*M* HCl	−0.64
Co(II) → Co	0.1*M* KCl	−1.20
Cu(II) → Cu	0.1*M* $HClO_4$	+0.01
Eu(III) → Eu(II)	0.1*M* KCl	−0.69
Fe(III) → Fe(II)	Citrate, pH 6	−0.23
$2H^+ \rightarrow H_2$	0.1*M* KCl	−1.58
In(III) → In	0.1*M* HCl	−0.56
K(I) → K	0.1*M* $(CH_3)_4NOH$	−2.33
Mn(II) → Mn	1.0*M* KCl	−1.51
Nb(V) → Nb(IV)	Citrate, pH 3	−0.95
Ni(II) → Ni	0.1*M* KCl	−1.1
$O_2 \rightarrow H_2O_2 \rightarrow H_2O$	0.1*M* KCl	−0.05, −0.9
Pb(II) → Pb	1.0*M* HCl	−0.44
Sb(III) → Sb	1.0*M* HCl	−0.15
Sn(IV) → Sn(II) → Sn	4*M* NH_4Cl, 1*M* HCl	−0.25, −0.52
Tl(I) → Tl	0.1*M* KCl	−0.48
Ti(IV) → Ti(III)	Oxalate, pH 4	−0.37
U(VI) → U(V)	0.1*M* HCl	−0.18
Zn(II) → Zn	1.0*M* KCl	−1.00

Reference:

[a] From the extensive compilation of *Handbook of Analytical Chemistry*, L. Meites, Ed., McGraw-Hill, New York, 1963, Sec. 5.

and europium, the rare earths are similar to the alkaline earths in polarographic behavior.

In acid solutions, several of the noble metals, Fe(III), and other strong oxidizing agents such as MnO_4^-, do not exhibit a wave at the DME because these substances have half-wave potentials more anodic than the potential of the oxidation of mercury. A diffusion current, which can be used for analytical purposes, is observed for these substances even at the most positive DME potential available. In the determination of strong oxidants it is preferable to select a complexing electrolyte which brings the wave within the domain of the DME, or to use an electroanalytical technique which employs a platinum or other inert electrode. The half-wave potentials of several such elements which can be determined at the platinum electrode are given in Table 4.

Table 4. Half-wave potentials of some inorganic substances at platinum electrodes

Reaction	Supporting electrolyte	Half-wave potential, V vs. SCE	Ref.
Ag (I) → Ag	0.1M KNO_3	+0.3	a
Au(III) → Au	0.1M HCl	+0.5	b
Ce(IV) → Ce(III)	1.0M H_2SO_4	+1.2	c
Fe(III) → Fe(II)	1.0M HCl	+0.46	d
$I_2 \rightarrow 2I^-$	0.05M H_2SO_4	+0.48	e
Ir(IV) → Ir(III)	1.0M HCl	+0.72	f
Pu(IV) → Pu(III)	1.0M HCl	+0.71	g

References:

[a] H. A. Laitinen and I. M. Kolthoff, *J. Phys. Chem.*, **45**, 1061 (1941).

[b] M. B. Bardin and V. S. Temyanko, *J. Anal. Chem.* (*USSR*), **14**, 751 (1959).

[c] K. J. Vetter, *Z. Physik. Chem.*, **196**, 360 (1951).

[d] O. H. Müller, *J. Am. Chem. Soc.*, **69**, 2992 (1947).

[e] E. Morgan, J. E. Harrar, and A. L. Crittenden, *Anal. Chem.*, **32**, 756 (1960).

[f] N. K. Pshenitsyn, N. A. Ezerskaya, and V. D. Ratnikova, *J. Inorg. Chem.* (*USSR*), **3**, 1791 (1958).

[g] F. A. Scott and R. M. Peekema, in *Proc. U. N. Intern. Conf. Peaceful Uses Atomic Energy, 2nd*, Geneva, 1958, **28**, pp. 573–578.

D. Potential-Sweep Chronoamperometry

Among the many techniques that are extensions of classical polarography, one which is receiving increasing application is the method in which the entire current–potential characteristic or polarogram is obtained at a single mercury drop by rapid scanning of the applied potential. The technique is known variously as *potential-sweep chronoamperometry*, *single-sweep oscillographic polarography*, or *cathode-ray polarography*.[28] The basic controlled potential circuitry of Figure 3 is employed, with E as a fast sweep generator, and the current is usually monitored with an oscilloscope.

The current–time transient that is observed when a linear potential sweep is applied to the electrolysis of a reversibly reduced species is shown in Figure 9. The initial applied potential is chosen at a point where there is no reduction, and as the potential is increased, the current increases until the electrode becomes completely polarized. The current then decreases with time, as in constant potential chronoamperometry. Analytical determina-

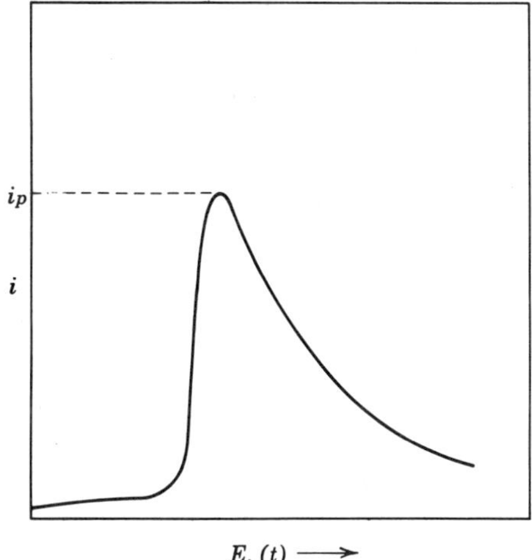

Fig. 9. Current–potential curve for potential-sweep chronoamperometry.

tions are based on measurements of the peak current, i_p, which in the case of a plane electrode is given by the Randles-Ševčik equation[28–30]:

$$i_p = (2.69 \times 10^5)n^{3/2}AD^{1/2}C^0v^{1/2} \tag{9}$$

where v is the rate of potential scan in V sec^{-1} and the other symbols have their usual significance. The potential at which the peak current occurs is $28.5/n$ mV cathodic of the classical polarographic half-wave potential.

The technique is more sensitive than classical polarography[31] and also has the advantage that a complete polarogram can be observed in a few seconds. As can be seen from equation 9 the sensitivity of the method can be increased by increasing the rate of potential scan. The peak current, however, varies only as the square root of v, while the charging current increases directly with v; thus practical rates of potential scan are limited to about 1 V/sec. A possible disadvantage for analytical work is the fact that, because of the relatively rapid rate of potential scan, polarographic waves due to irreversible reactions are much more drawn out and peak currents are less, than in the case of reversible systems.

Much of the utility of potential-sweep chronoamperometry as an analytical tool has come about through developments in the method and apparatus of *differential* cathode-ray polarography.[32] This technique involves the use of two electrolysis cells, two synchronized dropping mercury electrodes, and a differential circuit which allows subtraction of the signal of one cell from the other. Timing circuits control mechanical detachment of the drops at

7 sec intervals, and during the last 2 sec of drop life, the linear potential sweep is applied. Since the rate of change of drop area is low at the end of drop life, matching of the electrodes is less critical. The great advantage of the differential method is that automatic compensation for the residual current can be achieved by operating one cell with the supporting electrolyte alone. Accurate comparison of an unknown solution with a standard solution can also be performed. Differential cathode-ray polarographs are available commercially and examples of their application are given in Table 5. Some of these procedures involve separations prior to polarographic measurement.

Table 5. Analytical determinations by differential cathode-ray polarography

Elements determined	Lowest concentration	Matrix	Ref.
Al	4 ppm	Fe alloys	a
Cu, Pb, Al, Fe, Cr, Mn	0.2%	Ni alloys	b
In	1 ppm	Zn alloys	c
Nb	0.5%	Pb–Ti ceramics	d
Pb, Bi	0.02 ppm	Cast iron	e
Sn	10 ppm	Steels	f
Te	10 ppm	Brass, cast iron	g

References:

[a] R. C. Rooney, *Analyst*, **83**, 546 (1958).
[b] Southern Analytical Ltd. Application Note No. 483D.
[c] G. F. Reynolds and H. I. Shalgosky, *Anal. Chim. Acta*, **18**, 345 (1958).
[d] G. C. Goode, J. Herrington, and W. T. Jones, *Anal. Chim. Acta*, **35**, 91 (1966).
[e] R. C. Rooney, *Analyst*, **83**, 83 (1958).
[f] P. H. Scholes, *Analyst*, **86**, 392 (1961).
[g] E. J. Maienthal and J. K. Taylor, *Anal. Chem.*, **37**, 1516 (1965).

E. Stripping Analysis

In addition to the applications with the DME, potential sweep chronoamperometry has also been used extensively for measurements at single, suspended mercury drops,[33,34] for both analytical work and kinetic studies.[30] This hanging mercury drop electrode (HMDE) is also the basis of a very sensitive electroanalytical method, *stripping analysis*, in which the substance to be determined is first plated and concentrated on the electrode at constant potential, and then stripped off, usually by a linear potential sweep electrolysis.[33] The current–potential characteristic of the stripping process is similar to that of Figure 9. Conditions of the deposition and disssolution steps are chosen so that there is a considerable enhancement in the ratio of faradaic to residual currents. It has been possible to analyze solutions in the concentration range of 10^{-6} to $10^{-9}M$ and high purity materials for part per billion impurities by this technique. Both anodic and cathodic stripping methods have been employed, and solid electrodes have been used. Table 6 lists a few selected examples of applications of the technique.

Table 6. Analytical determinations by stripping analysis

Elements determined	Lowest concentration	Matrix	Electrode used	Ref.
Cd, Tl, Zn	$10^{-9}M$	Solutions	HMDE	a
Bi, Cd, Cu, In, Pb, Tl, Sn	0.01 ppm	Ge, Si, Semiconductors	HMDE	b
Bi, Cd, Cu, In, Sb	0.1 ppm	Zn	HMDE	c
Ni	$5 \times 10^{-8}M$	Solutions	Pt, Au	d
Cl^-	$10^{-6}M$	UO_2SO_4 solutions	Hg pool	e
Ga, Zn, Cd	0.01 ppm	Al	HMDE	f

References:

[a] R. D. DeMars and I. Shain, *Anal. Chem.*, **29**, 1825 (1957).

[b] P. F. Kane and K. R. Burson, in *Standard Methods of Chemical Analysis*, 6th ed., Vol. III, Part B, F. J. Welcher, Ed., Van Nostrand, New York, 1966, pp. 1802–1807.

[c] W. Kemula, Z. Kublick, and S. Glodowski, *J. Electroanal. Chem.*, **1**, 91 (1959).

[d] M. M. Nicholson, *Anal. Chem.*, **32**, 1058 (1960).

[e] R. G. Ball, D. L. Manning, and O. Menis, *Anal. Chem.*, **32**, 621 (1960).

[f] L. N. Vasileva and E. N. Vinogradova, *J. Anal. Chem.* (*USSR*), **18**, 397 (1963).

F. Derivative Polarography

Several of the techniques of voltammetry have been demonstrated to be improved by utilizing the derivative of the current–potential characteristics. Derivative signals have been shown to be especially useful in classical polarography, and an elegant instrument has been designed which continuously takes the derivative of the current–potential wave by means of analog computer circuitry.[35] Derivative polarography is capable of considerably better resolution of overlapping polarographic waves than is conventional polarography, and is less affected by residual current behavior.[17]

G. Alternating Current Polarography

In the technique of *alternating current polarography*,[36] a sine-wave ac voltage of about 10 mV amplitude and about 100 Hz frequency is superimposed on the regular, slow dc potential sweep of classical polarography, and the resulting ac component of the current is measured. The DME is usually employed and the polarogram of ac current versus dc potential yields peaks corresponding to the half-wave potentials of the electroactive species. An important feature of ac polarography is its extreme sensitivity to the reversibility of the electrode reaction, because of the rapid rates of potential change involved. Thus, signals for irreversible systems are suppressed and may be completely absent in the ac polarogram. The charging and discharging of the double layer contributes to the measured alternating current, and since the double-layer capacity is strongly affected by the adsorption of substances on the electrode, this has been made the basis of the determinatoin of these materials. Adsorption–desorption processes give rise to what are called *tensammetric* peaks in the ac polarogram, and these have been used for the analysis of many organic substances which are not otherwise electroactive. The sensitivity of ac polarography is somewhat better than that of classical polarography and has been extended by such refinements as phase-sensitive detection.[37]

H. Square-Wave, Pulse, and Tast Polarography

Efforts to minimize the influence of the charging current in polarography have led to the development of several techniques which take advantage of the fact that, when a potential is applied to an electrode system, the charging current generally decays more rapidly than the faradaic current. Sampling of the current signal after a suitable delay time therefore results in a better ratio of faradaic current to charging current. Three instrumental systems based on this principle have been devised, each utilizing the DME.

Square-wave polarography[38] involves the imposing of a continuous, ac square wave of about 200 Hz on the dc linear sweep, and measurement of the current flowing near the ends of successive half-cycles, during an interval near the end of drop life. The characteristics of square-wave polarography are similar to those of ac polarography: derivative-type peaks are obtained and the method is less sensitive for irreversible than for reversible electrode reactions. For reversible systems, the technique is useful at concentration levels as low as 10^{-6} to $10^{-7}M$, a definite improvement over ac polarography.

In *pulse polarography*,[39] a single, square-wave voltage pulse of about 40 msec duration is applied to each mercury drop at a predetermined time, and the electrolysis current is measured during the last 20 msec of the pulse. Either regular or derivative-type polarograms are obtained, depending upon whether the pulses are progressively increased in amplitude, or pulses of constant height are superimposed on the dc linear sweep. Pulse polarography is about equally sensitive to both irreversible and reversible systems, and has been used at the 10^{-7} to $10^{-8}M$ concentration level.

The experimental conditions of *Tast* or *strobe polarography*[40] are similar to those of classical polarography, except that the current is measured for only a brief interval near the end of drop life. This relatively simple refinement of classical polarography extends the analytical concentration range to $10^{-6}M$, because most of the double-layer charging current is not measured.

An advantage of the technique of derivative polarography and the methods of ac polarography, square-wave polarography, and derivative pulse polarography is that determinations of substances can easily be carried out in the presence of a large excess of a more easily reducible material, provided the half-wave potential separation is favorable. Unlike classical polarography, the rate-of-change response of the derivative-type techniques is unaffected by the presence of a large steady current; and a zinc wave, for example, can be distinguished "on top of" a large diffusion current due to cadmium. The polarographic-wave resolving power of each of the various derivative-type methods is the same,[17] and is significantly better than that of classical dc polarography.

I. Cyclic Techniques

Several electroanalytical techniques have been developed in which the electrode potential is rapidly scanned, first in one direction as in potential-sweep chronoamperometry, then immediately in the reverse direction. The cycle may be repeated once, several times, or until a steady state is achieved, and the resulting current is usually displayed on an oscilloscope. These techniques have been employed with various types of electrodes and in unstirred as well as stirred solutions. Examples are cyclic triangular-wave voltammetry[30,41] and oscillographic polarography.[42] The chief virtue of these methods is that electrode reaction intermediates which are formed during the first half-cycle of potential change may be detected on the reverse half-cycle and subsequent sweeps. The techniques thus have been extremely important in the elucidation of electrode reaction mechanisms, but have not been extensively applied to analytical problems.

J. Chronopotentiometry

Another electrochemical technique that has been a popular tool in the study of electrode reactions, without receiving wide application in analysis in preference to other methods, is *chronopotentiometry*. Chronopotentiometric experiments are carried out in unstirred solutions with the apparatus shown schematically in Figure 10. A constant current, i_0, is maintained between a working electrode, W, and a counter electrode, C; and the potential of the working electrode versus that of a reference electrode, R, is observed as a function of time. Electrolysis of the electroactive species establishes a diffusion gradient, and as the electrolysis proceeds the concentration of this species at the working electrode surface decreases to zero. In the case of a reversible process, the electrode potential is given by the Nernst equation, and when the concentration of the reactant at the electrode surface approaches zero there is a large inflection in potential. The behavior of the electrode potential is similar to that of the indicator electrode in a potentiometric titration (Figure 1), as the constant current "titrates" the electroactive material. The time from the initiation of the electrolysis to the potential inflection is called the *transition time*, τ, and this is related to the bulk concentration of the reactant by the Sand equation,[43]

$$\tau^{1/2} = (nFAC^0D^{1/2}\pi^{1/2})/(2i_0) \tag{10}$$

for linear diffusion, in which the symbols have their usual significance. Much of the interest in chronopotentiometry stems from the simplicity of the instrumentation compared with that required for three-electrode potentiostatic experiments. However, as an analytical technique compared with polarography, chronopotentiometry suffers from the disadvantages that (*1*) charging current corrections and the analysis of mixtures are less straightforward and (*2*) analyses at low concentrations are more limited because the chronopotentiograms are less well defined.[44]

K. Voltammetry with Forced Convection

Most of the analytical work performed with solid electrodes has involved movement of the electrode or stirring of the solution during the electrolysis. The determination of current–potential characteristics by voltammetry in stirred solutions is different from the analogous experiments in chronoamperometry or polarography because the electrolysis current is dependent on

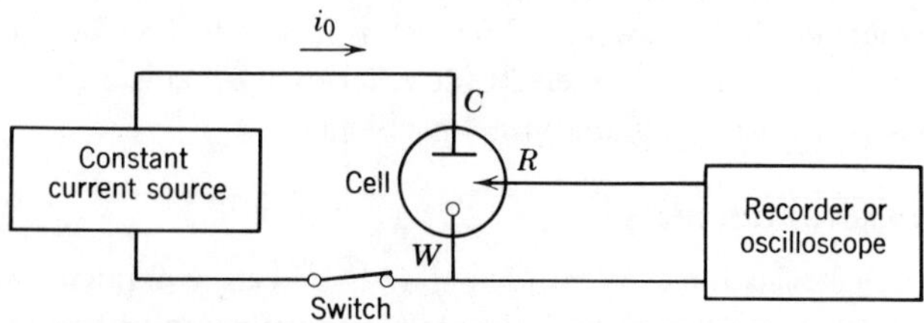

Fig. 10. Schematic diagram of apparatus for chronopotentiometry. *R*, reference electrode; *W*, working electrode; *C*, counter electrode.

both convection and diffusion. Conditions of continuous, uniform convection are employed so that a steady state is rapidly achieved; unlike diffusion-controlled methods, either the current or the potential can be controlled with equivalent results.

Voltammetric analyses at controlled potential utilize the basic circuits of Figure 3, and most commonly an electrode which can be rapidly rotated in the solution. A typical example of this type of electrode[45] is a short length of 1-mm diam platinum wire sealed in the side of a glass tube, perpendicular to the axis of rotation. These electrodes are usually rotated at speeds of approximately 600–1000 rpm and the electrolysis current is measured as a function of applied potential as in polarography.

The current–potential curves obtained in stirred solutions are sigmoid as in polarography (Fig. 4), and exhibit a plateau or *limiting current*, corresponding to potentials where the concentration of the electroactive species is zero at the electrode surface. Rigorous theoretical analysis of electrolyses in stirred solutions has been possible only with a few electrode configurations because of the complications of hydrodynamics.[46] Generally, the experiments are described in terms of a concept of Nernst of a diffusion layer at the electrode of thickness δ. According to this model, reacting species are brought up to this layer by stirring, and then transfer through the layer occurs by diffusion only. Based on these considerations the equation for the limiting current, i_l, in the usual notation, is:

$$i_l = (nFADC^0)/\delta \tag{11}$$

which indicates the analytically significant proportionality between the limiting current and the bulk concentration. The parameter δ really contains all the hydrodynamic factors whose exact influence is unknown. Increased rates of convection are envisioned as decreasing δ, which increases the current.

Half-wave potential equations analogous to equations 6 and 7 can be derived for these voltammetric waves. The higher rate of mass transfer embodied in most convection voltammetric procedures reveals the irreversibility and kinetic complications of many electrode reactions; however, the inexactness of most theoretical treatments renders the methods less useful for kinetic studies. A notable exception to this is the rotating disk electrode,[47] for which the mass transfer process is well defined. Voltammetric current measurements with rotating platinum electrodes have been applied to the determination of all the elements listed in Table 4 and a number of other substances whose half-wave potentials are out of the potential range of mercury electrodes.

IV. AMPEROMETRIC TITRATIONS

Most of the polarographic and voltammetric techniques described above have been used to monitor the concentration of a reactant in a titration; when the electrolysis current is the measured variable, the procedures are known as *amperometric titrations*. A very large number of amperometric titration methods have been published[48]; these are classified according to whether one or two polarized indicator electrodes are used in the titration. As in potentiometry, titration procedures based on precipitations, oxidation–reduction, acid–base, and complexometric reactions have been developed.

In determinations with one polarized electrode, the indicator electrode potential is selected with respect to that of a reference electrode so that the diffusion or limiting current of the electroactive species is registered. Various types of titration curves are obtained, some of which are shown in Figure 11, depending on whether the substance being titrated is electroactive (Fig. 11*A*), the titrant is electroactive (Fig. 11*B*), or both are electroactive (Fig. 11*C*). Curvature of the lines in the vicinity of the equivalence point is caused by a lack of completeness of the titration reaction and volume changes during the titration; in cases of excessive curvature, the equivalence point is determined by extrapolation of the linear segments of the curve.

In amperometric titrations with two polarized electrodes, or *biamperometric titrations*, a constant potential is applied between two similar electrodes, usually platinum, and the resulting current then depends on the current–potential characteristics of the couples involved at both electrodes. The principles of the technique are illustrated by the titration of iodine with thiosulfate. When each of the species of the reversible iodine–iodide system is present in the solution, an electrolysis current is passed because iodine can be reduced at the cathode and iodide can be oxidized at the anode. If the iodine is titrated with an irreversible couple such as thiosulfate, and the applied potential is not sufficient to cause the reduction of tetrathionate, the current decreases to zero as in Figure 11*A*. This "dead stop" titration,[49]

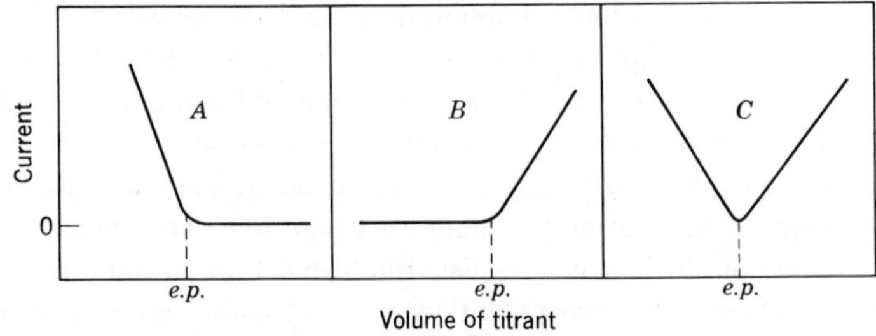

Fig. 11. Types of amperometric titration curves. e.p., equivalence point.

and the reverse titration with iodine, have been extensively applied in the determination of water by the Karl Fischer method,[50] and in the analysis of oxidants which liberate iodine from iodide solutions.

Amperometric titrations are especially suitable for determinations at millimolar concentration levels, where other end-point detection schemes may not be feasible. The accuracy and precision of the method in these dilute solutions is about 0.5% in many cases, and in more concentrated solutions approaches the 0.1% of good volumetric procedures. Some representative examples of the technique are given in Table 7.

Table 7. Analytical determinations by amperometric titration

Substance determined	Titrant	Electrodes	Titration curve type, Figure 11	Ref.
As(III)	$Br_2(KBrO_3, Br^-)$	Pt–SCE	B	a
Cr(VI), V(V)	Fe(II)	Pt–SCE	B	b
F^-	$La(NO_3)_3$	DME–SCE	B	c
Hg_2^{2+}	EDTA	DME–SCE	A	d
Pb(II)	$K_2Cr_2O_7$	DME–SCE	C	e
PO_4^{3-}	UO_2^{2+}	DME–Hg pool	B	f
Ru(IV)	Hydroquinone	Pt–SCE	A	g
SO_4^{2-}	$Pb(NO_3)_2$	DME–SCE	B	h
Zr(IV)	Cupferron	DME–SCE	B	i

References:

[a] Reference 45.

[b] T. D. Parks and E. J. Agazzi, *Anal. Chem.*, **22**, 1179 (1950).

[c] A. Langer, *Ind. Eng. Chem., Anal. Ed.*, **12**, 411 (1940).

[d] B. Matyska, J. Doležal, and D. Roubalová, *Chem. Listy*, **49**, 1012 (1955); B. C. Southworth, J. C. Hodecker, and K. D. Fleischer, *Anal. Chem.*, **30**, 1152 (1958).

[e] I. M. Kolthoff and Y. D. Pan, *J. Am. Chem. Soc.*, **61**, 3402 (1939).

[f] I. M. Kolthoff and G. Cohn, *Ind. Eng. Chem., Anal. Ed.*, **14**, 412 (1942).

[g] N. K. Pshenitsyn and N. A. Ezerskaya, *J. Anal. Chem. (USSR)*, **16**, 203 (1961).

[h] I. M. Kolthoff and Y. D. Pan, *J. Am. Chem. Soc.*, **62**, 3332 (1940).

[i] P. J. Elving and C. E. Olson, *Anal. Chem.*, **28**, 338 (1956).

V. COULOMETRY

In analysis by coulometry, the species which is determined is quantitatively electrolyzed in a portion of the sample solution, and the amount present is calculated from the total electric current consumed via Faraday's law:

$$\text{wt. of substance electrolyzed} = \frac{M}{nF}\int_0^t i\,dt \qquad (12)$$

where M is the molecular weight of the substance and t is the time of the electrolysis. The conditions of the electrolysis are chosen so that the current efficiency of the overall reaction is effectively 100%. The principal coulometric techniques are *constant-current coulometry* and *controlled-potential coulometry*. Coulometric methods are similar to titrations, but because the titrant is essentially an electric current, these techniques have the important advantage of not requiring standard solutions. The combination of quantitative reactions and purely electrical measurements results in very high levels of precision and accuracy.

A. Constant-Current Coulometry

Constant-current coulometry, or *coulometric titrations*, involve the *in situ* electrochemical generation of a suitable titrating species which then reacts with the substance determined. Direct electrode reaction of the substance determined may also take place but most of the transformation generally occurs via the intermediate; the exact nature of the reaction is unimportant to the analysis as long as the *overall* current efficiency is 100%. The electrolysis is carried out at constant current and the equivalence point of the titration is detected by a suitable electrochemical method such as potentiometry or amperometry, or by an optical method. The product of the constant current and the elapsed time to the equivalence point, $i_0 \times t$, yields the total coulombs of charge from which the weight of the substance determined can be calculated.

Figure 12 shows the basic equipment required for constant-current coulometry. Electrolysis at the generator electrode, G, produces the titrating species which then reacts with the substance determined. Many titrants which are unstable as volumetric solutions can be generated in this way. Such powerful reducing agents as Sn(II) and Ti(III) can be produced cathodically, and good oxidants such as Ag(II), Br_2, and Cl_2 can be produced anodically. The counter electrode, C, is isolated from the titration solution with a semiporous membrane such as fritted glass to prevent reversal of the generation reaction at this electrode. The detection system shown in Figure 12 with an indicator electrode, I, and a reference electrode, R, might be either potentiometric or amperometric. This detection system can be linked to the timer/current control circuitry to effect automatic termination of the titration; a number of commercial coulometric titrators have this feature.

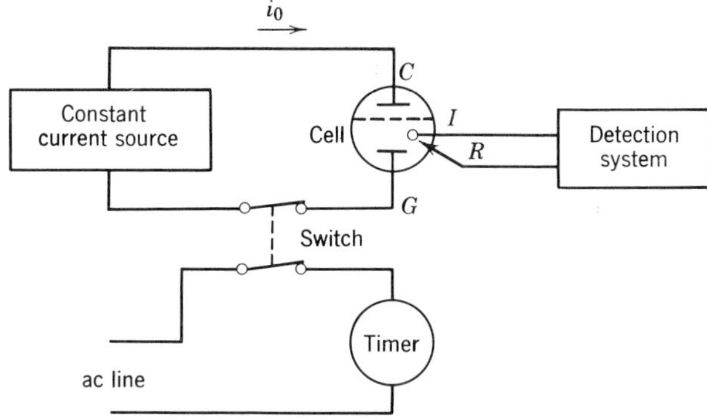

Fig. 12. Schematic diagram of apparatus for constant-current coulometry. G, generator electrode; C, counter electrode; *I*, indicator electrode; *R*, reference electrode.

The routine accuracy and precision of constant-current coulometry is very good, with relative errors and standard deviations of 0.1%, even at the submilligram level in a few procedures. Very careful assay work on larger quantities has achieved two orders of magnitude better accuracy and precision. For example, $K_2Cr_2O_7$ has been standardized with a relative standard deviation of 0.003% by titration with electrogenerated Fe(II).[51] and acids have assayed to ±0.004%.[52] Examples of other determinations by coulometric titration are given in Table 8. In addition to its use as a conventional analytical technique, constant-current coulometry has been combined with stripping analysis for the determination of plating and metal oxide layer thicknesses.

B. Controlled-Potential Coulometry

In controlled-potential coulometry, the substance determined is electrolyzed at a working electrode whose potential is controlled or maintained constant throughout the electrolysis. As the concentration of the electrolyzed substance decreases during the electrolysis, the current decays exponentially according to the relationship:

$$i = i_* e^{-kt} \tag{13}$$

where t is the elapsed time of the electrolysis, i_* is the initial current at $t = 0$, and k is a factor which depends on the conditions of the electrolysis. The current is integrated and the electrolysis is continued until the current reaches a suitably low value, indicating that transformation of the electroactive species is complete.

Table 8. Analytical determinations by constant-current coulometry

Substance determined	Titrant generated	E. P. detection method	Ref.
As(III), Sb(III)	Br_2	Biamperometric	a
Ce(III)	Ag(II)	Potentiometric	b
Ca(II), Cu(II), Zn(II), Pb(II)	EDTA	Potentiometric	c
Cl^-, Br^-, I^-	Ag(I)	Potentiometric	d
Fe(III)	Ti(III)	Potentiometric	e
Pt(IV)	Sn(II)	Potentiometric or spectrophotometric	f
V(V)	Fe(II)	Potentiometric	g
Zn(II)	$Fe(CN)_6^{4-}$	Potentiometric	h

References:

[a] R. J. Myers and E. H. Swift, *J. Am. Chem. Soc.*, **70**, 1047 (1948); R. A. Brown and E. H. Swift, *Ibid.*, **71**, 2717 (1949).

[b] D. G. Davis and J. J. Lingane, *Anal. Chim. Acta*, **18**, 245 (1958).

[c] C. N. Reilley and W. W. Porterfield, *Anal. Chem.*, **28**, 443 (1956).

[d] J. J. Lingane, *Ibid.*, **26**, 622 (1954).

[e] P. Arthur and J. F. Donahue, *Ibid.*, **24**, 1612 (1952).

[f] A. J. Bard, *Ibid.*, **32**, 623 (1960).

[g] W. Oelsen and P. Göbbels, *Stahl u. Eisen*, **69**, 33 (1949); N. H. Furman, C. N. Reilley, and W. D. Cooke, *Anal. Chem.*, **23**, 1665 (1951).

[h] J. J. Lingane and A. M. Hartley, *Anal. Chim. Acta*, **11**, 475 (1954).

The successful controlled-potential coulometric determination depends on maximizing the value of k so that electrolyses can be carried out in a minimum length of time, usually 15 min or less. These rapid electrolyses require very efficient stirring of the solution and a large ratio of electrode area to solution volume. Electrolysis currents of several hundred milliamperes are generated even for the typical milligram-size samples so that three-electrode potentiostatic circuits (such as that of Fig. 3*b*) are required to compensate ohmic losses. Integration of the current is accomplished most conveniently by one of the two electronic schemes depicted in Figure 13; these systems can be operated at accuracies of 0.05%.

Control of the electrode potential in this type of coulometry affords a high degree of selectivity, compared with constant-current coulometric titrations, because it is fundamentally the electrode potential that determines which reactions will occur. Half-wave potential and standard potential data can be consulted regarding the practicality of proposed separations, and the

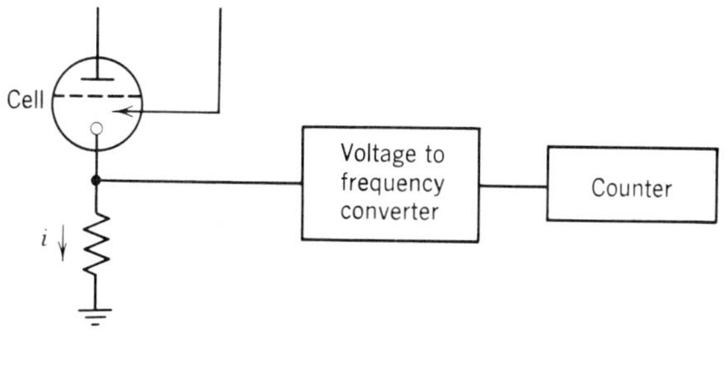

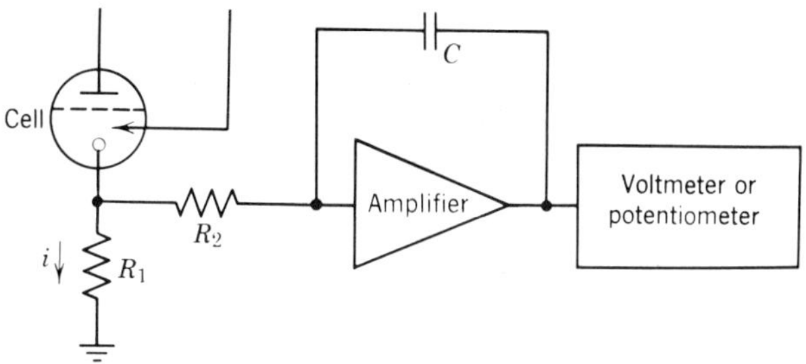

Fig. 13. Types of electronic integrators for controlled-potential coulometry. (*a*) voltage to frequency converter type[53]; Counts $\propto \int_0^t idt$; (*b*) operational amplifier type[54]; output voltage $= (R_1/R_2C)\int_0^t idt$.

Nernst equation can be used to calculate separation factors for reversible systems. For example, to achieve 99.9% transformation of a reversibly reduced species, the electrode potential must be $-(RT/n_1F)\log 10^{-3}$ or $0.177/n_1$ volt cathodic of the half-wave potential, where n_1 is the number of electrons involved in the reaction. At the same time, the electrode potential must be $0.177/n_2$ volt anodic of the half-wave potential of a reversibly reduced second species, to insure less than 0.1% reduction. Thus half-wave potential differences of[55]

$$E_{1/2}{}^1 - E_{1/2}{}^2 = 0.177/n_1 + 0.177/n_2 \tag{14}$$

are required for essentially complete separations of equal quantities of reversibly electrolyzed substances. Due to irreversibility in some electrode reactions and n values greater than 1, some determinations can be carried out with as little as 0.2 V differences in potentials.

The accuracy and precision of controlled-potential coulometry is similar to that of constant current coulometry, being limited chiefly by the quality of the integration and difficulties with residual current corrections at the low levels. The technique is frequently useful for the determination of the quantity of an element in a specific oxidation state in a mixture of several oxidation states, and several components of mixtures can often be determined in a single sample by sequential adjustment of the electrode potential. Table 9 lists a number of examples of determinations by controlled-potential coulometry. In several procedures two electrolysis steps are involved: an oxidation followed by a reduction, or vice versa. The first step serves to transform all of the substance determined into a single oxidation state and may be employed to avoid a high residual current which is not present in the second measurement step.

Table 9. Analytical determinations by controlled-potential coulometry

Substance determined	Reactions	Working electrode	Ref.
Ag	Ag(I) → Ag	Pt	a
Au	Au(III) → Au	Pt	b
Co, Ni	Ni(II) → Ni	Hg	c
	Co(II) → Co		
Cu	Cu(II) → Cu	Hg	d
Eu	Eu(III) → Eu(II)	Hg	e
	Eu(II) → Eu(III)		
Ir	Ir(IV) → Ir(III)	Pt	f
Pb	Pb(II) → Pb	Hg	d, g
Pu	Pu(IV) → Pu(III)	Pt	h
	Pu(III) → Pu(IV)		
U	U(VI) → U(IV)	Hg	i

References:

[a] L. L. Merritt, E. L. Martin, and R. D. Bedi, *Anal. Chem.*, **30**, 487 (1958).

[b] J. E. Harrar and F. B. Stephens, *J. Electroanal. Chem.*, **3**, 112 (1962); L. Duncan, U. S. At. Energy Comm. Rept. HW–SA–2455 (1962).

[c] J. J. Lingane and J. A. Page, *Anal. Chim. Acta*, **13**, 281 (1955).

[d] J. J. Lingane, *J. Am. Chem. Soc.*, **67**, 1916 (1945).

[e] W. D. Shults, *Anal. Chem.*, **31**, 1095 (1959).

[f] J. A. Page, *Talanta*, **9**, 365 (1962).

[g] P. R. Segatto, *J. Am. Ceram. Soc.*, **45**, 102 (1962).

[h] F. A. Scott and R. M. Peekema, in *Proc. U. N. Intern. Conf. Peaceful Uses Atomic Energy, 2nd, Geneva, 1958*, **28**, 573–578.

[i] G. L. Booman, W. B. Holbrook, and J. E. Rein, *Anal. Chem.*, **29**, 219 (1957).

VI. ELECTROGRAVIMETRY AND ELECTROSEPARATIONS

Electrogravimetry applies the techniques of electroplating to chemical analysis and is the oldest quantitative electroanalytical method.[56] The solution analyzed is exhaustively electrolyzed, and the element or compound determined is deposited on an electrode in a weighable form. In the absence of interfering substances which would codeposit, and when there is no need for high current efficiency, electrogravimetric determinations are carried out with the simple two-electrode constant applied voltage system of Figure 3*a*. To obtain a satisfactory deposit the voltage is adjusted manually to maintain a specific electrolysis current. The controlled-potential three-electrode system (Fig. 3*b*) is used for the more precise potential control required in the analysis of mixtures. Except for the fact that the quantities of substance electrolyzed are about two orders of magnitude larger, the methodology and apparatus of electrogravimetry are similar to that of controlled-potential coulometry. The working electrode is usually a platinum gauze cylinder and the counter electrode a concentric platinum spiral.

Many metals such as Cu, Cd, Pb, Sn, Zn, Bi, Au, Ag, and the platinum metals have been determined by electrogravimetry.[57] The routine accuracy and precision of the method is about 0.1–0.5%, which is comparable to classical gravimetric methods. Results within 0.01% can be obtained with little difficulty for some metals, particularly Cu.[58]

Both two-electrode and controlled-potential methods are employed frequently for the removal of the more easily reduced components from a sample, prior to the determination of its constituents by other analytical techniques. Large platinum or mercury pool cathodes are used, and quantities of the order of several grams are deposited. For example, iron can be removed from solutions of steel by mercury cathode electrolysis to facilitate the determination of Al, V, Zr, Ce, and La.[59] An important separation in radiochemical investigations is the electrolytic deposition of the lanthanide and actinide metals on mercury and solid metal cathodes. Separation factors for reversible systems can be calculated as described in Section V-B for coulometry.

VII. ELECTROGRAPHY AND ELECTROSPOT TESTING

The method of electrographic testing of metals involves the process of electrolytic dissolution of the material into an aqueous test medium, usually filter paper wetted with an appropriate reagent solution. The basic arrangement for electrography is shown in Figure 14. The specimen is made the anode in a circuit with an aluminum cathode, and the filter paper is pressed in between the two electrodes. A potential difference of a few volts is applied

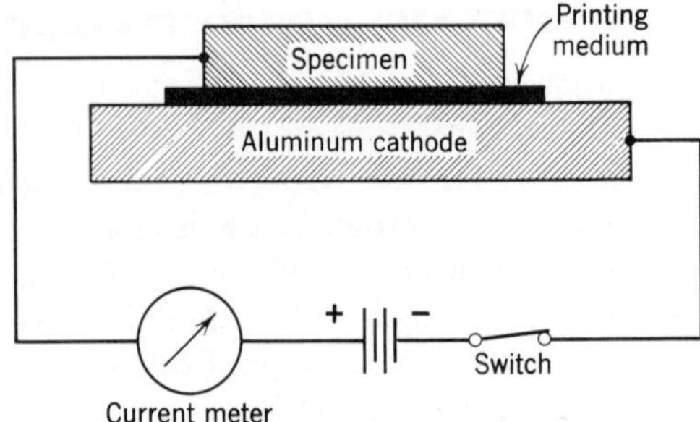

Fig. 14. Schematic diagram of apparatus for electrography (see reference 60).

and electrolysis is carried out until sufficient material, as little as a few micrograms, is transferred to the paper. The paper is then treated with specific spot test reagents to identify the elements in the specimen by their color reactions. Semiquantitative estimations can also be performed under carefully controlled conditions; and impressions of the specimen surfaces, for example for the examination of imperfections in protective coatings, can be obtained with special techniques.[60]

VIII. CONDUCTOMETRY AND CONDUCTOMETRIC TITRATIONS

Conductometry, the measurement of the *specific conductivity* (or its reciprocal the *resistivity*) of electrolytic solutions, is the basis for both the direct determination of concentration and the measurement of concentration during a titration. Analytical conductivity techniques involve the measurement of ac voltages and currents, and the techniques are classified according to the ac frequency employed. Low frequency conductometry utilizes an ac frequency of about 60–1000 Hz, and the experimental conditions are chosen so that the bulk conductivity of the solution is measured. High frequency conductometry (sometimes called *oscillometry*) is based on responses to both conductivity and dielectric constant or capacitance at frequencies of the order of 1–100 MHz.

In low frequency conductometry, the usual electrode arrangement consists of two identical, planar, platinized platinum electrodes immersed in the analysis solution. Platinization of the electrodes increases the surface area, and hence the double layer capacity, at each electrode, with the result that the double-layer capacitive *reactance* is insignificant compared with the solution resistance. Moreover, because the charge transported at the electrode–solution interface is conveyed chiefly by the large double-layer capacitance, faradaic processes are unimportant. Under these conditions the voltage–current properties of the cell reflect only changes in solution conductivity.

Various types of direct-reading and bridge-type solution conductivity instruments are available, and measurements can also be made with commercial audiofrequency impedance bridges.

High frequency conductometry is unique in that the metal electrodes are not in direct contact with the solution; indeed, there often are no electrodes in the conventional sense. The cell containing the solution which is analyzed is located within a mechanical assembly which constitutes a capacitor or inductor connected to the measuring circuit. The measuring circuit is thus coupled to the solution through the walls of the vessel. High frequency conductivity instruments[61] are based on the measurement of (*1*) the rf current passed by the cell as a result of an impressed rf voltage; (*2*) the resistive loading of an oscillator, of which the cell is a part; or (*3*) the shift in frequency of the oscillator.

Both low and high frequency conductometry have been extensively used in process stream monitoring applications. Both types of conductometric techniques have also been employed in acid–base, precipitation, and complexometric titrations, and in some cases are more accurate than the corresponding potentiometric titrations. However, because the conductivity of a solution depends on all of the ionic species present, conductometry is the least selective electroanalytical technique. High frequency conductometry offers a distinct advantage in cases where fouling of the electrodes may be a problem, as in precipitation titrations and in other heterogeneous systems.

General References

Electroanalytical Chemistry—General

J. J. Lingane, *Electroanalytical Chemistry*, 2nd ed., Interscience New York, 1958.

P. Delahay, *New Instrumental Methods in Electrochemistry*, Interscience, New York, 1954.

Potentiometry and Potentiometric Titrations

I. M. Kolthoff and N. H. Furman, *Potentiometric Titrations*, 2nd ed., Wiley, New York, 1931.

D. G. Davis, in *Comprehensive Analytical Chemistry*, C. L. Wilson, D. W. Wilson, and C. R. N. Strouts, Eds., Elsevier, New York, 1964, Vol. IIA, Chapt. III.

Voltammetry, Polarography, and Related Techniques

L. Meites, *Polarographic Techniques*, Interscience, New York, 1965, 2nd ed.

J. Heyrovský and J. Kutå, *Principles of Polarography*, Academic, New York, 1966.

I. M. Kolthoff and J. J. Lingane, *Polarography*, Interscience, New York, 1952, Vols. I and II.

H. Schmidt and M. von Stackelberg, *Modern Polarographic Methods*, Academic, New York, 1963.

B. Breyer and H. H. Bauer, *Alternating Current Polarography* and *Tensammetry*, Interscience, New York, 1963.

Treatise on Analytical Chemistry, I. M. Kolthoff and P. J. Elving, Eds., Interscience, New York, 1963, Part I, Vol. 4, Chapters 42–44, 46, 47, and 50.

D. E. Smith, in *Electroanalytical Chemistry*, Vol. 1, A. J. Bard, Ed., Marcel Dekker, New York, 1966, pp. 1–155.

Amperometric Titrations

J. T. Stock, *Amperometric Titrations*, Interscience, New York, 1965.

Coulometry

K. Abresch and I. Claassen, *Coulometric Analysis*, Franklin, Englewood, N. J., 1966.

G. A. Rechnitz, *Controlled-Potential Analysis*, MacMillan, New York, 1963.

W. D. Shults, in *Standard Methods of Chemical Analysis*, F. J. Welcher, Ed., Van Nostrand, New York, 1966, 6th ed., Vol. III, Part A, Chapt. 23.

Electrogravimetry and Electroseparations

A. J. Lindsey, in *Comprehensive Analytical Chemistry*, C. L. Wilson, D. W. Wilson, and C. R. N. Strouts, Eds., Elsevier, New York, 1964, Vol. IIA, Chapt. II.

N. Tanaka, in *Treatise on Analytical Chemistry*, I. M. Kolthoff and P. J. Elving, Eds., Interscience, New York, 1963, Chapt. 48.

Electrography and Electrospot Testing

H. W. Hermance and H. V. Wadlow, in *Standard Methods of Chemical Analysis*, F. J. Welcher, Ed., Van Nostrand, New York, 1966, 6th ed., Vol. III, Part A, Chapt. 25.

Conductometry and Conductometric Titrations

J. W. Loveland, in *Treatise on Analytical Chemistry*, I. M. Kolthoff and P. J. Elving, Eds., Interscience, New York, 1963, Chapt. 51.

T. S. Burkhalter, in *Comprehensive Analytical Chemistry*, C. L. Wilson, D. W. Wilson, and C. R. N. Strouts, Eds., Elsevier, New York, 1964, Vol. IIA, Chapt. V.

References

1. R. G. Bates, *Determination of pH*, Wiley, New York, 1964, p. 278.
2. *Chem. Eng. News*, May 30, 1966, p. 50.
3. N. H. Furman, in *Treatise on Analytical Chemistry*, Part I, Vol. 4, I. M. Kolthoff and P. J. Elving, Eds., Interscience, New York, 1963, p. 2286.
4. S. Siggia, D. W. Eichlin, and R. C. Rheinhart, *Anal. Chem.*, **27**, 1745 (1955).
5. C. N. Reilley and W. W. Porterfield, *Anal. Chem.*, **28**, 443 (1956); C. N. Reilley and R. W. Schmid, *Anal. Chem.*, **30**, 947 (1958).
6. C. N. Reilley, R. W. Schmid, and D. W. Lamson, *Anal. Chem.*, **30**, 953 (1958).

7. J. P. Phillips, *Automatic Titrators*, Academic Press, New York, 1959.
8. F. C. Anson and J. J. Lingane, *J. Am. Chem. Soc.*, **79**, 4901 (1957); H. A. Laitinen and C. G. Enke, *J. Electrochem. Soc.*, **107**, 773 (1960).
9. D. G. Peters and J. J. Lingane, *J. Electroanal. Chem.*, **4**, 193 (1962).
10. H. A. Laitinen and M. S. Chao, *J. Electrochem. Soc.*, **108**, 726 (1961).
11. R. E. Wilson and M. A. Youtz, *Ind. Eng. Chem.*, **15**, 603 (1923).
12. V. F. Gaylor, A. L. Conrad, and J. H. Landerl, *Anal. Chem.*, **29**, 224 (1957).
13. T. R. Mueller and R. N. Adams, *Anal. Chim. Acta*, **23**, 467 (1960).
14. A. L. Beilby, W. Brooks, and G. L. Lawrence, *Anal. Chem.*, **36**, 22 (1964); H. E. Zittel and F. J. Miller, *ibid.*, **36**, 45 (1964).
15. H. E. Zittel and F. J. Miller, *Anal. Chem.*, **37**, 200 (1965).
16. P. Delahay, *Double Layer and Electrode Kinetics*, Interscience, New York, 1965.
17. D. J. Fisher, W. L. Belew, and M. T. Kelley, in *Polarography 1964*, G. J. Hills, Ed., Interscience, New York, 1966, p. 89.
18. P. Delahay, *New Instrumental Methods in Electrochemistry*, Interscience, New York, 1954, Chapt. 3.
19. F. G. Cottrell, *Z. Physik, Chem. (Leipzig)*, **42**, 385 (1902).
20. J. Heyrovský, *Chem. Listy*, **16**, 256 (1922).
21. L. Meites, *Polarographic Techniques*, 2nd ed., Interscience, New York, 1965, p. 35.
22. J. J. Lingane and H. A. Laitinen, *Ind. Eng. Chem., Anal. Ed.*, **11**, 504 (1939).
23. D. Ilkovič, *Collection Czech. Chem. Commun.*, **6**, 498 (1934); D. Ilkovič, *J. Chim. Phys.*, **35**, 129 (1938).
24. L. Meites, *Polarographic Techniques*, 2nd ed., Interscience, New York, 1965, p. 111·
25. L. Meites, *Polarographic Techniques*, 2nd ed., Interscience, New York, 1965, p. 150·
26. L. Meites, *Polarographic Techniques*, 2nd ed., Interscience, New York, 1965, p. 391.
27. H. H. Willard and J. A. Dean, *Anal. Chem.*, **22**, 1264 (1950).
28. L. A. Matheson and N. Nichols, *Trans. Electrochem. Soc.*, **73**, 193 (1938); J. E. B. Randles, *Trans. Faraday Soc.*, **44**, 322, 327 (1948).
29. A. Ševčik, *Collection Czech. Chem. Commun.*, **13**, 349 (1948).
30. R. S. Nicholson and I. Shain, *Anal. Chem.*, **36**, 706 (1964).
31. R. C. Rooney, in *Analytical Chemistry 1962*, P. W. West, A. M. G. MacDonald, and T. S. West, Eds., Elsevier, New York, 1963, p. 231.
32. H. M. Davis and J. E. Seaborn, *Electronics Eng.*, **25**, 314 (1953); G. F. Reynolds and H. M. Davis, *Analyst*, **78**, 314 (1953).
33. I. Shain, in *Treatise on Analytical Chemistry*, Part I, Vol. 4, I. M. Kolthoff and P. J. Elving, Eds., Interscience, New York, 1963, Chapt. 50.
34. W. Kemula and Z. Kublik, in *Advances in Analytical Chemistry and Instrumentation*, C. N. Reilley, Ed., Interscience, New York, 1963, p. 123.
35. M. T. Kelley, H. C. Jones, and D. J. Fisher, *Anal. Chem.*, **31**, 1475 (1959).
36. B. Breyer and H. H. Bauer, *Alternating Current Polarography and Tensammetry*, Interscience, New York, 1963.
37. G. Jessop, British Patent No. 640,768 (1950).
38. G. C. Barker and I. L. Jenkins, *Analyst*, **77**, 685 (1952).
39. G. C. Barker and A. W. Gardner, *Z. Anal. Chem.*, **173**, 79 (1960).
40. E. Wahlin and A. Bresle, *Acta Chem. Scand.*, **10**, 935 (1956).
41. P. Delahay, *New Instrumental Methods in Electrochemistry*, Interscience, New York, 1954, p. 135.
42. R. Kalvoda, *Techniques of Oscillographic Polarography*, 2nd ed., Elsevier, New York, 1965.

43. H. J. S. Sand, *Phil. Mag.*, **1**, 45 (1901).
44. J. J. Lingane, *Analyst*, **91**, 1 (1966).
45. H. A. Laitinen and I. M. Kolthoff, *J. Phys. Chem.*, **45**, 1079 (1941).
46. P. Delahay, *New Instrumental Methods in Electrochemistry*, Interscience, New York, 1954, Chapt. 9.
47. B. Levich, *Acta Physicochim.* (*USSR*), **17**, 257 (1942); K. F. Blurton and A. C. Riddiford, *J. Electroanal. Chem.*, **10**, 457 (1965).
48. J. T. Stock, *Amperometric Titrations*, Interscience, New York, 1965.
49. C. W. Foulk and A. T. Bawden, *J. Am. Chem. Soc.*, **48**, 2045 (1926).
50. J. Mitchell and D. M. Smith, *Aquametry*, Interscience, New York, 1948, p. 86.
51. G. Marinenko and J. K. Taylor, *J. Res. Natl. Bur. Std.* (*U. S.*), **67A**, 31 (1963).
52. J. K. Taylor and S. W. Smith, *J. Res. Natl. Bur. Std.* (*U. S.*), **63A**, 153 (1959); E. L. Eckfeldt and E. W. Schaffer, Jr., *Anal. Chem.*, **37**, 1534 (1965).
53. R. Ammann and J. Desbarres, *J. Electroanal. Chem.*, **4**, 121 (1962); E. N. Wise, *Anal. Chem.*, **34**, 1181 (1962); A. J. Bard and E. Solon, *Ibid.*, **34**, 1181 (1962).
54. G. L. Booman, *Anal. Chem.*, **29**, 213 (1957).
55. W. D. Shults, in *Standard Methods of Chemical Analysis*, 6th ed., Vol. III, Part A, F. J. Welcher, Ed., Van Nostrand, New York, 1966, p. 467.
56. W. Cruickshank, *Ann. Physik*, **7**, 105 (1801).
57. J. A. Page, in *Handbook of Analytical Chemistry*, L. Meites, Ed., McGraw-Hill, New York, 1963, p. 5–170 to 5–182.
58. *ASTM Methods for Chemical Analysis of Metals*, American Society for Testing Matter., Philadelphia, 1960, p. 422; T. J. Murphy and J. K. Taylor, *Anal. Chem.*, *37*, 929 (1965)
59. F. J. Welcher, *Standard Methods of Chemical Analysis*, 6th ed., Vol. III, Part A, Van Nostrand, New York, 1966, p. 448.
60. H. W. Hermance and H. V. Wadlow, in *Standard Methods of Chemical Analysis*, 6th ed., Vol. III, Part A, F. J. Welcher, Ed., Van Nostrand, New York, 1966, Chapt. 25.
61. T. S. Burkhalter, in *Comprehensive Analytical Chemistry*, Vol. IIA, C. L. Wilson, D. W. Wilson, and C. R. N. Strouts, Eds., Elsevier, New York, 1964, p. 232.

Chapter 6
ANALYTICAL FLAME SPECTROSCOPY

ROMAN I. BYSTROFF, Lawrence Radiation Laboratory, Livermore, California

I. Introduction 186
II. History of Analytical Flame Spectroscopy 187
  A. Flame Emission 187
  B. Atomic Absorption 188
III. Principles and Description of the Techniques 188
  A. Introduction 188
  B. The Sample 189
    1. Dissolution 189
    2. Pretreatment 191
    3. Standards 192
  C. Introduction of the Sample 193
    1. Ideal Conditions 193
    2. The Pneumatic Nebulizer 193
    3. Other Nebulizers 194
    4. Nonflame Sampling 194
  D. The Burner and the Flame 195
    1. Diffusion Flames 195
    2. Premix Flames 195
    3. Spectral Emission of Flames 198
    4. Long Path Burners 199
  E. Principles of Flame Emission 199
    1. Line Emission 199
    2. Band Emission 201
    3. Continuous Emission 202
  F. Principles of Atomic Absorption 202
  G. Sources for Atomic Absorption Spectroscopy 204
    1. Hollow Cathode Lamps 205
    2. Discharge Lamps 205
    3. Other Sources 206

H. Processes in the Flame Affecting the Analyte 206
1. Introductory Remarks 206
2. Compound Formation 207
3. Ionization 209
4. Departures from Thermodynamic Equilibrium 211
5. Physical Effects 212
I. The Spectrometer 212
1. Common Terms 212
2. Filters 213
3. Monochromators 214
4. Resonance Monochromators 214
5. Detectors 214
6. Readout 215
IV. Scope and Sensitivity 215
A. Detection Limits 215
B. Applications to Metallurgical Materials 218
C. Trends 218
References 220
Selected Readings 223

I. INTRODUCTION

Analytical flame spectroscopy is a specialization of the general field of optical spectroscopy. It concerns the use of excitation phenomena found in flames, which are relatively low energy sources compared to the electric field devices normally used in emission spectroscopy. The analysis involves the measurement of either the emission from or absorption in flames. The newer discipline of atomic absorption spectroscopy, while dealing with the inverse of the emission process, with few exceptions utilizes the flame as a sampling device. The same relatively inexpensive spectrometers and the characteristics of flame sampling and chemistry are common to discussions of both techniques.

It is the purpose of this chapter to introduce the materials scientist to the principles, capabilities, and limitations of the techniques without plunging into an excess of spectroscopic theory or the specific practical subtleties to which analytical chemists devote their interests. For such information beyond the scope of this chapter, a number of excellent articles and books are listed at the end of this chapter.

The desired impression is of the existence of a theoretically, instrumentally, and procedurelly simple, inexpensive technique capable of the analysis of major and trace constituents with a precision and accuracy comparable to and frequently better than colorimetric methods, which is to say, about 1% relative standard deviation. The scope of these methods include virtually all metals and some nonmetallic elements as well.

Atomic absorption spectroscopy is perhaps more emphasized, since the popularity of the technique has surpassed that of flame emission spectroscopy in the last few years.

II. HISTORY OF ANALYTICAL FLAME SPECTROSCOPY

A. Flame Emission

The history of analytical flame spectroscopy dates to the original works of Kirchoff and Bunsen,[1] who in 1860–61 published a number of papers on qualitative observations using a spectroscope and a Bunsen burner. A number of pioneers pursued the problem of introducing samples into the flame in a reproducible fashion, so as to obtain quantitative results. Gouy, in 1876[2,3] developed the pneumatic nebulizer, which as a means of introducing liquids as aerosols into the flame, is still the simplest and most used today. It is interesting to note that interest in the flame declined as advances were made in arc and spark excitation over subsequent decades. Not until the 1920's did a revival occur in Europe, exemplified by the classic work of Lundegardh,[4] who developed flame spectroscopy to its present level and demonstrated the determination of more than 30 elements. In the 1930's commercial flame photometers became available from the firm of Carl Zeiss. The elegant simplicity of the instrument, which comprised only a colored filter, a flame atomizer, and a photodetector, together with the capability of determining parts per million of specific alkali elements, stimulated widespread interest. In the United States the growth of the applied technique began with the interest of a number of workers[5–7] in the Lundegardh spectrographic technique during the period 1939–1946, and with the subsequent introduction of a flame attachment for the well-known Beckman Instruments spectrophotometer, and a simple photometer by Perkin-Elmer Corp. Since then the literature of flame emission spectroscopy has grown to impressive proportions. Mavrodineanu, who has compiled a comprehensive bibliography in this field[8,9] through the year 1959, cites 1700 articles, a number which has since easily trebled. Thorough summaries of developments and of the principles of flame emission spectroscopy are found in books by Dean,[10] Herrmann and Alkemade,[11] Poluektov,[12] Mavrodineanu and Boiteux,[13] and articles by Gilbert[14,15] and Dean.[16] In addition, Margoshes and Scribner[17,18] have reviewed most of the significant work in this field for the years 1962–1967.

B. Atomic Absorption

While the phenomenon of the resonance absorption of radiation was first observed by Wollaston[19] in 1802 as the "Fraunhofer lines" in the solar spectrum, and the theory and principles were thoroughly understood at least by the time of the first printing of Mitchell and Zemansky's definitive volume on resonance radiation in 1934,[20] the credit for the present day analytical application belongs to Walsh.[21] In 1955 he demonstrated the use of hollow cathode lamps with an air–acetylene premixed flame. The lamps were similar to those first described by Schüler,[22] and commonly used by spectroscopists for hyperfine structure determinations because of their extraordinarily sharp line spectrum. Walsh was able to distinguish the lamp radiation from the flame radiation by modulating the source radiation with a mechanical chopper, while tuning the detector amplifier to the frequency of the chopper.

Within a few years commercial versions of the Walsh system became available, often as simple modifications of equipment used for flame emission spectroscopy.[23] The technique of manufacturing long-lived and more intense hollow cathode lamps improved,[24–27] extending the practicability and scope of the method. The evolution of the technique is still in progress and is characterized by new flame systems,[28] new burner systems,[29,30] nonflame systems,[31–35] different sources of radiation,[36,37] new methods of detection,[26] and even the inception and growth of the allied technique of atomic fluorescence spectroscopy.[38–41]

Journal articles dealing with the analytical aspects of atomic absorption now number more than 1000. Bibliographies and review articles have appeared regularly.[17,18,42–52] There have been four books on the subject, several chapters in general treatises, and in books on flame methods. These are included in Reference list at the end of this chapter.[11–16,53–55] In addition, commercial instrument manufacturers have made available detailed technical notes and articles on a regular basis.[56–58]

III. PRINCIPLES AND DESCRIPTION OF THE TECHNIQUES

A. Introduction

In principle flame emission and absorption photometry are relatively simple. The sample is prepared as a solution. The solution is converted into a mist, in a process referred to as nebulization. The mist is introduced into a flame, where it is dried and converted to a solid particle. The particle evaporates into the atomic or molecular species stable at the temperature of the flame. In flame emission spectrometry, some fraction of the light emitted by excited atomic or molecular species is optically focused on an energy discriminator—the monochromator. The intensity of the light from the particular atomic or molecular transition isolated by the monochromator is measured, usually with a phototube or photomultiplier. In atomic absorption, a separate source of light is focused on the flame. Free unexcited atoms in the flame will absorb some specific wavelength of the source light, and the light flux will be attenuated. The attentuation, which is proportional to the free atom population in the flame, is measured with a photomultiplier. A monochromator is usually placed between the photomultiplier and flame to isolate the source radiation of interest and to reject most of the flame radiation.

In the ensuing discussion, each of the functional aspects in the gross description above are discussed separately to illustrate where in the analytical system the variabilities in sensitivity and selectivity exist. In Figure 1, a schematic representation illustrates the functional parts of both systems.

Another approach, which space forbids here, would be the theoretical derivation of the detection limits for either system of analysis. These equations have a pedagogical value of illustrating the kinds of factors and interrelationships involved. The reader is referred to the work of Winefordner et al.[59–61]

B. The Sample

1. DISSOLUTION

Since all but two variations of the techniques involve solutions, the reader should be aware of the limitations and advantages involved. The choice of solvent, for the most part, can be dictated by the expediency of the dissolution. In general, water-based solvents are conveninent to use. The smallest amount of mineral acid to affect the dissolution of metals and alloys and still prevent hydrolysis is desirable. If fused-salt dissolution is necessary, minimization of the amount of salt, and regard for its purity is important. It should be obvious that the solvents must be free of the element sought. This is particularly binding when the techniques are used for trace analysis,[55] and imposes a *blank limitation*. Normal reagent grade solvents are usually satisfactory since the predominant range of sensitivities lies in the parts per million. Blank problems arise more frequently when preconcentration methods are used.

Under the topic of processes in the flame (Section III-H), it will be seen that a variety of interferences do occur. Some of these are attributable to specific anionic species that form compounds with the analyte which are difficult to dissociate in the flame. A judicious choice of the solvent acid

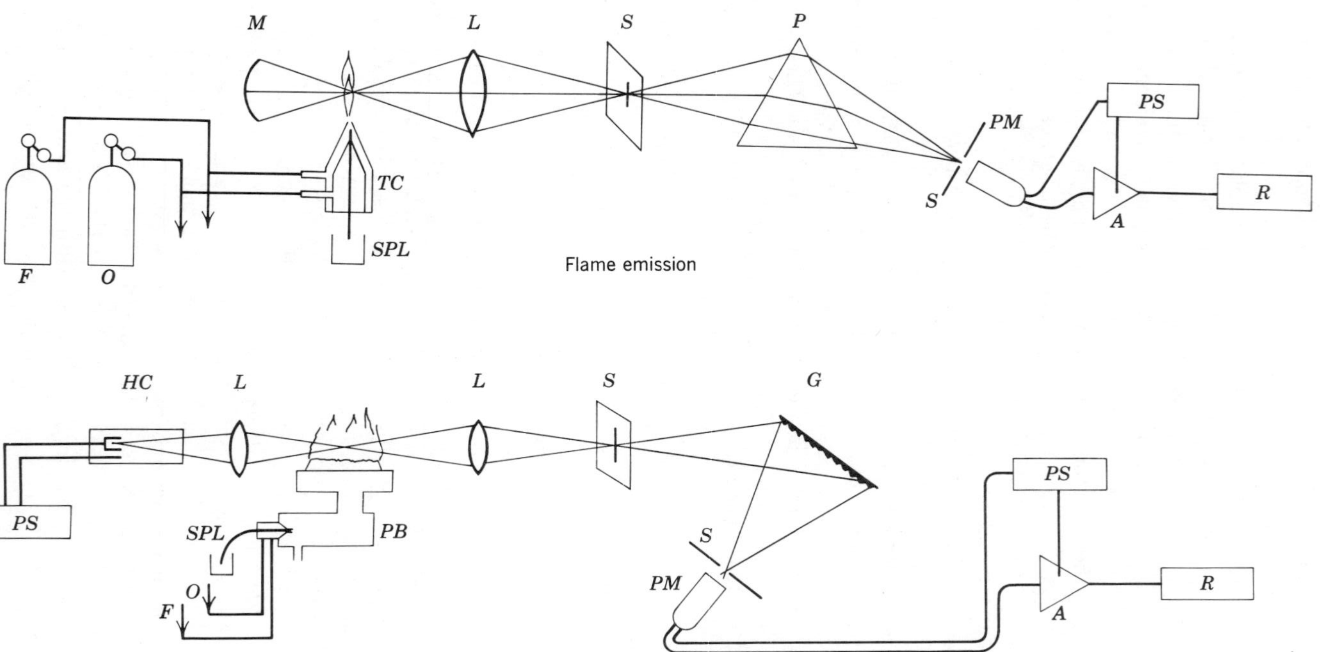

Fig. 1. Schematic systems for flame emission and atomic absorption analysis. Symbols are: *A*, amplifier; *F*, fuel supply; *G*, grating; *HC*, hollow cathode lamp; *L*, lens; *M*, mirror; O, oxidant supply; *P*, prism; *BP*, premix burner; *PM*, photomultiplier; *PS*, power supply; *R*, readout; *S*, slit; *SPL*, sample.

would be called for. Another type of interference results from the presenee of easily ionized metals in the flame. The analyte signal may be enhanccd or depressed if it was found necessary to use a basic or acid salt flux in the simple dissolution.

Organic solvents are even more desirable than water, particularly those of low viscosity, low boiling points, and of an easily combustible nature. Ethanol, acetone, methyl isobutyl ketone, and gasoline are burned directly, with a significant enhancement of emission or absorption signals.

2. PRETREATMENT

It is of course preferable that, other than the dissolution, dilution to volume, and perhaps addition of some beneficial additives, no further treatment of the sample would be necessary. This is not always possible. Pretreatment is sometimes necessary for the removal of anions and metals that cause spectral or chemical interference, particularly in flame emission work. The addition of chemical steps, unless simple themselves, detracts from the great speed of flame analysis. Still, if matrix effects must be bypassed, or additional sensitivity is needed, a few general separations can add enormously to the power of the technique. The extraction separations listed in Table 1 are relatively nonselective. Note that a large number of metals are

Table 1. Some extraction systems used in flame emission and absorption analysis.

Chelating agent	Solvent	Metals extracted	Conditions	Ref.
APDC[a]	MIBK[b]	34	pH 2–14	c,d
		28	pH 4	c,d
		not Groups I, II, III		
Oxine	MIBK or $CHCl_3$	30	pH 1–11	e
Cupferron	MIBK	V, Ti, Cu,	pH 1	d
		Cu, Fe, Ni, Mn	pH 7	f
Dithizone	MIBK	Cd, Pb, Zn	pH 7	e
	or $CHCl_3$	15	pH 0–14	e

[a] APDC is ammonium pyrrolidine dithiocarbamate.

[b] MIBK is methyl isobutyl ketone.

[c] H. Mallisa and E. Schoffman, *Mikrochim. Acta*, **1**, 187 (1955).

[d] C. E. Mulford, *Atomic Absorption Newsletter*, **5**, 88 (1966), Perkin-Elmer Corp., Norwalk, Conn.

[e] J. Starý, *The Solvent Extraction of Metal Chelates*, (MacMillan, New York, 1964), p. 80, 139.

[f] B. Delaughter, *Atomic Absorption Newsletter*, **4**, 273 (1966), Perkin-Elmer Corp., Norwalk, Conn.

separated from anions and Group I, II, and III metals which are the principal interferants in flames. A fixed and convenient chemical procedure may be adopted by relying on the selectivity of the spectrometry to distinguish among the extracted metals.

Inhomogeneities of trace elements on the microscale in the solid should always be considered. Sampling problems of this sort can be handled by pretreatment of the solid sample by grinding, crushing, and mixing, as is common in X-ray and emission spectroscopy. Such methods are exacting and subject the sample to contamination. The dissolution of sufficiently large samples is a preferable approach, and is the normal procedure for flame methods.

3. STANDARDS

Both flame methods are relative, as opposed to absolute techniques such as coulometry and methods based on stoichiometric reactions, and therefore require standards. In the majority of cases, standards of the analyte in water suffice for atomic absorption methods. If any kind of matrix effect is found, however, it may be necessary (in lieu of an additive with buffering properties) to match the unknown solution composition as closely as possible to obtain quantitative results. The method of standard additions is the simplest and most straightforward way of assuring this. Briefly, known incremental amounts of the standard analyte are added to equal portions of the sample, and the solutions are brought to the same volume with the solvent used in preparing the standard. The emission or absorption signals from this series of solutions establish the sensitivity, defined here as the change in concentration for a unit change in signal. The sensitivity times the magnitude of the signal from the diluted but "unspiked" sample gives the analyte concentration. If the sensitivity is not constant with varying concentration, the method fails. Mostyn and Cunningham[62] have pointed out another limitation for this method. If complex matrix effects occur, such that the *ratio* of the analyte to some other component of the solution affects the signal, the method will be biased, even though a constant sensitivity is apparent. The ratio is not easy to control. Clearly, the existence and methods of suppressing such effects must be recognized, and the analyst must be wary of the infrequent exception to the rule.

The addition of internal standards is a technique used most often in flame emission methods and is discussed in detail by Dean.[11] By taking the ratio of the analyte signal to that of the internal standard, a number of uncontrolled fluctuations, or mismatches between successive solutions, can be compensated, thus improving the precision of the measurement.

C. Introduction of the Sample

1. IDEAL CONDITIONS

The ideal conditions for the introduction of the sample into the flame would include: (*1*) The sample, converted to a dry powder, would possess a uniform particle size in the submicron range so that the vaporization rate of the particle would not be significant. (*2*) The number of particles introduced into the flame would be very uniform in the time scale of the measurement, and this rate should be as high as commensurate with the availability of energy in the flame to totally vaporize the sample. In other words, the flame temperature should not be significantly lowered by sample introduction.

In practice the solution containing the sample is converted to a mist by means of one of a variety of nebulization techniques. The most common in use today is the pneumatic nebulizer.

2. THE PNEUMATIC NEBULIZER

In its simplest form the pneumatic nebulizer (sometimes called an atomizer; a slight misnomer) consists of a capillary tube to carry the sample solution and a gas jet perpendicular or concentric at the end of this tube. The Venturi effect produces a vacuum at the end of the sample tube sucking more sample continuously. Typical rates of sample consumtion range from 0.1 to 6 ml/min. The high velocity of the jet breaks the liquid into small droplets. Mavrodineanu[13] devotes an entire chapter to a review of the theory and practice in nebulizer design. It will suffice here to point out that the rate and efficiency of nebulization will depend critically on the design, particularly on the orifice dimensions, the spacing between the sample tube and the gas orifice, the gas pressure, the velocity and direction of flow, and on physical properties of the solution such as viscosity, surface tension, and density. Because it is desirable to nebulize solutions of different viscosities, densities, and surface tension, not only have nebulizers with a variety of sizes of sample intake capillaries been found useful, but some workers resort to constant feed pumps, such as syringe drives, peristaltic,[63] and gas pumps.[64] A measure of the efficiency of nebulization is the mean droplet size, and the distribution about the mean. Generally speaking, droplets below 10 μ in size will become airborne, and will evaporate very quickly in the heat of the flame. The larger droplets, representing the tail of the particle size distribution curve, are lost, either because they fall out of the gas stream, or in integral burner–atomizers, because they evaporate too slowly and pass through the analytical region of the flame. For the most part, commercial nebulizers are capable of producing mean droplet sizes in the 10–20 μ range for water, but a significant volume is lost in the fewer droplets of large size. Much of the increased sensitivity observed in with organic solvents has to do with the finer droplet size produced because of more favorable physical properties.

Burners which are integral with the nebulizer, also termed "total consumption" burners, for the most part produce the diffusion flames to be discussed later. Here, the inefficient nebulization pattern has an important effect on the fluctuations of background and analyte radiation, which ultimately appears as flame "noise." In contrast with total consumption burner–atomizers, premix burners employ a spray trap which is simply a chamber in which the larger droplets are permitted to fall out onto the walls, while the smaller, airborne droplets continue to the flame port. This chamber has been designed in many ways, with particular consideration given to intimately mixing the combustible gases and subsequently providing a laminar flow of gases to the flame port.

A recent significant advance in burner design takes advantage of the fact that the droplet cloud can be processed in the premix chamber.[65] The solvent is evaporated with IR heaters, leaving airborne dry particles containing the analyte. The solvent is condensed and removed, and only the dry particles of sample reach the flame. This sytem would seem to come very close to the ideal, and indeed the analytical sensitivities for atomic absorption some of which appear in Section IV, Table 5, would indicate close to total sample utilization.

3. OTHER NEBULIZERS

Recently there have been interesting developments in nebulization using ultrasonic transducers.[66,67] The droplet mean diameters are in the range 0.8–1 μ; thus not only is nebulization efficient, but the sample may be introduced at a rate independent of the gas flow. Other types of nebulizers (spinning disks, electrostatic, etc.) are infrequently used, and will not be discussed here.

Where samples are already in the powdered form, burners have been designed so that the sample can be introduced conveniently into the flame. Shipitsyn[68] has described a pulsating hammer device for introducing soil samples for a potassium analysis by flame emission photometry. Solid samples can also be introduced as slurries.[69]

4. NONFLAME SAMPLING

In passing, mention should be made that the flame is not the only energy cell into which samples have been introduced for measurement by atomic absorption. Early work by Nelson[70] utilized samples deposited on a graphite grid, and vaporized by induction heating from capacitor discharge through a coil. Recent work by Mossotti[71] has been reported, where atomic absorption is observed in the plume produced by laser excitation of metallic samples.

D. The Burner and the Flame

The study of the flame is a scientific discipline in its own right, and is usually covered in the annual proceedings of the Combustion Institute and in a number of excellent books.[72–75] The characteristics of flames are important to burner design and to attempts by analytical chemists to understand and control the excitation phenomena. This discussion will be limited to illustrations of characteristics that are of some importance to the analysis and the flame systems most commonly used.

1. DIFFUSION FLAMES

Two general classes of flames are used—diffusion and premixed flames. In the diffusion flame, pure flowing fuel is allowed to contact the flowing oxidant. The rate of combustion is determined by the rate of diffusion of one into the other. In practice, in order to achieve atomization, the gases are introduced through ports concentric with the sample inlet tube. The gas velocities to achieve efficient atomization are usually so high that the Reynolds number is exceeded and no stable combustion surface is possible. The presence of vortices and pockets of gas–oxidant mixtures literally exploding cause such flames to be acoustically noisy. The flame itself, termed a turbulent diffusion flame, is stable over wide ranges of flow and of fuel to oxidant ratios. The primary analytical interest is temporal stability of flame position, flame size, shape, composition, and temperature. These factors depend ultimately on the degree of gas flow and pressure regulation, and the changes in effective dimensions of the gas and sample ports due to corrosion, thermal effects, and encrustation of particulate matter from the flame (carbonization) or sample (excess solids content) or dust carried in the gas stream. Turbulence, particularly that due to entrainment of room air, can be controlled by means of a gas sheath about the flame. Such burners have been shown to result in increased precision for analysis.[76]

2. PREMIX FLAMES

Premixed gases result in a flame in which there is a well-defined combustion surface. The combustion rate, or flame speed, must be at least matched by the velocity of cold gas passing the flame port. If the gas velocity is increased beyond the ability of the burner port to "anchor" the flame, "blow off" of the flame occurs. The stability of a flame at a port is diagramatically illustrated in Figure 2 by a plot of composition vs. gas velocity at constant

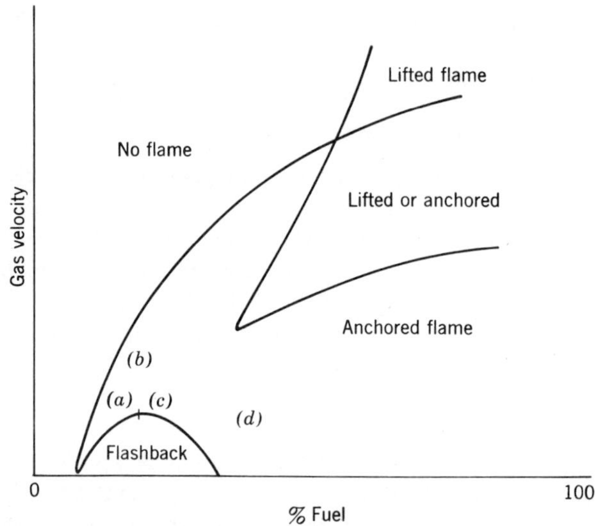

Fig. 2. Schematic of flame stability at a port. *(a)* Lean flame. *(b)* Stoichiometric composition. *(c)* Rich flame. *(d)* Incandescent carbon flame.

port geometry. The flame speed, more precisely the velocity of the flame front, is inversely related to the quenching distance,[77] the minimum size channel which will permit flame propagation. There is an analytical preference for slower flames. Small flame ports necessary to quench fast flames have a proclivity to clog when solutions containing much solid (1–10%) are used. Larger ports with fast flames increase the possibility of flashback, which can be serious for gas mixtures which develop large detonation velocities and pressures. Thus, for very fast flames, many workers have dealt only with diffusion flames, which, of course, will not flashback.

A number of flame components combinations are used in analytical work, offering a variety of excitation conditions. Some, but not all, of these appear in Table 2, together with maximum temperatures and features which make them attractive. The presence of solvent will generally lower the actual temperature. In the case of the diffusion oxygen–acetylene flame, Winefordner[78] has measured temperature decreases of 200°K.

Premixed flames are particularly advantageous for atomic absorption where it is desirable to have a long optical path in the flame. This is simply achieved with flame ports designed as slots. An additional advantage arises from the absence of turbulence, and therefore minimization of background emission fluctuations. A premix flame system is composed of two or more

Table 2. Commonly used flames for emission and absorption spectroscopy

Oxidant	Fuel	Max. temp., °K	Mac. flame speed (cm sec^{-1})	Suitability
Air	Propane	2200[b]	82[b]	Least ionization.
Air	Hydrogen	2320[b,c]	320[b,c]	Chemiluminescence; low background.
Air	Acetylene	2570[c]	160[c]	Convenient for a wide variety of elements.
Oxygen	Hydrogen	2970[d]	900[d]	Low background; some refractory elements.
Oxygen	Propane	3070[d]	500[d]	Convenient for a wide variety.
Oxygen	Acetylene	3330[c]	1130[c]	Ref. elements.
Nitrous Oxide	Acetylene	3230[a]	180[a]	Ref. elements.
Oxygen	Cyanogen	4830[d]	176[d]	Ref. elements.

[a] W. G. Parker and H. G. Wolfhard, *Proc. Fourth Intern. Symp. Combustion*, 420 (1952); *Proc. Fifth Intern. Symp. Combustion*, 718 (1954).

[b] See reference 13, p. 17.

[c] E. Bartholomé, *Z. Electrochem.*, **54**, 169 (1950).

[d] See reference 14, p. 74.

zones. The inner, or primary reaction zone is usually the most luminous. The outer cone is a secondary reaction zone which represents a diffusion flame between the surrounding air and incompletely oxidized products of preceding reaction zones. Between these, additional secondary reaction zones may occur as an interconal region. This is illustrated in Figure 3. The interconal region can be sizable under fuel-rich conditions, and in the case of oxy-acetylene or nitrous oxide–acetylene, a very significant analytical zone. Strongly reducing conditions exist in this region due to a significant vapor pressure of free carbon atoms which react with oxygen. Indeed, the atomic species of even the most refractory elements are observed. This subject will be taken up again, but it should be pointed out that because of the extreme incandescence of fuel-rich turbulent diffusion flames, the radiation of atomic species is quite difficult to observe. The premix flame offers a strong advantage by offering a clear-cut separation of zones which can be optically segregated.

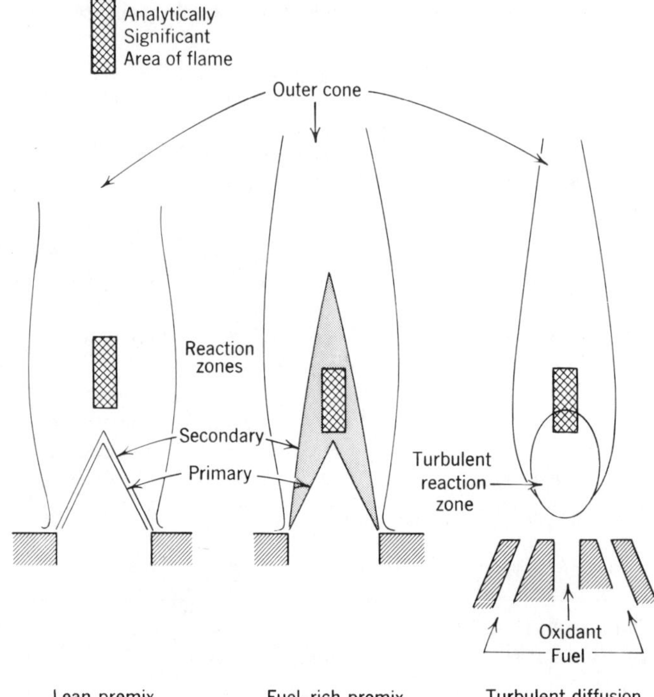

Fig. 3. Schematic appearance of flames used in emission and atomic absorption spectroscopy.

3. SPECTRAL EMISSION OF FLAMES

The presence of molecular bands, continuous or discrete radiation due to species other than the analyte can make the measurement of the analytical line intensity difficult. All oxygen flames exhibit strong OH band systems, and hydrocarbon flames have C_2 and CH band systems of lesser magnitude throughout the visible and UV spectrum. In emission work, such background radiation may constitute an uncertain correction, or may have to be reduced by decreasing the bandpass of the spectrometer and thus reducing also the sensitivity for the analyte. A judicious choice of the flame system sometimes offers a means of avoiding the problem. The air–hydrogen flame is notable for the absence of visible radiation and is favored for flame emission analysis.

Considerations of flame background radiation are avoided in atomic absorption when the source and detector are modulated. The amplifiers then will not respond to the emission signal from the flame. Excessive emission can saturate the photodetector and prevent any response, however. Even with less emission, the random fluctuations of the flame emission intensity appear as unwanted noise on the absorption signal. These difficulties are easily overcome if the source intensity can be increased, or if the bandpass of the monochromator can be reduced. The latter reduces the intensity of continuous radiation faster than that of monochromatic radiation from the source.

Some workers[79,80] have applied the modulation principle to flame emission to distinguish between the flame and analyte emission. For instance, this can be done by rapid pulsing of the solution into the flame.[81] The noise considerations discussed above would still apply, of course.

4. LONG PATH BURNERS

Mention should be made of burners which have been designed to increase the residence time of atoms in the beam for atomic absorption and therefore the sensitivity. Fuwa and Vallee[30] have injected the flame into long tubes through which absorption was observed. Although the tube is insulated, the flame gases entrain air and cool enough so that only the most long-lived metals (with respect to recombination) show the expected increased sensitivity. Zinc, for instance, can be determined to less than 1 ppb of solution in this manner. Long tubes have also been used in the form of furnaces.[82] Progress in the nonflame approach is still in the experimental stage.[83,84,33–35]

E. Principles of Flame Emission

The emission spectrum of the analyte in a flame can consist of a number of lines, intensity maxima with half-band widths in the order of less than 0.1 Å, and of bands of undefined spectral width, normally greater than a few angstroms. The latter arise from molecular transitions—the former from atomic transitions.

1. LINE EMISSION

In contrast to processes in an arc or spark, where excitation arises from electron impact, flames possess excitation which is purely thermal in character, at least outside the primary reaction front. With a realization that other processes may also occur, thermodynamic equilibrium is a fair assumption in the analytical region of the flame. The famous Maxwell-Boltzman equation then describes the distribution of atoms between two energy states:

$$N_u = \left(\frac{N_0 g_u}{g_0}\right) \exp\left[-\Delta E/kT\right] \tag{1}$$

where

N_u = the number of excited atoms per cm^3; u designates upper state

N_0 = the number of atoms in the less energetic state per cm^3

g_u, g_0 = the statistical weights for each state, $g = 2J + 1$, where J is the inner quantum number describing the state

ΔE = the excitation energy, $\Delta E = E_u - E_0 = h\nu$, usually expressed in eV

k = the Boltzman constant

T = the absolute temperature

The intensity of a line depends on the population of the upper state and the rate of transfer between the two states. This is expressed in the following equation which represents the basis of all spectrochemical quantitative analysis using line spectra:

$$I = \frac{10^{-7} h\nu_0 g_u A_t}{4\pi k T B(T)} \cdot PL \exp\left[-\Delta E/kT\right] \qquad (2)$$

where

I = the steradiancy, the radiant energy of a single integrated line in watts/cm^2-ster

P = the partial pressure of the analyte present in the flame as atoms, expressed in atm

h = Planck's constant

A_t = the Einstein coefficient, the probability in sec^{-1} that an energy state will radiate

L = the thickness of the region of the flame focused on the spectrometer slit

$B(T)$ = the partition function, representing the distribution of atoms among all states. For flames,

$$B(T) = \sum_i g_i \exp -(E_i - E_0)/kT \cong g_0$$

since most atoms are in the ground state

ν_0 = the frequency of the center of a radiating line

$k, T, g_u, \Delta E$ have been defined.

The equation illustrates the following points:

1. The intensity is proportional to the partial pressure of analyte, and therefore proportional to the concentration.

2. The intensity of a line will increase exponentially with T until $B(T)$ increases significantly, representing the depopulation of the ground state, or until some other process (like ionization) occurs.

3. For a given value of A_t, lower excitation potentials favor emission intensity.

Those lines representing transitions to the ground state are termed resonance lines. The emission from these lines are capable of being reabsorbed by ground state atoms. Now the concept of the Doppler width becomes important and is given by the equation

$$\Delta\nu = \frac{2(2R \ln 2)^{1/2}}{c} \nu_0 (T/M)^{1/2} \tag{3}$$

where

$\Delta\nu$ = the frequency width of the band at half maximum in sec^{-1}
M = the atomic weight of the emitter-absorber
R = the gas constant
c = the speed of light
ν_0, T have been defined

Because of the temperature dependence of the Doppler width, the emission at higher temperatures has a greater handwidth than the absorption-reemission of atoms in a cooler region of the flame. Reemission, however, is random in direction. Thus the net appearance of the spectral distribution of a resonance line viewed from one direction would show a reversal due to self-absorption in the center of the line. This is illustrated in Figure 4. The low-resolution spectrometers used in flame spectroscopy would not be able to resolve such spectral effects, but the total integrated intensity of the line would show a depressed dependence on concentration, ultimately approaching an $I = C^{1/2}$ relation, where C is the analyte concentration. Self-absorption itself is concentration dependent, thus equation 2 represents the limiting concentration dependence at low concentration. Nonresonance lines do not show self-absorption since the lower states are negligibly populated. Ionization effects will be discussed later.

2. BAND EMISSION

Molecular emission is too complex in principle to discuss here; the reader is referred to Mavrodineanu[13] or Gaydon.[85] Analysis of an analyte by use of molecular emission has generally been limited to those elements which are difficult to dissociate to atoms at flame temperatures. Not only do bands tend to be weakly emitting, but their inherent spectral width makes them much more subject to radiation interference from other flame species and matrix components. Nevertheless, prior to the more recent work on fuel-rich flames, band spectra were successfully used, particularly in the analysis of Sc, Y, and the rare earths,[86] boron,[87] and a few other elements. The nonmetallics P, C, S, and N all have analytically useful narrow bands.[88]

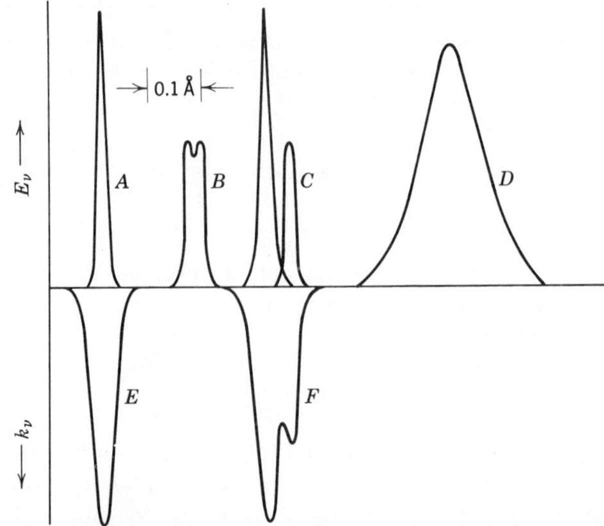

Fig. 4. Spectral distributions of source emission and flame absorption lines. *(A)* Hollow cathode resonance line. *(B)* Self-absorption in a resonance line. *(C)* Resonance line with an adjacent isotope line; or with an adjacent ion line. *(D)* Bandpass of a spectrometer used with a continuous source. *(E)* Absorption coefficient of a resonance transition in the flame, illustrating Doppler broadening. *(F)* Same as *(E)*, with an adjacent isotope line.

3. CONTINUOUS EMISSION

Continuous emission is less well understood. It may arise from the superposition of many bands, or spectrally wide bands. Solid particles radiate continua and exist in incandescent flames. Only rarely would analytical use be made of continuous radiation due to an analyte; more frequently this radiation acts as a spectral interference.

F. Principles of Atomic Absorption

Any discussion of principles of the absorption process in the flame should elucidate factors affecting the sensitivity, the shape of the analytical curve, and the detection limit. Detection limits are for the most part dependent on the noise limits set by the flame emission, and instrumental design. Currently obtained detection limits can be found in Section IV of this chapter. Some of the factors affecting the shape and sensitivity of the response curve involve processes in the flame, and are deferred.

Within the flame the sensitivity for absorption depends on (*1*) the number of free ground state atoms present in the flame, (*2*) factors affecting the absorption coefficient, (3) the source energy distribution, and (4) the relationship of source image in the flame to the distribution of the atom population in the flame.

The fraction of atoms in excited states is generally very small. The normal flame temperatures are such that only very low-lying atomic states can be significantly populated, and for most elements $N_0 = N_T$, the total number of unionized atoms/cc, applies in equation 1. Even for the worst case, cesium, the 8521 Å resonance line arises from a state which contains only $4.4 \times 10^{-4} N_T$ atoms/cc at 2000°K, and $3.0 \times 10^{-2} N_T$ atoms/cc at 4000°K. In contrast to flame emission, thermal fluctuations in the flame would have little effect on the significant atom population. Of course the same cannot be said if compound dissociation or ionization processes occur.

The absorption coefficient for a single transition, k_ν, can be shown to obey equation 4.

$$\int k_\nu d\nu = \left(\frac{c}{\nu}\right)^2 \frac{g_u A_t N_T}{8\pi g_0} \tag{4}$$

The terms in the equation have been defined under equations 1 and 2. Figure 4 shows the appearance of k_ν with Doppler broadening, a gaussian distribution. The source may also have a similar distribution, but to be general, several possible distributions are illustrated in Figure 4, and designated as the function E_ν.

For monochromatic light the well-known Beer's Law is known to hold

$$A_\nu = 1 - (I/I_0) = 1 - \exp\left[-k_\nu l\right] \tag{5}$$

where A_ν is the absorption, I is intensity, and l is path length. A perfectly general relationship is given in equation 6 for the absorption observed under nonmonochromatic conditions.

$$A = \int E_\nu (1 - \exp\left[-k_\nu l\right]) d\nu / \int E_\nu d\nu \tag{6}$$

In the limit, where the spectral width of the source become much less than the Doppler width of the analyte in the flame, k_ν can be approximated by $k_{\nu 0}$, the absorption coefficient of the line maximum; this equation reduces to a linear form of Beer's Law:

$$\ln (1 - A)^{-1} = k_{\nu_0} l \tag{7}$$

The constant $k_{\nu 0}$, which depends on $T^{1/2}$, implies a highly linear response. Such response is the usual case in practical atomic absorption with hollow cathode sources, and is a particularly desirable feature since only one standard (and the zero point) will establish the calibration curve. If the source radiation is broad or nonideal as in Figure 4*d*, linearity will still be the rule at low-absorption values, and curvature toward the concentration axis will occur as the absorption increases.

It should be realized that the source beam will be presented an atom concentration profile in the flame. Such a profile is illustrated in Figure 5*a*, and can vary for each element. The effective portion of the beam is that which passes by the slit of the monochromator. This portion of beam, focused in the flame is illustrated in Figure 5*b*. It is evident that regions in the flame of differing populations (and perhaps differing emission and chemistry) are traversed by the beam in a flame of some length. This concept is important to instrument and burner design. For instance, Boling[96] by designing a three-parallel-slot burner achieved a wider flame, thus obtaining a greater sensitivity and linearity than was possible with a single-slot 10 cmburner on the same instrument. Amos and Willis[97] noted increased sensitivities with decrease in length and an increase in width of a slot-type burner. The efficacy of passing the same beam through the flame several times is limited in flames where large vertical atom population gradients exist.

G. Sources for Atomic Absorption Spectroscopy

The requirements for the spectral output of a source to achieve resonance absorption in a flame are straightforward: The intensity must be high

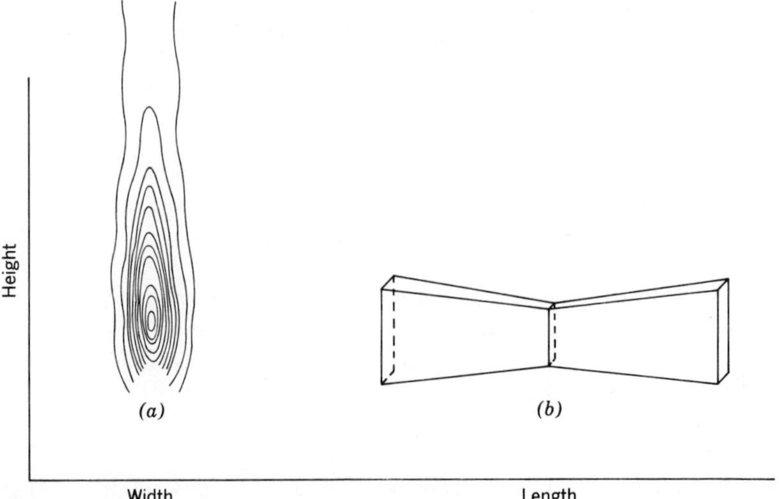

Fig. 5. Flame atom profiles and optical sample space in atomic absorption spectroscopy. *(a)* Equal concentration contours of neutral atoms in a flame. Barium in lean air–acetylene, taken from C. S. Rann and A. N. Hambly, *Anal. Chem.*, (*37*, 879 (1965). *(b)* An illustration of optical divergence of a spectrometer slit image along the length of a flame.

enough to override that of the flame; the light must be monochromatic at a resonance line of the analyte with little or no background continuum; and the intensity must not show self-absorption and must be very constant in time. In practice one or more of these must be compromised.

1. HOLLOW CATHODE LAMPS

In its simplest form, the hollow cathode lamp consists of a cup-like cathode and a rod anode sealed in an atmosphere of a few mm of noble gas. The cathode is constructed of the metal whose spectra is desired, or of a metal possessing a simple spectrum and high-excitation potential relative to the analyte metal. The metal is evenly deposited inside the cup as a foil or ring by electrodeposition, sputtering, or spinning of a melt. The phenomenon of sputtering by ionized rare gas atoms accelerated at the cathode surface is responsible for metal atoms appearing in the cup. This process is discussed in more detail in books on gas ionization phenomena.[89] The atoms are excited by electron bombardment in the plasma present in the cup. The spectra emitted includes resonance lines and ion lines of both the metal and the rare gas, and are notably narrow because of the low-thermal activity of the sputtered atoms. The relative intensity of the lines is limited by potential melting of cathode material at high currents and by self-absorption by the lost atoms in front of the cup. Improvements in the technology of lamps[26,90,91] have increased the available intensity, reduced self-absorption, increased the life of the lamp (current lamps exceed 2000 hr in life at normal currents), and reduced the presence of ion lines. The stability of the intensity after a 10–20 min warm-up time is very good, especially with current regulated power supplies. Only the cost (roughly $100–200), the consideration that one lamp is required for each analyte element, and the unavoidable presence of noble gas lines make these lamps less than ideal. Multielement lamps are available for up to five elements. Demountable lamps have been used at the loss of convenience and sophistication. The presence of other lines places a premium on monochromators of high dispersion to isolate the resonance lines; but this requirement exists only for some elements and affects the linearity of the response rather than the response itself.

2. DISCHARGE LAMPS

Discharge lamps such as the Osram lamps are suitable for the alkali metals and yield higher intensities than the corresponding hollow cathode lamps. There is a greater tendency to self absorption than in hollow cathode lamps.

Microwave excited "electrodeless" discharge lamps[92] can be made for any element which has a reasonably volatile compound—like the iodides. Like other discharge lamps, their intensity is high, self-absorption may be serious, and close control of their stability is an undeveloped art.

3. OTHER SOURCES

Flames have been used as sources. With extreme care and a null point technique, Malmstadt[93] succeeded in determining sodium by absorption with a precision of better than 0.5% using the simplest possible arrangement: two flames; an optical filter; and a photodetector. Lithium isotope ratios[94] and the rare earths[95] have been determined using flames as sources and for sampling.

Continuous sources can be used for atomic absorption if the monochromator has sufficient resolution to pass a band of radiation not very much greater than that of the resonance line in the flame. Light passed from the source whose energy does not fall within the analyte Doppler width remains unattenuated so that the absorption represents a smaller fraction of the total integrated intensity. So long as the overall noise in the system is low, a loss in sensitivity may be tolerated. Indeed continuous sources, at least in the visible region of the spectrum, are sufficiently intense and stable so that only instrumental noise is limiting. There is, of course, a great economic and practical advantage in being able to use only one source for all elements provided the savings is not lost in a more elaborate spectrometer. Fassel et al.[36] have reported considerable success with a tungsten lamp or xenon arc and a table model spectrometer.

H. Processes in the Flame Affecting the Analyte

1. INTRODUCTORY REMARKS

It is unavoidable that in a chemically dynamic system such as the flame, processes must occur which affect the analyte, particularly as different matrices are compared. The effects are numerous and attempts to classify them, while commendable, have let to problems of semantics. Part of the confusion arises naturally, for the effects seldom arise singly, and the multiplicity of variables compounds the difficulty of confirmation. The analytical art consists of finding ways to control the effects. It is the simplicity and universality of the control measures which determine the popularity of the analytical technique.

Chemical effects may arise from any species in the environment of the analyte, usually anions. The magnitude of the effect depends on how close and how much of the species there is, how fast a response there is to the changing environment, and how stable the analyte forms are in the flame. Ionization might be considered in the same sense. Some physical effects peculiar to the flame will also appear in the ensuing discussion.

2. COMPOUND FORMATION

In the rapid process of drying of the aerosol, selective precipitation of stable compounds occurs. To a great degree, the relative volatility of these compounds will control the rate of atomization in the flame.

Volatile compounds can become gases and subsequently atomize most rapidly on entry into the flame reaction zone. Atoms are therefore produced in the hottest portion of the flame. This is particularly noticeable with refractory metals, where dissociation of the oxides is quite temperature dependent. Examples exist of larger signals than normal for the rare earths in the presence of NH_4Cl,[12] and Ti, Zr, Hf, Ta with HF.[97] In Figure 6, two effects are present. Ammonium chloride enhances the signal by a constant amount while a linear depression of signal occurs with increasing chloride concentration. The first effect is attributable to a more volatile ammonium compound, while the second effect may be due to the mass action of chloride producing some recombination in the flame.

The formation of relatively involatile compounds effects a decrease in signal. Other than the effect of recombination present in the flame, the rate of volatilization of the compound plays a significant part. This rate may be such that the maximum atom population appears higher in the flame and in a region of lower effective temperature. To various degrees a depression of

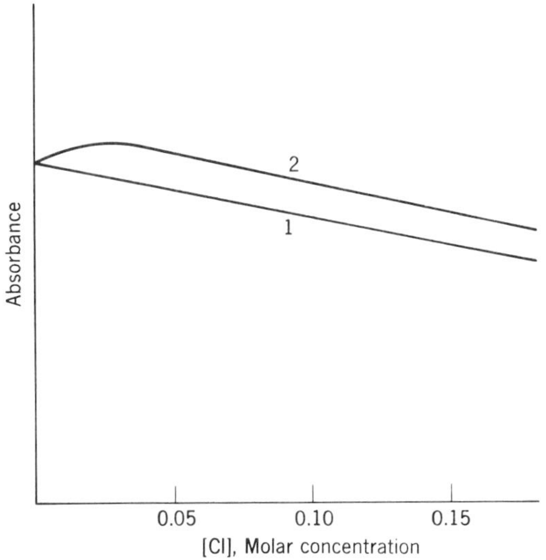

Fig. 6. Effect of chloride concentration on aluminum by atomic absorption spectroscopy. *(1)* Effect of HCl, NaCl, or NaCl with EDTA. *(2)* Effect of NH_4Cl.

signal is a common observation in determinations of alkaline earth metals where the refractory compounds of P, Si, S, Al, Cr, Ti, B, Zr, Hf, V, Mo, W, U and others are present. The effect is most pronounced in cooler flames, where the vaporized compound may not even thermally dissociate. Often, the decrease in signal with increasing concentration of anion is linear and stoichiometric. This fact is the basis for numerous indirect determinations.[98] The effect is illustrated in Figure 7 for calcium absorption in the presence of phosphate. An excess of phosphate has little additional effect in depressing the signal. This is termed saturation.

The control of anionic effects of this sort has been accomplished in a variety of ways. The most common are the addition of a complexing agent for the analyte or addition of a competitor for interfering species. An example would be the addition of EDTA (ethylenediamine tetraacetate) in the former case and excess La or Sr in the latter to determine Ca in the presence of phosphate. If the signal is not strongly suppressed, addition of an excess of the interfering anion may buffer the effect of variable amounts

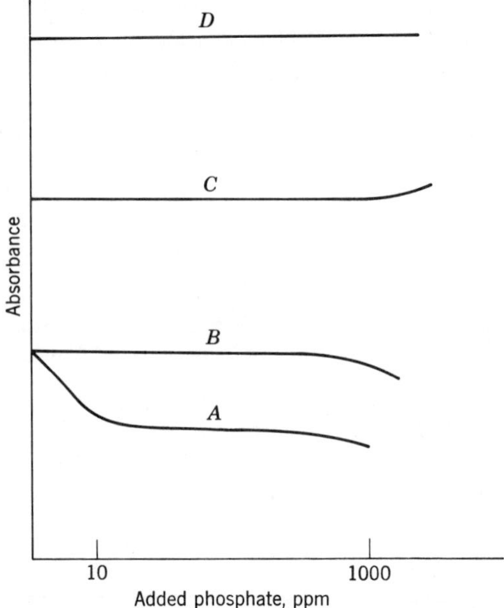

Fig. 7. Various effects on calcium absorption. *(a)* Depression and saturation of signal with added phosphate in air–acetylene flames. *(b)* Masking of phosphate depression with Sr, La, or EDTA additives in air–acetylene flames. *(c)* Absence of phosphate depression in fuel-rich oxyacetylene; ionization enhancement at high-phosphate concentrations. *(d)* Ionization suppression with 1000 ppm potassium.

in the sample. A particular height in the flame may be found where the effect is minimal, and is generally above the point of maximum sensitivity. The most satisfactory technique will be discussed shortly.

Cation interferences, aside from ionization, may arise from compound formation by the same mechanism discussed above. Distinguishing cation and anion effects in this sense are primarily semantic, as the cation in solution may become the oxyanion of an involatile salt or compound.

In the case of refractory elements, several factors minimize the chances of observing free atoms. At most flame temperatures, the rate of volatilization of the oxide may be slow and the free energy for thermal dissociation may be marginal. Even if dissociation occurs, the high partial pressure of the primary flame components O and OH assures rapid recombination. Thus only band emission spectra are observed for most commonly used flames.

The most striking advance in flame emission and atomic absorption in the last few years has been the development of the hot premixed fuel-rich flames for practical analysis. Specifically, the rich oxyhydrogen, oxyacetylene and nitrous oxide–acetylene systems have been studied, but the latter has the advantage of a low-flame speed, and has seen the most development. Within the secondary reaction zone of the fuel-rich acetylene flames, the concentration of the O species is so reduced by the reaction of $C + O \rightarrow CO$ ($\Delta H^\circ = -11.1$ eV), that metal oxygen recombination is negligible. Even boron and uranium resonance lines have been observed. The significance for analytical work is as follows: (*1*) the scope of elements detectable in resonance is broadened; (*2*) spectral interferences in flame emission are reduced where resonance lines are available; (*3*) interferences due to compound formation are minimized and the use of additives (which might add contamination) made unnecessary for this purpose. The latter effect is illustrated for the calcium phosphate system in Figure 7.

Occasionally synergistic effects are observed due to compound formation involving the analyte and two concomitant species. Amos and Willis[97] describe a Ti—Fe—F effect which enhances the Ti absorption signal in the nitrous oxide–acetylene flame. Mostyn and Cunningham[62] observed effects with Mo—Mn—Fe—Al—NO_3^- which were undoubtedly due to compounds.

3. IONIZATION

Flame emission has been shown to increase exponentially with temperature. Atomic absorption sensitivities increase with temperature if compound dissociation is important. The upper limit of gain from increases in temperature is usually limited by the eventual loss of an electron by the analyte. In flames, the ionization process can be treated as an equilibrium process

$$K = [M^+][e^-]/[M] \tag{8}$$

The constant for the dissociation is related to the ionization potential, U, by Saha's equation which has the form

$$\log K = [(-UC_1)/T] + \tfrac{5}{2}\log T - C_2 \tag{9}$$

where C_1 and C_2 are constants.

Examples of the magnitude of ionization for different elements and temperatures is illustrated in Table 3. The importance of ionization is apparent if one considers mass action. Not only is the degree of ionization dependent on the concentration of the analyte, but it depends on the degree of ionization of other atoms present and their concentration. The elements with low ionization potentials will enhibit the strongest effects.

Table 3. Degree of ionization of metals as a function of ionization potential, partial pressure, and temperature

Element	Ionization potential, eV.	Temperature, °K, at 10^{-4} atm[b] 2000	Temperature, °K, at 10^{-4} atm[b] 3500	Temperature, °K, at 10^{-6} atm[b] 2000	Temperature, °K, at 10^{-6} atm[b] 3500
Cs	3.893	0.01	0.86	0.11	>0.99
Rb	4.176	0.004	0.74	0.04	0.99
K	4.339	0.003	0.66	0.03	0.99
Na	5.138	0.000,3	0.26	0.003	0.90
Li	5.390	0.000,1	0.18	0.001	0.82
Ba	5.210	0.000,6	0.41	0.006	0.95
Sr	5.692	0.000,1	0.21	0.001	0.87
Ca	6.111	0.000,03	0.11	0.000,3	0.67
Mg	7.644	0.000,000,4	0.01	0.000,004	0.09

[a] B. L. Vallee and R. E. Thiers, in *Treatise on Analytical Chemistry*, Part I, Vol. 6, I. M. Kolthoff and P. J. Elving, Eds. Interscience, New York, 1965, p. 3500.
[b] Partial pressure of added metal.

In flame emission, ionization causes a concave response curve at low concentrations of analyte. The curve may be complicated by the presence of electrons from the flame species like oxygen ion, O^-, particularly at temperatures over 2500°K. Ionization of molecular species can occur also. The reduction of ionization effects on the response curve and the sensitivity can be affected by the addition of an element of low ionization potential. This raises the electron population in the flame, and by equation 8, decreases the analyte cation population. In Figure 8, an increase in signal is noted for Ca, Sr, and Ba in fuel-rich nitrous oxide–acetylene at high concentrations of potassium.

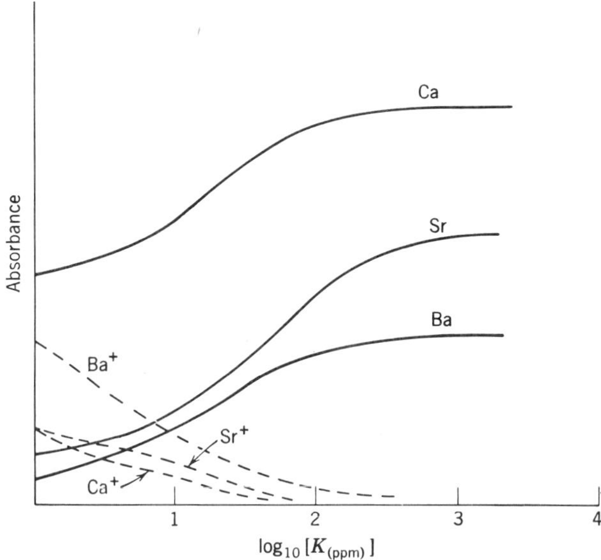

Fig. 8. Effect of increasing amounts of potassium on calcium, strontium, and barium absorbance in a nitrous oxide–acetylene flame. From Amos and Willis.[97]

Since the nitrous oxide–acetyle flame is hot enough to ionize a great number of elements to some extent, the addition of large amounts of potassium (1000 ppm) serves to suppress ionization effects and increase sensitivity in this flame system.

4. DEPARTURES FROM THERMODYNAMIC EQUILIBRIUM

The reaction zone of a flame is a source of a bewildering variety of atomic and molecular free radicals and ionized active species. The lifetimes and diffusion of these species define the flame front characteristics. Under specific flame conditions free radicals may be present in significant quantities in the analytical zone of the flame, and their reactions with the analyte can produce departures from thermodynamic equilibrium. The phenomena is particularly significant for flame emission and is called chemiluminescence. Increased excitation by this process can result in a many-fold increase in sensitivity. Gilbert[76] and Gibson et al.[99] discuss chemiluminescence in the air–hydrogen flame with a sheath burner. Both authors note the part played by the addition of organic solvents, and the especially notable sensitivity for Sn in this flame. Fristrom[100] notes the excess of hydrogen radical in air–hydrogen flames.

5. PHYSICAL EFFECTS

Physical effects in the flame play a minor role, but must be dealt with in analyses. Heating of the burner by hot flames can lead to changes in flame character and sensitivity. Acoustical coupling with extraneous objects can occur with premix burners, resulting in loud "singing flames." Carbonization of the burner port with acetylene flames can lead to clogging. These effects can be avoided. Scattering by flame particles is more significant, particularly for atomic absorption and fluorescence spectroscopy. A common means of obtaining corrections for scatter is to observe the attentuation of an ion or rare gas line from the hollow cathode lamp. Such a line will of course not be absorbed by the analyte. Caution must be observed, for molecular absorption can occur, and is characteristically of the same magnitude as scatter. The correction for molecular absorption must be obtained by use of a continuous source at wavelengths quite close to that of the analyte resonance line, or in an analyte-free blank. Molecular absorption has been observed for metal oxides,[101] and various organic compounds in the ultraviolet.[102]

I. The Spectrometer

1. COMMON TERMS

There are a number of terms used in specifying the capability of spectrometers which can be elucidated as follows:

The dispersing element is usually a grating or prism.

Speed, or relative luminosity, is the ratio of the focal distance f to the effective diameter, or aperature, A, of the beam falling on the dispersing element. The speed can be specified by the f/number or by giving both f and A. A low f/number, such as $f/6$, corresponds to a "fast" instrument, because the solid acceptance angle of the spectrometer, A/f^2, is large.

Dispersion is usually specified inversely as reciprocal linear dispersion, in units Å/mm. It is measured at the slit or focal plane.

The bandpass or bandwidth is the increment of the spectrum passed by the spectrometer, and is the product of the slit width in mm, and the dispersion.

The resolution of a spectrometer is a function of dispersion and speed. Fast instruments of low dispersion generally have poor resolution. For a grating, the relation is $\lambda/\delta\lambda$ = R. P. = Resolving Power = NA, where N is the number of rulings per mm of grating surface, and $\delta\lambda$ (in Å) is the number generally given to represent resolution. Actually, $\delta\lambda$ is a theoretical number; the true resolution depends on the quality of the grating, on the finite width of the slit, and on the optical quality of the spectrometer.

The effective resolution cannot generally be specified. It is the product of the resolution of the spectrometer and the spectral output of the source for an atomic absorption system. Thus, the effective resolution for atomic absorption with a line source is determined by the linewidth rather than the spectrometer bandwidth or resolution. Fassel et al.[111] have demonstrated that linewidths of 0.1 Å and less do not guarantee atomic absorption to be free from spectral interference. (See Table 4 for a comparision of these factors in several instruments.)

Table 4. Specifications of some instruments for atomic emission and absorption determinations

	Instrument			
Characteristic	JACO[a]	Beckman[b]	Parkin-Elmer[c]	Techtron[d]
Dispersing element	Grating	Grating	Grating	Grating
Resolution, Å	0.2	0.2	0.2	3.3
Dispersion, Å/mm	16	25	6.5 UV 13 VIS	33
Focal length, mm	500	350	400	500
Speed, f/	8.6	10	9	10
Lines/mm (grating)	1182	1200	2880 UV 1440 VIS	600
Detector	1P28 R136	1P28 R136	9592 B	1P28

[a] Jarrell-Ash Co., Model 82-500 Series.
[b] Beckman Instruments, Model 979.
[c] Perkin-Elmer Corp., Model 303 (flame emission as an accessory).
[d] Techtron Pty. Ltd. (Aztec Instruments, Inc.), Model AA-3.

2. FILTERS

The most economical optical descriminators are filters. Colored gelatin or glass filters used singly or in combination can provide an effective bandpass of 350–450 Å.[10] This is sufficient spectral isolation for the determination of the alkali metals and calcium by flame photometry, but some degree of spectral interference is encountered as the price for economy. The use of multilayer interference filters with effective bandwidths of less than 100 Å are much more effective in reducing background radiation interference. A filter photometer for flame emission is not a suitable choice for metallurgical analyses in which the presence of many interfering lines and generally complex spectra are likely to be the rule. The use of filter photometers for atomic absorption is subject to less serious limitations. Interference filter bandwidths are adequate to isolate the resonance lines from nonabsorbing lines in the hollow cathode output for most of the common elements. An imperfect isolation of lines is usually fatal only to the linearity at higher concentrations of analyte, and may compromise the sensitivity somewhat.

3. MONOCHROMATORS

For complete freedom in the choice of spectral line, and for relative freedom from background radiation and nearby line spectra, the medium dispersion monochromator is the spectrometer of choice. Both prisms and gratings are used as dispersion elements. The dispersion of a prism is best in the ultraviolet region (for example, 10 Å/mm) and decreases in merit with increasing wavelength (about 100 Å/mm in the visible region). Gratings exhibit constant dispersion which depends on the number of rulings per mm, N and the focal length, f.

$$\frac{d\lambda}{dx} = \frac{1}{fN} \tag{10}$$

While prism instruments are common, most manufacturers now produce grating instruments because for a given cost replica gratings produce better dispersion than prisms.

4. RESONANCE MONOCHROMATORS

Resonance monochromators have been described in the literature[104] and proposed by Walsh[26] as element specific detectors for atomic absorption. In principle, a cold atomic gas will fluoresce is proportion to the intensity of resonance radiation falling on it. If the source is a hollow cathode lamp and the detector is a hollow cathode lamp with a mask to prevent its emission from affecting a photomultiplier used to detect the fluorescence, then this combination will serve as a complete atomic absorption instrument for one element. Instruments utilizing the resonance monochromator are available commercially for applications which call for the determination of one or two elements.[113]

5. DETECTORS

Photomultipliers are now almost universally used for detection of radiation in the 1800–8500 Å region. The spectral response will vary with the composition of the cathode. The inexpensive 1P28 is adequate for all but the 1800–2000 Å and 6500–8500 Å region. This would limit the determination of elements such as As, Se, K, Rb, and Cs. Photomultipliers with broader spectral response, such as the EMI 9592B and Hamamatsu R136, are also being used in commercial instruments.

The use of photomultipliers at the limit of their performance can be the determining factor on the limit of detection, particularly in flame emission spectroscopy. The determination of the alkali elements at one part in 10^{-12} depends on as much on the care in the selection and use of a high-gain photomultiplier, as in the design of high sensitivity amplifiers.[105]

6. READOUT

Much of the ultimate detection ability achieved by flame methods depends on the sophistication employed in the processing and amplification of the signal. While this subject is also beyond the scope of this chapter, a few observations are possible.

Because there is merit in electronically rejecting the emission signal from the flame in absorption measurements, AC amplification is now commonly used in instruments capable of both modes of measurement. This kind of amplification has the property of rejecting long period or DC instability such as dark current drift in photomultipliers, which is of importance at the limit of detection for emission measurements. For atomic absorption measurements, the use of the double-beam technique (by providing an optical AC reference signal from the hollow cathode lamp) serves to compensate for DC drift in the lamp emission. This kind of attention to systematic sources of noise makes scale expansion techniques possible. The limiting noise is then that inherent in the burner and flame system. It is common to provide the readout with some means of damping the remaining fluctuations. In addition to long time constants provided by simple RC filters, analog integration and statistical averaging readout attachments[106] are in vogue in commercial instruments. The ability to read an output response linear in concentration is a convenience that is also provided instrumentally by some manufacturers.

IV. SCOPE AND SENSITIVITY

A. Detection Limits

The materials scientist is of necessity concerned with a variety of questions concerning the definition of a material. Questions regarding oxidation states, dissolved gases, nonmetallic interstitials, chemical structure, and the microdistribution of elements are best answered by other techniques. The concern here is for elemental composition analysis. Flame techniques at present offer an intermediate capability between survey techniques such as emission and mass spectroscopy which are fast, exacting, and somewhat inprecise, and chemical analysis techniques, which are precise but slow. Experience in the field indicates atomic absorption spectroscopy is replacing chemical elemental analysis when the analyte constitutes from 0.0001–5% of the matrix material because of its speed, specificity, and comparable precision and accuracy. It is too soon to say there is competition for the survey techniques in speed, even with automation, but optimism is not unrealistic. The flame methods are certainly preferable if only a few elements are sought.

These statements must be taken in light of the detection limits and scope actually achieved in flame spectroscopy. In Table 5 the detection limits are listed graphically on a logarithmic scale. Lists of detection limits such as this must necessarily be considered illustrative and temporary; the following cautionary and explanatory comments are included to protect the analyst and better gauge reality.

1. The detection limits presented are the best achieved with commercially available instruments and a variety of burner and nonflame systems. They represent signals two or three times the random fluctuations or noise. Noise is a difficult parameter to reproduce. Thus, the precision with which some limits are quoted is deemphasized by the graphical representation.

2. The limits are those obtained by direct methods. While some indirect methods exist for anions, these methods are less specific and involve additional chemistry, so that they are less attractive.

3. A great many of the detection limits (particularly those for flame emission) are optimistic in the sense that they are obtained with only the analyte present. Applicability can be limited by the presence of matrix interference. Some of the atomic absorption limits are optimistic also because of the dependence on the availability and quality of source lamps.

4. The normal range for determinations would be at least one to two orders of magnitude greater concentration than that quoted for the detection limit.

The scope of flame techniques is evident from the fact that all but hydrogen, the noble gases, and the rare radioactive elements are included, although some are barely detectable.

The arrangement of the elements by groups in the periodic chart serves to emphasize the dependency on chemical properties, notably the electronic excitation of the elements. Thus elements in Group IA, with low excitation potentials and few resonance lines, have the highest sensivitity in emission, while Group IIB elements are much more sensitive in absorption because of their high excitation potentials. The refractory elements, particularly the rare earths, are best done in absorption; the detection limits in emission are often for band spectra, and therefore lack specificity. Occasionally there is little choice. Boron, for instance, is much more sensitive when the oxide band is used than by observing the resonance line.

The graphical representation of upper bounds for the detection limits are taken primarily from the survey work of Amos and Willis[97] on the nitrous oxide–acetylene flame, the work of Ramirez-Muñoz[57,113] with a furnace–

Table 5. Best reported detection limits for solutions of elements taken alone (free of interfering effects). Concentrations given in ppm, weight to volume. Solid line: Flame emission spectroscopy. Broken line: Atomic absorption spectroscopy.

Concentration (ppm, w/v) 10^3 10^2 10^1 1 10^{-1} 10^{-2} 10^{-3} 10^{-4} 10^{-5}

Group	Element	Reference Em.	Reference Abs.
IA	Li	e	e
	Na	e	a
	K	e	a
	Rb	e	a
	Cs	e	e
IB	Cu	e	a
	Ag	e	f
	Au	e	a
IIA	Be	e	c
	Mg	e	a
	Ca	e	a
	Sr	e	a
	Ba	e	c
IIB	Zn	e	a
	Cd	e	f
	Hg	e	f
IIIA	B	e	c
	Al	e	c
	Sc	e	c
	Y	e	c
	La	e	c
IIIB	Ga	e	a
	In	e	a
	Tl	e	a
IVB	C	e	—
	Si	e	b
	Ge	d	e, c
	Sn	e	c
	Pb	e	e
IVA	Ti	e	b
	Zr	e	c
	Hf	d	c
VB	N	e	—
	P	e	e
	As	e	e
	Sb	e	e
	Bi	e	e
VA	V	e	b
	Nb	e	c
	Ta	e	c
VIB	S	e	—
	Se	—	e
	Te	e	a
	Cr	e	a
VIIB	Mo	d	a
	W	d	c
	F	e	—
	Cl	e	—
	Br	e	—
	I	e	—
VIIA	Mn	e	a
	Re	e	c
VIII	Fe	e	a
	Co	e	e
	Ni	e	e, a
	Ru	e	e
	Rh	e	f
	Pd	e	f
	Os	d	a
	Ir	d	a
	Pt	e	a
R.E.	Ce	e	—
	Pr	e	c
	Nd	e	g
	Sm	e	g
	Eu	d	c
	Gd	e	c
	Tb	e	g
	Dy	d	c
	Ho	d	c
	Er	e	g
	Tm	e	g
	Yb	e	g
	Lu	e	c
Act.	Th	d	c
	U	d	—

[a] See reference 57, Vol. 1, Nos. 1, 2 and 4, Vol. 2, No. 1; also reference 114.

[b] See reference 27. Also J. S. Cartwright, C. Sebens, and D. C. Manning, *Atomic Absorption Newsletter*, **5**, 91 (1966).

[c] See reference 97. Estimated detection limits at 0.1% absorption signals.

[d] D. Golightly and V. A. Fassel, IS–1200 (TID–4500), p. 56, July 1965.

[e] See reference 15.

[f] See reference 30. Also I. Rubeška and J. Štupar, *Atomic Absorption Newsletter*, **5**, 69 (1966), or J. Štupar, *Mikrochim. Acta*, **1966**, 722.

[g] See reference 108.

condenser system[107] and the presentation by Anderson[108] of work with the less common elements.

B. Applications to Metallurgical Materials

A wide variety of industrial metallurgical materials are now being analyzed by atomic absorption procedures. The validity and ease of analysis in comparison to wet chemical techniques have been systematically demonstrated for both ferrous and nonferrous alloys. Sattur[109] discusses extensive work on the analysis of sixteen elements in standard nonferrous alloys from the National Bureau of Standards and presents atomic absorption analysis results which compare very satisfactorily with those certified or determined chemically. The concentration range of impurities in the alloys is 0.005–0.1%. Bell[110] gives similar data for ten common alloying elements in aluminum alloys. Statistical variations are typically better than 1% relative, and very few interferences are noted. Scott et al.[116] report similarly on analyses of ferrous alloys. Comparative analyses of uranium base alloys[114] show that there is no loss in accuracy and 4-fold increase in speed (30 analyses/hr) over the usual spectrophotometric methods employed in routine industrial applications.

Most of the reports dealing with metallurgical samples emphasize the dissolution and additive procedures, as these steps are somewhat of an art, while the actual determination is straightforward. The most useful compilation of applications reports have been assembled by Ramirez-Muñoz[113] and by instrument manufacturers.

The use of extractions to remove the bulk or the matrix or enhance the sensitivity of flame methods has been summarized by Ramirez-Muñoz[115] and in more general source books.

The ASTM[117] has surveyed the 1968 levels of industrial and government laboratory uses of atomic absorption and found that 60% of the usage is on routine analyses, 35% on methods development and special problems, and 80% of the laboratories use the technique to establish in-house calibration standards. The high level of usage seems to indicate a very pronounced conversion of most laboratories from chemical to flame methods. It is significant that the ASTM is considering adoption of many atomic absorption procedures for use as standard methods for interlaboratory use. To indicate the scope of routine applications, the summary from the ASTM report plus additional developments are reproduced in Table 6.

C. Trends

The field of analytical flame spectroscopy has been marked by unusual growth and a rapid series of innovations. This condition is likely to continue

Table 6. Metallurgical materials analyzed routinely by atomic absorption spectrophotometry

Material	Elements analyzed
Copper-base alloys	Ag,Al,Fe,Mn,Ni,Pb,Sb,Sn,Zn
Lead-base alloys	Ag,Bi,Ca,Cu,Fe,Ni,Sb,Sn
Aluminum-base alloys	Ag,Al,Ca,Cd,Cr,Cu,Fe,Li,Mg,Mn,Ni,Pb,Sn,Zn
Zinc-base alloys	Al,Cd,Cu,Mg,Pb
Magnesium-base alloys	Al,Cu,Mn,Pb,Zn
Tin-base alloys	Ag,Bi,Cu,Fe,Ni,Pb,Sb
Copper alloys	Ag,Au,Bi,Ca,Cd,Co,Cr,Ni,Pb,Sb,Se,Zn
Copper, refined	Ag,As,Cd,Fe,Ni,Pb,Sb,Te,Zn
Lead, refined	Ag,Bi,Cd,Cu,Fe,Zn
Uranium	Cr,Fe,Mg,Mn,Mo,Ni,Si
Silicon	Al,Ca,Cu,Fe,Mn
Pyrites	Ca,Cu,Mg,Pb,Zn
Titanium alloys	Al
Gold–silver–palladium alloy	Au,Ag,Pd
Cesium	Na,K,Rb
Rubidium	Cs,Na,K
Brines	Ca,Na,K,Mg
Nickel-base alloys	Bi,Cr,Mg,Pb,Sb,Sn,Zn
Tantalum	Fe,Mo
Niobium	Fe
Molybdenum	Pb
Zirconium alloys	Zn
Sinter	Cd,Na,K,Mg,Pb
Steel	Al,Co,Cr,Cu,Mg,Mn,Mo,Ni,Pb,V
High-temperature alloy	Al,Ca,Mg,Pb
Ferrosilicon	Al,Ca,Mg,Mn
Zinc spelter	Cd,Pb,Sb
Raw materials	Al,Ca,Cu,Mg,Pb,Si,Zn
Cast iron pigs	Cu,Mn
Low-alloy steel	Cu,Mn
Stainless steel	Co,Cr,Cu,Mn,Mo,Ni
Slags	Al,Ba,Ca,Cr,Cu,Fe,Mg,Mn,Mo,Ni,Pb,Si,Sr,Zn

for some time. Several possible areas of expanding capability can be forseen. There are significant efforts being made to simplify the operation of the instruments (dial an element), reduce the cost (filters for monochromators), increase the reliability of the results (automatic internal standardization), and expand the multielement capabilities (multielement hollow cathode lamps) and allow for simultaneous determinations[118] (seven elements per sampling). Sources (electrodeless discharge) may undergo change, because the cost of hollow cathodes remains high. With the advent of mini-computers it is easy to contemplate radical changes in the overall system which would allow for control of the variables.

Atomic fluorescence spectroscopy, a flame technique of great sensitivity and promise, but slow to gain acceptance, will in all likelihood blossom, particularly if brighter sources (like tunable lasers) become available. As the material scientist comes to require economical alternatives to mass spectroscopy and neutron activation for trace impurities, flame techniques have the potential to rise to the challenge.

References

1. G. Kirchoff and R. Bunsen, "Chemical Analyses By Spectrum-observations," Phil. Mag. **20**, 89 (1860).
2. G. L. Gouy, "Photometric Research on Colored Flames," *Compt. Rend.*, **83**, 269 (1876).
3. G. L. Gouy, "Report on Colored Flames," *Ann. Chim. Phys.* (Vesérie) **18**, 5 (1879).
4. H. Lundegradh, *Die Quantitative Spektralanalyses der Elemente*, Vols. 1 and 2, Fischer, Jena, 1929, 1934.
5. R. B. Barnes, D. Richardson, J. W. Berry, and R. L. Hood, *Ind. Eng. Chem., Anal. Ed.*, **17**, 605 (1945).
6. V. R. Ells, *J. Opt. Soc. Am.*, **31**, 534 (1941).
7. J. Cholak and D. M. Hubbard, *Ind. Eng. Chem., Anal., Ed.*, **16**, 728 (1944).
8. R. Mavrodineanu, "Bibliography on Analytical Flame Spectroscopy 1846–1956," Part I and II, *Appl. Spectry.*, **10**, 51, 137 (1956).
9. R. Mavrodineanu, "Bibliography on Analytical Flame Spectroscopy 1956–1959," Part I, II, and III, *Appl. Spectry.*, **13**, 132, 150 (1959); ibid., **14**, 17 (1960).
10. J. A. Dean, *Flame Photometry*, McGraw-Hill, New York, 1960.
11. R. Herrmann and C. T. J. Alkemade, *Chemical Analysis by Flame Photometry*, (*Chem. Anal.*, Vol. 14), 2nd ed., translated by P. T. Gilbert, Jr., Interscience, New 1963.
12. N. S. Poluektov, *Techniques in Flame Photometric Analysis*, translated by C. N. Burton and T. I. Burton, Consultants Bureau, New York, 1961.
13. R. Mavrodineanu and H. Boiteux, *Flame Spectroscopy*, Wiley, New York, 1965.
14. P. T. Gilbert, Jr., Special Technical Publication No. 269, Am. Soc. Testing Mater., Philadelphia, 1960, pp. 73–156.
15. P. T. Gilbert, Jr., *Analysis Instrumentation—1964*, L. Fowler, R. J. Harmon, and D. K. Roe, Eds., Plenum Press, New York, 1965, pp. 193–233.
16. J. A. Dean, *Develop. Appl. Spectry.*, **3**, 207 (1963).
17. M. Margoshes and B. F. Scribner, *Anal. Chem.*, **38**, 297R (1966); **40**, 223R (1968).
18. M. Margoshes and B. F. Scribner, *Anal. Chem.*, **36**, 329R (1964).
19. W. H. Wollaston, *Phil. Trans. Roy. Soc. London, Ser. A*, **92**, 365 (1802).
20. A. C. G. Mitchell and M. W. Zemansky, *Resonance Radiation and Excited Atoms*, Cambridge Univ. Press, New York, 1934.
21. A. Walsh, *Spectrochim. Acta*, **7**, 108 (1955).
22. H. Schüller, *Z. Physik.*, **35**, 323 (1926); **59**, 149 (1930).

23. G. F. Box and A. Walsh, *Spectrochim. Acta*, **16**, 255 (1960).
24. W. G. Jones and A. Walsh, *Spectrochim. Acta*, **16**, 249 (1960).
25. J. C. Burger, W. Gillies, and G. K. Yamasaki, ETD-6403, Westinghouse Electric Corp., Elmira, New York, Sept. 1964.
26. J. V. Sullivan and A. Walsh, *Spectrochim. Acta*, **21**, 721 (1965).
27. J. S. Cartwright, C. Sebens, and W. Slavin, *Atomic Absorption Newsletter*, **5**, 35 (1966).
28. J. B. Willis, *Nature*, **207**, 715 (1965).
29. R. N. Kniseley, A. P. D'Silva, and V. A. Fassel, *Anal. Chem.*, **35**, 910 (1963).
30. K. Fuwa and B. L. Vallee, *Anal. Chem.*, **35**, 942 (1963).
31. R. H. Wendt and V. A. Fassel, *Anal. Chem.*, **38**, 337 (1966).
32. J. P. Mislan, Atomic Energy of Canada, Ltd., AECL-1941, Chalk River, Apr. 1964.
33. R. Woodriff and J. Ramelow, *Spectrochim. Acta*, **23B**, 665 (1968).
34. J. A. Goleb, *Anal. Chem.*, **35**, 1978 (1963).
35. J. A. Goleb and Y. Yokoyama, *Anal. Chim. Acta*, **30**, 213 (1964).
36. V. A. Fassel, V. G. Mossotti, W. E. L. Grossman, and R. N. Kniseley, *Spectrochim. Acta*, **22**, 347 (1966).
37. C. W. Frank, W. G. Schrenk, and C. E. Meloan, *Anal. Chem.*, **38**, 1005 (1966).
38. J. D. Winefordner and T. J. Vickers, *Anal. Chem.*, **36**, 161 (1964).
39. J. D. Winefordner and R. A. Staab, *Anal. Chem.*, **36**, 165 (1964).
40. J. D. Winefordner and R. A. Staab, *Anal. Chem.*, **36**, 1367 (1964).
41. D. M. Dagnall, T. S. West, and P. Young, *Talanta*, **13**, 803 (1966).
42. W. Slavin, *Atomic Absorption Newsletter*, **1**, March 1962; **1**, Sept. 1962; **2**, Oct. 1963; **3**, Sept. 1964; **4**, Feb. 1965; **5**, May 1966.
43. B. Bermijo Martinez, *Quim. Ind.* (Bilbao), **10**, 39 (1963).
44. D. J. David, *Spectrochim. Acta*, **20**, 1185 (1964).
45. R. Lockyer, *Advances in Analytical Chemical Instrumentation*, Vol. 3, C. N. Reilly, Ed., Interscience, New York, 1964, pp. 1–29.
46. H. L. Kahn, *J. Chem. Educ.*, **43**, 72, 103A (1966).
47. N. S. Poluektov and Yu. V. Zelyukova, *Zavodsk. Lab.*, **30**, 33 (1964).
48. I. Rubeska and I. Velicka, *Chem. Listy*, **59**, 769 (1965).
49. H. Schleser, *Z. Instrumentenk*, **73**, 25 (1965).
50. S. Tardon, *Chem. Listy*, **58**, 417 (1964).
51. G. Thilliez, *Chim. Anal. (Paris)*, **46**, 3 (1964).
52. W. Slavin, *Appl. Spectry.*, **20**, 281 (1966).
53. W. T. Elwell and J. A. F. Gidley, *Atomic Absorption Spectrophotometry*, Pergamon Press, 1961.
54. J. W. Robinson, *Atomic Absorption Spectroscopy*, Dekker, New York, 1966.
55. R. P. Weberling and J. F. Cosgrove, *Trace Analysis*, G. H. Morrison, Ed., Wiley, New York, 1965, Chap. 7, pp. 245–269.
56. *Atomic Absorption Newsletter*, (Perkin-Elmer, Norwalk, Conn.).
57. *Flame Notes*, Application Engineering (S. I. O.), Beckman Instruments, Inc., Fullerton, Calif.
58. The Element, Aztec Instruments, Inc., Wesport, Conn.
59. J. D. Winefordner and T. J. Vickers, *Anal. Chem.*, **36**, 1939, 1947 (1964).
60. J. D. Winefordner and C. Veillon, *Anal. Chem.*, **37**, 416 (1965).
61. P. J. T. Zeegers, R. Smith, and J. D. Winefordner, *Anal. Chem.*, **40**, No. 13, 26A (1968).
62. R. A. Mostyn and A. F. Cunningham, *Anal. Chem.*, **38**, 121 (1966).

63. J. Isreeli, M. Pelavin, and G. Kessler, *Ann. N. Y. Acad. Sci.*, **87**, 636 (1960).
64. See reference 14, p. 88.
65. Laminar Flow Burner Assembly, No. 105250, Beckman Instruments, Fullerton, Calif.
66. W. J. Kirsten and G. O. B. Bertilsson, *Anal. Chem.*, **38**, 648 (1966).
67. H. Dunken, G. Pforr, W. Mikkeleit, and K. Geller, *Spectrochim. Acta*, **20**, 1531 (1965).
68. S. A. Shipitsyn, V. V. Kiryushkin, and A. A. Ermolaev, *Zavodsk. Lab.*, **31**, 557 (1965).
69. P. T. Gilbert, Jr., *Anal. Chem.*, **34**, 1025 (1962).
70. L. S. Nelson and N. A. Kuebler, *Spectrochim. Acta*, **19**, 781 (1963).
71. V. G. Mossotti, K. Laqua, and W. D. Hagenah, *Spectrochim. Acta*, **23B**, 197 (1967).
72. A. G. Gaydon and H. G. Wolfhard, *Flames—Their Structure, Radiation, and Temperature*, Chapman and Hall, London, 1953.
73. B. Lewis, and G. von Elbe, *Combustion, Flames, and Explosions of Gases*, Academic Press, New York, 1961.
74. R. M. Fristrom and A. A. Westenberg, *Flame Structure*, McGraw-Hill, New York, 1965.
75. A. G. Gaydon, *Spectroscopy and Combustion Theory*, Chapman and Hall, London, 1948.
76. P. T. Gilbert, Jr., *Proceedings of the Xth Colloquium Spectroscopicum Internationale*, E. R. Lippincott and M. Margoshes, Eds., Spartan Books, Washington, D. C., 1963, pp. 171–215.
77. See reference 74, p. 27.
78. J. D. Winefordner, C. T. Mansfield, and T. J. Vickers, *Anal. Chem.*, **35**, 1611 (1963).
79. R. Herrmann and W. Lang, *Z. Anal. Chem.*, **213**, 1 (1964).
80. R. Herrmann, W. Lang, and K. Ruediger, *Z. Anal. Chem.*, **206**, 241 (1964).
81. See reference 13, p. 23.
82. G. L. Vidale, ASTIA AD-237.918, Office Tech. Service, P. B. Dept. 148.206, 44 pp., 1960.
83. B. M. Gatehouse and A. Walsh, *Spectrochim. Acta*, **16**, 602 (1960).
84. B. V. L'vov, *Spectrochim. Acta*, **17**, 761 (1961).
85. A. G. Gaydon, *The Spectroscopy of Flames*, Wiley, New York, 1957, pp. 48–72.
86. T. C. Rains, H. P. House, and O. Menis, *Anal. Chim. Acta*, **22**, 315 (1960).
87. J. A. Dean and C. Thompson, *Anal. Chem.*, **27**, 42 (1955).
88. See reference 76, pp. 187–190.
89. F. Llewellyn-Jones, *The Glow Discharge*, Wiley, New York, 1966, Chap. 5.3.
90. J. Vollmer, *Atomic Absorption Newsletter*, **5**, 35 (1966).
91. W. Gilles, G. Yamasaki, and J. C. Burger, *Soc. Appl. Spectry. Meeting, Chicago, June 1966*, paper No. 89.
92. W. F. Meggers and F. O. Westfall, *J. Res. Natl. Bur. Stands.*, **44**, 447 (1950).
93. H. V. Malmstadt and W. E. Chambers, *Anal. Chem.*, **32**, 225 (1960).
94. D. C. Manning and W. Slavin, *Atomic Absorption Newsletter*, **1**, No. 8 (1962).
95. R. K. Skogerboe and R. A. Woodriff, *Anal. Chem.*, **35**, 1977 (1963).
96. E. A. Boling, *Spectrochim. Acta*, **22**, 425 (1966).
97. M. D. Amos and J. B. Willis, *Spectrochim. Acta*, **22**, 1325 (1966).
98. See reference 14, p. 112.
99. J. H. Gibson, W. E. L. Grossman, and W. D. Cooke, *Anal. Chem.*, **35**, 266 (1963).
100. See reference 74, pp. 330–334.
101. S. R. Koirtyohann and E. E. Pickett, *Anal. Chem.*, **38**, 585 (1966).

102. See reference 75, p. 218.
103. R. P. Bauman, *Absorption Spectroscopy*, Wiley, New York, 1962.
104. See reference 13, pp. 124–136.
105. M. T. Kelley, D. J. Fisher, and H. C. Jones, *Anal. Chem.*, **31**, 178 (1959).
106. E. A. Boling, *Anal. Chem.*, **37**, 482 (1965).
107. A. Hell, W. F. Ulrich, N. Shifrin, and J. Ramirez-Muñoz, *Appl. Opt.*, **7**, 1317 (1968).
108. J. W. Anderson, "Application of Atomic Absorption Spectroscopy to Rare Earth Analysis," Paper 98, *Soc. Appl. Spectry, Meeting, Chicago, Ill., June 1966.*
109. T. W. Sattur, *Atomic Absorption Newsletter*, **5**, 37 (1966).
110. G. F. Bell, *Atomic Absorption Newsletter*, **5**, 73 (1966).
111. V. A. Fassel, J. O. Rasmuson, and T. G. Cowley, *Spectrochim. Acta*, **23B**, 579 (1968).
112. Model AR 200 Spectrophotometer, Varian Techtron Pty. Ltd., Melbourne, Australia; Cary Instruments, Monrovia, Calif.
113. J. Ramirez-Muñoz, *Atomic-Absorption Spectroscopy*, Elsevier, New York, 1968.
114. M. L. Jursik, *Atomic Absorption Newsletter*, **6**, 21 (1967).
115. J. Ramirez-Muñoz, *Revista del Instituto del Hierro y del Acero*, **89**, 44 (1964) (in Spanish).
116. T. C. Scott, E. D. Roberts, and D. A. Cain, *Atomic Absorption Newsletter*, **6**, 1 (1967).
117. L. L. Lewis, *Anal. Chem.*, **40**, No. 12, 28A (1968).
118. R. Mavrodineanu and R. C. Hughes, *Appl. Opt.*, **7**, 1281 (1968).

Selected Readings

W. Slavin, *Atomic Absorption Spectroscopy*, Interscience, New York, 1968.
J. Ramirez-Muñoz, *Atomic Absorption Spectroscopy and Analysis by Atomic Absorption Flame Photometry*, Elsevier, New York, 1968.
J. Dean, *Flame Photometry*, McGram-Hill, New York, 1960.
J. W. Robinson, *Atomic Absorption Spectroscopy*, Dekker, New York, 1966.
W. Slavin, *Appl. Spectry.*, **20**, 281 (1966).
P. T. Gilbert, Jr., Special Technical Publication No. 269 (Am. Soc. Testing Mater., Philadelhpia, 1960).-
R. Mavrodineanu and H. Boiteux, *Flame Spectroscopy*, Wiley, New York, 1965.
B. L. Vallee and R. E. Thiers, "Flame Photometry" in *Treatise on Analytical Chemistry.*
I. M. Kolthoff and P. J. Elving, Eds., Interscience, New York, 1965.
H. L. Kahn, *J. Metals*, **18**, 1101 (1966).

Chapter 7

SPECTROPHOTOMETRY AND SPECTROFLUORIMETRY

JAMES A. HOWELL, Western Michigan University, Kalamazoo, Michigan
DAVID F. BOLTZ, Wayne State University, Detroit, Michigan

I. Introduction 226
A. Delineation of Radiative Processes 228
B. Nature of the Interaction of Ultraviolet and Visible Radiant Energy with Matter 228

II. Spectrophotometric Analysis 236
A. Scope of Spectrophotometric Analysis 236
B. Instrumentation 236
1. Essential Components 236
2. Spectrophotometers 236
3. Filter Photometers 239
4. Practical Considerations in Selection of Instruments 240
C. General Methodology 243
1. Preparation of Sample 243
2. Elimination of Interferences 243
3. Measurement of the System 244
4. Sources of Error 246
5. Presentation and Evaluation of Spectral Data 248
6. Special Techniques 248
D. Applications 253
1. Metals 253
2. Nonmetals 253
3. Selected Methods 254

III. Spectrofluorimetric Analysis 254
A. Scope of Spectrofluorimetry 254
B. Instrumentation 254
1. Essential Components 255
2. Spectrofluorimeters 257

3. Filter Fluorimeters 260
4. Practical Considerations in Selection of Instrument 263
C. General Methodology 263
1. Preparation of Sample 264
2. Factors Affecting Fluorescence 266
3. Measurement of the System 268
4. Sources of Error 269
5. Presentation and Evaluation of Data 269
D. Applications 269
1. Metals 270
2. Nonmetals 270
3. Selected Methods 271
References 272

I. INTRODUCTION

The development of spectrophotometry and spectrofluorimetry has extended over a period of many years with numerous contributions from many disciplines which have found these methods of measurement applicable. Spectrophotometric analysis has at this time attained such a state of sophistication and maturity that it is one of the most widely used methods of chemical analysis. Spectrofluorimetric analysis, although developed to a reliable and useful stage, is a much newer method of analysis whose potential analytical capabilities are still being extensively investigated. The main objectives of this chapter are (*1*) fundamental theoretical concepts underlying these methods, (*2*) essential features of instrumentation, (*3*) the general methodology involved, and (*4*) representative applications to metallurgical analysis.

This discussion will be limited primarily to those analytical methods involving absorption and fluorescence in the ultraviolet and visible regions and to instruments which employ photoelectric systems. Analytical methods involving absorption or fluorescence of X-rays are discussed in Chapter 8 and flame emission, absorption, and fluorescence methods are found in Chapter 6. Instruments are classified into two general categories: namely, those instruments which employ prisms or gratings in their dispersing systems are denoted as spectrophotometers or spectrofluorimeters and those instruments using filters to isolate a specific wavelength region are designated filter photometers or filter fluorimeters. The recommendations on nomenclature and terminology as outlined in *Analytical Chemistry* have been followed.[1]

Table 1. Recommended terminology

Term	Symbol	Definition
Radiant power	P	The rate of transfer of radiant energy, i.e., radiant flux
Transmittance	T	The ratio of the radiant power transmitted by the sample (P) to the radiant power incident on the sample (P_0) both being measured at the same wavelength and with the same slit width; $T = P/P_0$
Absorbance	A	The logarithm to the base 10 of the reciprocal of the transmittance; $A = \log 1/T = -\log T$
Absorptivity	a	A constant characteristic of the absorptivity capacity of a specific absorber at a particular wavelength. The absorbance divided by the product of the concentration (mg/ml or μg/ml) and sample path length (in cm); $a = A/bc$
Molar absorptivity	ϵ	The absorptivity expressed in liter/mole cm when concentration is in moles/liter and sample path length is in cm. (If a is calculated using a concentration expressed in the form of g/liter, is equal to $a \times$ mol wt of absorber.)
Path length	b	Internal length of absorption cell, in cm
Concentration	c	Amount of absorber per unit volume moles/liter gram/liter = mg/ml mg/l. = μg/ml. = ppm[a]
Resolution		The ratio of the average wavelength of two spectral lines which can just be identified as a doublet, to the difference in their wavelengths
Slit widths	SW	The slit width is the mechanical distance, in mm, between the sides of the narrow aperture which permits radiant energy to enter and leave the monochromator
	ESW	The width of the image of exit slit, along the wavelength scale, at which the radiant power is half of the maximum.
	SSW SRI	The width of the image of the exit slit along the wavelength scale, i.e., the spectral region isolated
Bandwidth	OBW	Observed band width is the width at one-half of the absorbance maximum as measured with a spectrophotometer
	NBW	Natural band width is the width of one half of the absorbance maximum due to a specific absorber. It is characteristic of absorber and not of instruments

[a] For solvents with a density of 1.000 g/cm^3.

A. Delineation of Radiative Processes

Absorption is a process by which molecular or ionic species reduce the radiant power of an incident beam of radiant energy by conversion of the radiant energy into other forms, e.g., thermal energy. *Spectrophotometry* is a term used to describe the methodology by which spectral transmittance or reflectance is measured as a function of wavelength. *Luminescence* is the phenomenon wherein certain molecular or ionic species upon irradiation by a particular wavelength of radiant energy absorb energy and then subsequently emit radiant energy of a somewhat longer wavelength, regardless of the mechanism or the time required for this emission to occur. Hence, Rayleigh scattering and Mie scattering which may be involved in turbidimetric and nephelometric methods and Raman scattering which is involved in Raman spectrometry are excluded as luminescence processes. Luminescence should also not be confused with chemiluminescence, which is a process in which there is an emission of light resulting from a chemical reaction. Two separate and distinct luminescence processes are *fluorescence* and *phosphorescence*. Some authors delineate these two processes on the basis of the time lag between irradiation and reemission; the fluorescence process exhibiting reemission within 10^{-8} sec while the reemission of radiant energy takes place from 10^{-8} sec to many seconds or minutes later in the case of phosphorescence. We prefer to differentiate between these two processes by defining fluorescence as the emission of radiant energy resulting from an electron making the transition from an excited singlet energy state to its ground state and phosphorescence as due to a similar electronic transition from the metastable excited triplet state to the ground state. Furthermore, by limitation of luminescence processes to those in which ultraviolet or visible radiant energy is used for irradiation, electroluminescence, a topic beyond the scope of this chapter, has also been excluded from consideration. *Spectrofluorimetry* or *spectrophotofluorimetry* and *fluorescimetry*, *fluorimetry*, or *fluorometry* are equivalent terms used to designate the methodology of measuring fluorescence. An analogous term, *phosphorimetry*, is applicable to the measurement of phosphorescence.

B. Nature of the Interaction of Ultraviolet and Visible Radiant Energy with Matter

The interaction of electromagnetic radiant energy with matter is a highly complex phenomenon which is still being studied. In restricting this discussion to the ultraviolet and visible regions, the main interactions will be those involving the excitation of outer orbital electrons to higher energy levels with simultaneous vibrational, rotational, and translational excitations. In principle, any electron associated with an atom, ion, or molecule is capable of undergoing one or more electronic transitions until such time as the electron has been completely dissociated from its parent species. In practice, however, ultraviolet–visible energy and instrumental limitations permit the observation of transitions of only the more easily excited electrons. These electronic transitions generally fall into one of four classifications: (*1*) transitions among split *d* and *f* energy levels, (*2*) transitions of the charge-transfer type, (*3*) transitions of nonbonding electrons, and (*4*) transitions from sites of unsaturation, e.g., double and triple bonds.

Transitions arising from split *d* and *f* energy levels occur with the elements of the transition, lanthanum, and actinium series. All of the *d* or *f* levels of the free ions of these metals are of equivalent energy; however, when these ions are complexed by some ligand, a coordinating agent, their orbitals may undergo some distortion due to the influence of the ligand. The degree of distortion will be different for some of the orbitals and consequently they will no longer all possess the same energy. If these orbitals are not completely occupied with electrons, then it is possible for the ion or molecule whose electrons reside in the lower energy levels to absorb a certain wavelength of radiant energy and elevate some of the electrons to one of the vacant higher energy levels.

Transitions of the charge transfer type result, as the name implies, from excitation of an electron by its transfer from one atom to another. A typical example of this behavior might be illustrated by the following:

$$[(Fe^{+3})\leftarrow(SCN^{-})]^{+2} + h\nu \rightarrow [(Fe^{+2})\leftarrow(SCN)]^{+2}$$

Typically, these complexes exhibit very intense absorption bands even though they may be regarded as weakly bound complexes. Orgel[2] discusses in much detail the characterization of many of these complexes.

In many molecular and ionic species there are electrons which do not appear to participate in bonding and reside in nonbonding molecular orbitals. Molecules with electrons residing in these nonbonding orbitals often are capable of absorbing ultraviolet light resulting in the excitation of these electrons to unoccupied higher energy levels known as antibonding orbitals. Transitions of nonbonding electrons are generally classified into two types depending upon the type of antibonding orbital to which they are excited. These transitions are designated as $n \rightarrow \sigma^*$ and $n \rightarrow \pi^*$ where σ orbitals are associated with single bonds, π orbitals are associated with double and triple bonds, and the asterisks denote antibonding orbitals. Ketones frequently exhibit $n \rightarrow \pi^*$ transitions as evidenced by a weak absorption band in the vicinity of 285 mμ. Certain alcohols in the vapor state exhibit $n \rightarrow \sigma^*$ transitions with weak absorption bands around 200 mμ

and below. The excitation of nonbonding electrons does not have a high statistical probability and consequently strong absorption bands should not be anticipated.

Transitions from sites of unsaturation are of the $\pi \longrightarrow \pi^*$ type and are consequently associated with the presence of either double or triple bonds in the ion or molecule. In addition to the $n \longrightarrow \pi^*$ transitions, many of the ketones exhibit $\pi \longrightarrow \pi^*$ transitions which are generally characterized by a relatively strong absorption band in the region of 180 mμ. A more detailed discussion of these transitions has been given by Dyer[3] and Bauman.[4]

From the previous discussion it is apparent that certain molecular structures, or groups of atoms, can be characterized by certain types of electronic transitions. For example, carbonyls, $>$C=O, should exhibit an absorption band near 180 mμ corresponding to $\pi \longrightarrow \pi^*$ transitions and another band near 285 mμ corresponding to the $n \longrightarrow \pi^*$ transitions. Likewise, functional groups such as $>$C=C$<$, —C≡C—, —C≡N, and —N=N— should behave in a somewhat predictable manner. These characteristic functional groups are called chromophores. Certain functional groups which do not exhibit an appreciable selective absorption in the ultraviolet and visible regions can produce very pronounced effects upon the absorption due to the chromophoric groups when they are attached to the chromophoric group. These groups are known as auxochromes and may produce the following spectral effects: (*1*) bathochromic shift, a shift of the absorption band to a longer wavelength; (*2*) hypsochromic shift, a shift of the absorption band to a shorter wavelength; (*3*) hyperchromic effect, an increase in the absorptivity; and (*4*) hypochromic effect, a decrease in the absorptivity.

Thus far we have only considered electronic excitation with no mention of spin, nor the effects of vibrational, rotational, and translational energies. In any electronic state where the electron spins are coupled, e.g., for every electron with positive spin there exists an electron with a corresponding negative spin, we have a singlet state which is indicative of the multiplicity of the molecule. If the molecule possesses two unpaired spins, it exists in a triplet state. Spectroscopic selection rules forbid transitions involving changes in multiplicity, implying that they are of low probability. In Figure 1, the various energy states which a molecule or ion might possess are shown graphically by plotting potential energy vs. the interatomic distance between the atoms. The lowest energy level is designated the ground state G while the higher energy levels will be designated S or T to indicate singlet or triplet excited states. In a diagram such as Figure 1 it is also possible to denote the vibrational energy levels superimposed upon any particular electronic state by means of horizontal lines within the energy wells. As

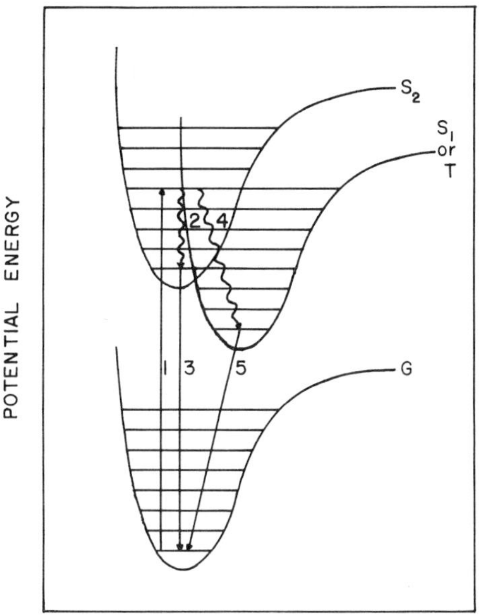

Fig. 1. Potential energy diagram.

shown in Figure 1 (line 1) the absorption of a photon could raise the molecule from whatever ground state vibrational level it might be in at the instant of absorption to one of the vibrational levels of the excited state singlet. In order to raise those electrons residing in a particular ground state vibrational level to a specific excited state singlet vibrational level a photon of energy* exactly equal to the potential energy difference between the two electronic-vibrational states would be required. At any given instant we should have a distribution of molecules residing in various ground state vibrational levels and consequently one should expect to find absorption over a specific wavelength range resulting in an absorption band. The wavelength of maximum absorbance should correspond to excitation from the most populated ground state vibrational level. Thus, an absorption band is comprised of many discrete energy transitions, but current instrumentation is incapable of resolving these individual hyperfine absorption lines.

In spectrophotometry only the process of excitation, a phenomenon which occurs in approximately 10^{-15} sec is considered. In fluorescence and phos-

*Energy may be related to frequency or wavelength by the equation $E = h\nu = hc\lambda$ where h is Planck's constant (6.6256×10^{-27} erg sec), ν is the frequency, c is the speed of light (2.9979×10^{10} cm sec^{-1}), and λ is the wavelength.

phorescence processes the molecule or ion deexcites itself from approximately 10^{-8} sec up to many seconds. The modes of deexcitation may be classified into two general groups: (*1*) radiative processes resulting in fluorescence and phosphorescence, and (*2*) nonradiative processes. Let us consider a molecule in the fifth highest vibrational level of the excited state S_2 (see Fig. 1). Deexcitation by collisions with neighboring molecules could lower the molecule's energy to correspond with the lowest, the first vibrational level of S_2 (via line 2). However, the molecule must return to one of the ground state vibrational levels in order to complete its deexcitation process. It might seem that a collision of sufficient magnitude could do this; however, collisions of this nature are highly improbable. It is possible with some molecules that a singlet or triplet state of intermediate energy between S_2 and G (not shown in Fig. 1) could exist so that its upper vibrational levels coincided with the lower vibrational levels of S_2, and its lower vibrational levels could coincide with the upper vibrational levels of the ground state G. In this case, the molecule could undergo complete deexcitation by means of collision. Collisional deexcitation by this process is known as internal conversion when the intermediate state is a singlet and intersystem crossing if it is a triplet. The time required for these processes is generally less than 10^{-9} sec even when the intermediate state is a triplet. If such an intermediate state does not exist, excluding the possibility of chemical reaction or dissociation, a radiative process of deexcitation will occur. This could occur directly as shown by line 3 in Figure 1 resulting in fluorescence emission. Another possibility for deexcitation might be a nonradiative decay by internal conversion (see line 4 of Fig. 1) to the lowest vibration level of the intermediate singlet S_1 and subsequently a radiative deexcitation (line 5) again resulting in fluorescence emission. Processes of this nature are generally believed to occur between 10^{-8} and 10^{-4} sec. Finally, another mode of deexcitation is possible when there exists an intermediate level triplet so that intersystem crossing can occur (line 4) followed by a phosphorescent radiative deexcitation (line 5). We have previously indicated that selection rules forbid changes of multiplicity for excitation processes. These rules are equally valid for radiative deexcitation and therefore we find the probability of phosphorescence emission to be low. Consequently, the time required for phosphorescence deexcitation is between 10^{-4} sec up to many seconds for complete deexcitation.

There are two fundamental laws describing the absorption of monochromatic radiant energy by homogeneous, transparent media. The Bouguer-Lambert law states that each layer of equal thickness will absorb an equal fraction of the beam of radiant energy which traverses it. Beer's law states that the fraction of the monochromatic radiant energy absorbed on passing through a solution is directly proportional to the concentration of the absorber. A complete derivation of the combined law which illustrates the theoretical aspects of these relationships will be presented.

A parallel beam of monochromatic radiant energy, whose dimensions of height and width are given by h and w, respectively, irradiates an area σ on the face of the hypothetical cell (Fig. 2). As the beam of radiant energy traverses the cell there will be a decrease in radiant power (P_0) due to the absorptivity characteristics of the small incremental volume σ (db). Attenuation of the beam may be described by

$$-(dP_\lambda/P_\lambda) = (d\phi_\lambda/\sigma) \tag{1}$$

where the subscript λ denotes a fixed wavelength and $d\phi_\lambda$ is the photon capture cross section at this wavelength and is given by

$$d\phi_\lambda = (\partial\phi_\lambda/\partial n_1)_\lambda \, dn_{1,n_2,n_3,\ldots n_n} + (\partial\phi_\lambda/\partial n_2)_\lambda \, dn_{2,n_1,n_3,\ldots n_n} + \cdots + (\partial\phi_\lambda/\partial n_n)_\lambda \, dn_{n,n_1,n_2,\ldots n_{n-1}} \tag{2}$$

where n is the number of molecules of a given absorbing species. If each of the species is an independent moiety, free from any interaction with any of its neighboring species, e.g., a condition realized at high dilution; then each partial term may be evaluated as follows:

$$(\partial\phi_\lambda/\partial n_1)_{\lambda,n_2,n_3,\ldots n_n} = k_1$$

$$(\partial\phi_\lambda/\partial n_2)_{\lambda,n_1,n_3,\ldots n_n} = k_2$$

$$(\partial\phi_\lambda/\partial n_n)_{\lambda,n_1,n_2,\ldots n_{n-1}} = k_n$$

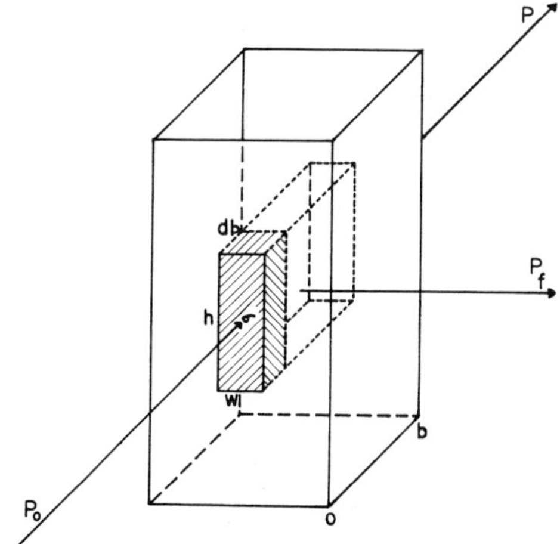

Fig. 2. Hypothetical absorption cell.

Substitution of these values into equation 2 gives

$$d\phi_\lambda = k_1dn_1 + k_2dn_2 + \ldots + k_ndn_n \tag{3}$$

By substitution of equation 3 into equation 1, the following expression results:

$$-(dP_\lambda/P_\lambda) = (1/\sigma)\,[k_1dn_1 + k_2dn_2 + \ldots + k_ndn_n] \tag{4}$$

Integration of this relationship over the limits of P_0 to P and from 0 to n_j for each absorbing species we obtain the following expression. (for simplicity, the subscript λ is omitted):

$$\ln\,(P_0/P) = (k_1n_1/\sigma) + (k_2n_2/\sigma) + \ldots + (k_nn_n/\sigma) \tag{5}$$

If transmittance is defined as follows: $T = P/P_0$, then the following equation can be written:

$$\ln\,(P_0/P) = -\ln T \tag{6}$$

Absorbance is defined by the following equation:

$$A = -\log T = \log\,(P_0/P) \tag{7}$$

If $a_j = k_j/2.303$ = absorptivity for a specific absorber, then the following equation defines the absorbance for the incremental volume σ (db).

$$A = (a_1n_1/\sigma) + (a_2n_2/\sigma) + \ldots + (a_nn_n/\sigma) \tag{8}$$

Summation of the increments for the entire cell gives:

$$\int_0^b \sigma db = \sigma b = v = \text{volume of the solution in the optical path}$$

where b is the length of the hypothetical cell. Inasmuch as $1/\sigma = b/v$, substitution of this relationship into equation 8 gives the following:

$$A = (a_1bn_1/v) + (a_2bn_2/v) + \ldots + (a_nbn_n/v) \tag{9}$$

However, $n_j/v = c_j$ when c_j is the concentration of a specific absorber. Substitution of c into equation 9 gives equation 10:

$$A = a_1bc_1 + a_2bc_2 + \ldots + a_nbc_n \tag{10}$$

For a single absorber:

$$A = abc \tag{11}$$

Equation 10 is the combined Beer-Bouguer law of absorption and imposes all of the conditions and restrictions indicated in its derivation.

If the absorbing molecules are capable of exhibiting fluorescence, or phosphorescence, attention is focused upon the emission of fluorescence and phosphorescence radiant energy, denoted by P_f in Figure 2. If there is only a single absorbing species in the optical beam, the absorbed radiant energy (not to be confused with absorbance) may be given by

$$\bar{A} = P_0 - P \tag{12}$$

Consequently, the radiant power of the observed fluorescent or phosphorescent radiant energy will be proportional to the absorbed radiant energy, i.e.,

$$P_f = \phi k' \bar{A} = \phi k'(P_0 - P) \tag{13}$$

where ϕ is the quantum efficiency and k' is a proportionality constant related to the design characteristics of a particular instrument. From equations 7 and 11 the following expressions are obtained:

$$\log P_0/P = abc$$

or

$$P = P_0\, e^{-2.303abc} \tag{14}$$

By substitution of this expression into equation 13, the following simplified equation results:

$$P_f = \phi k' P_0(1 - e^{-2.303abc}) \tag{15}$$

Further simplification of this relationship is permissible for small values of c by approximating the expansion of the exponential term to give:

$$P_f \approx KP_0abc \tag{16}$$

where $K = 2.303\phi k'$. If the radiant flux of the lamp, P_0, remains constant, then a linear function of concentration vs. radiant flux will exist for low concentrations (Fig. 4). The radiant flux approaches asymptotically the value KP_0ab with increasing concentration because of the approximation made for small values of c.

Frequently it is desired to observe the rate of decay for phosphorescence emission. Such studies can often give information related to the nature of the excited species, the quantum yield, and the average lifetime of the excited state. Phosphorescence decay rates are described by means of a first order kinetic decay expression

$$P_{f_t} = P_0\, e^{-t}/\tau \tag{17}$$

where P_{f_t} is the radiant power of the phosphorescence emission at time t, P_0 is the incident radiant flux at time 0, t is the time and τ is the average lifetime of the excited state.

II. SPECTROPHOTOMETRIC ANALYSIS

A. Scope of Spectrophotometric analysis

Spectrophotometric methods are used extensively in modern metallurgical analysis because they are relatively rapid, suitable for the determination of small amounts of metals and nonmetals, and require minimal expenditures for instrumentation and operation. In general, spectrophotometric methods are sensitive to low concentrations of metals and nonmetals (0.1–100 μg/ml) and are applicable to about a hundredfold concentration range.

B. Instrumentation

Although visual color comparisons can be used to obtain satisfactory results in many analytical determinations, the superiority and virtually universal use of photoelectric instrumentation in present day analytical practice dictates that only photoelectric spectrophotometers and filter photometers be considered in this chapter.

1. ESSENTIAL COMPONENTS

There are four integral components of instruments used for the relative measurement of radiant energy as a function of wavelength, namely: (*1*) a source of radiant energy, (*2*) a monochromator or an optical filter system for isolating a narrow wavelength region, (*3*) an absorption cell assembly which provides for alternate examination of the reference and sample solution, and (*4*) the photometer with which the ratio of radiant power transmitted by the sample solution to that of the reference solution is measured. These basic components will now be considered in more detail in reference to the design and operation of spectrophotometers and filter photometers.

a. Spectrophotometers.

The first essential component is a source of continuous radiant energy at constant and sufficient intensity for the region of the spectrum in which the intrinsically characteristic absorption bands of the sample are found. A tungsten filament lamp is widely used for the visible region and near infrared region (350–3,000 mμ). A continuous spectrum characteristic of a blackbody radiator at a radiation temperature of about 2800°K is simulated when the filament is heated to incandescence. The radiant emittance of a tungsten bulb depends upon the absolute temperature of the filament and upon the wavelength of the radiant energy being emitted. The glass envelope precludes use of a tungsten lamp source for the ultraviolet region. It is necessary that the power supply for heating the tungsten filament be well regulated to ensure constancy of the incident radiant energy.

The hydrogen (or deuterium) discharge lamp has been found to be the most suitable source of a continuous spectrum in the ultraviolet region (200–400 mμ). Special ultraviolet transmitting windows of fused silica enable these sources to be used in the far ultraviolet region to about 170 mμ. Measurements below 195 mμ require purging of the optical system with dry nitrogen to eliminate absorption due to oxygen and water vapor. These lamps are moderately expensive and have a limited life expectancy. A collimating lens or mirror is often attached to the source in order that radiant energy of all wavelengths shall be transmitted in a parallel path to the entrance slit of the monochromator.

The second essential component of a spectrophotometer is a monochromator, which permits the selection of radiant energy of the desired wavelength. A monochromator consists of entrance and exit slits and a dispersive device to separate a continuous spectrum into its constituent wavelengths. Either prisms or diffraction gratings are employed as dispersive devices. The capability of either a prism or a grating to separate two adjacent wavelengths is termed *resolving power*, R. The resolving power of a prism depends on its effective thickness, t, and the slope of the dispersion plot of the optical material of which it is made.

$$R = t(\partial n/\partial \lambda) \cong \lambda/\Delta\lambda$$

For a grating, the resolving power depends on the number of rulings, N, and the order of the spectrum, m

$$R = mN$$

The widths of the entrance and exit slits control the spectral slit width of the emergent radiant energy which is to be incident upon the absorbing system. Thus, the nominal wavelength which one purports to use is in reality slightly contaminated with radiant energy of slightly higher and slightly lower wavelengths. The optimum spectral slit width depends on radiant power of the source and the sensitivity of photoelectric detector. It is important that slit width settings be accurately reproducible.

The third essential component of a spectrophotometer is the assembly which holds the absorption cells and provides for the selection of the absorption cell to be irradiated. It is very important that the cells are positioned in a reproducible manner.

The absorption cells are constructed of glass for measurements in the 350–1000 mμ region, while quartz windows, or fused silica cells, are used for measurement in the ultraviolet region. For precision work it is preferable

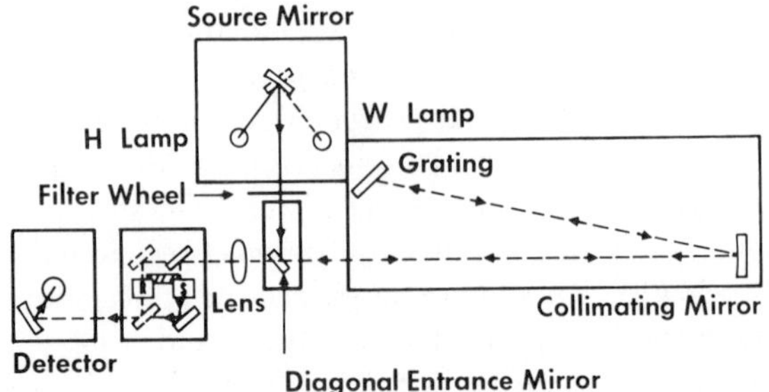

Fig. 3. Schematic diagram of Beckman DB Spectrophotometer. (Courtesy Beckman Instruments, Inc.)

that entrance and exit sides be plane parallel surfaces. The thickness of the cell is extremely important as an examination of the combined absorption law expression shows. It is very important that the cells used in a series of measurements be matched in respect to spectral transmission characteristics. Although absorption cells 1 cm in optical path lengths are widely used, 2, 5, and 10 cm cells are also available. In several instruments cylindrical cells, or cuvets are used. The positioning of the cuvets in the optical path is quite critical if reliable results are to be obtained.[5] Some cell compartments have provisions for thermostating.

The fourth essential component of a spectrophotometer is the transducer or photometer which detects the transmitted radiant energy and produces an electrical signal proportional to the radiant power of the beam. There are three types of photosensitive detectors commonly used: (*1*) barrier (photovoltaic) layer, (*2*) photoemissive, and (*3*) electron multiplier phototubes.

The photovoltaic or barrier layer cell consists essentially of a conductor in close contact with a semiconductor. Electrons are released at the interface of these two layers when the semiconductor, e.g., selenium, is irradiated. The electron flow is proportional to the radiant power of the incident radiant energy and the area being irradiated. The current is a linear function of radiant power provided the internal and load resistances are low. This type of detector has a spectral response somewhat similar to the eye and is suitable for photometric measurements in the visible region. However, the low impedance characteristic of the barrier layer cell restricts the use of ordinary amplifiers in measuring its output current. Hence, rather wide spectral slit widths are to be expected when this type of detector is employed. This detector is used primarily in filter photometers.

The photoemissive cells consist basically of a photosensitive cathode and an anode mounted in a glass envelope which is evacuated. The spectral response of the cell depends on the material used for the photocathode surface. Hence, by proper selection of a photocathode it is possible to sensitize the cell to specific ranges of the visible and ultraviolet region. Radiant energy striking the photocathode causes photoelectrons to be emitted. The number of electrons per microwatt is directly proportional to the radiant power of the incident radiant energy, and the maximum velocity of the electrons is directly proportional to the frequency of the incident radiant energy. Because of the low radiant power involved and magnitude of these photocurrents it is necessary to amplify the output current.

The electron multiplier phototube (photomultiplier) is especially versatile in respect to sensitivity and response time. This tube has a series of photosensitive surfaces each charged at a successively higher potential and the photoelectrons emitted by the first photocathode surface are accelerated from one dynode to the next with the current being increased in each step by the secondary emission of electrons. The extreme sensitivity of this tube to low radiant power and its fast response time makes it particularly suitable for high resolution, recording spectrophotometers.

Deflection-type meters or calibrated slidewires permit satisfactory readouts of photometric measurements at a single wavelength. However, the incorporation of a recorder with which the photometric reading is recorded as a function of the wavelength of the incident radiant energy is a valuable accessory. Especially in developing a spectrophotometric method and evaluating the effects of solution variables a recorder is a time saver and provides a permanent record of the spectral data obtained.

b. Filter Photometers

For routine determinations it is possible to use instruments of simplified construction. These instruments which use filters to isolate portions of the spectrum are commonly called filter photometers. Barrier layer photocells are frequently used because the relatively broad transmittance bands of the filters provide sufficient radiant energy to give a measurable output current. The proper selection of filters is important in isolating a specific region of the incident continuous spectrum. Table 2 can be used as a guide in selecting the appropriate filter. The wavelength of maximum transmittance for the filter should correspond closely to the wavelength of maximum absorbance of the colored solution. Hence, the color of the filter should be complementary to the color of the solution to be measured.

Filter photometers have either a direct reading, deflection-type meter or a calibrated potentiometer with a null point indicator. In general, the instrument is adjusted by attenuating the incident light beam or by changing the electrical resistance so that the meter or potentiometer reads 100% transmittance when the absorption cell is filled with a blank solution. The 0% T

Table 2. Complementary filter colors

Color of solution	Color of filter.
Orange	Blue-green
Yellow	Blue
Purple	Green
Red	Blue
Violet	Yellow-green
Green	Purple
Blue-green	Red-orange
Blue	Yellow

adjustment is made with no light reaching the detector. When the absorption cell is filled with colored solution, the corresponding $\% T$ value is obtained. The spectral bandwidth of most filter photometers is rather large, often 50 mμ, so that the preparation of a calibration graph of instrument readings vs. concentration is recommended.

4. PRACTICAL CONSIDERATIONS IN SELECTION OF INSTRUMENTS

There is such a variety of commercially available spectrophotometers that the selection of an appropriate instrument often poses a difficult problem to the chemist or technician of limited experience. The choice of an instrument depends largely on the function and service demands of a given laboratory. For a few routine determinations of a highly repetitive nature an inexpensive filter photometer may be sufficient. However, if new procedures are to be developed a spectrophotometer is highly desirable. The cost of a spectrophotometer is closely associated with its resolution, photometric accuracy, and wavelength range. High-resolution instruments usually have more expensive optical components and a sensitive detection system with its inherently more complex electronic components. Most spectrophotometric methods for metals are based on systems having moderately broad absorbance maxima and therefore do not require high resolution. However, many of the tervalent ions of the rare earth elements have very narrow absorbance maxima and high-resolution spectrophotometers are necessary for spectrophotometric analysis of these elements. Likewise, if other substances present in the sample have absorbance maxima close to the wavelength of maximum absorbance of the desired constituent, resolution is a valuable feature. A reasonable guide in matching a specific absorbance maximum with resolution is to have the effective spectral slit width to be one-tenth the half-bandwidth of the absorbance maximum.

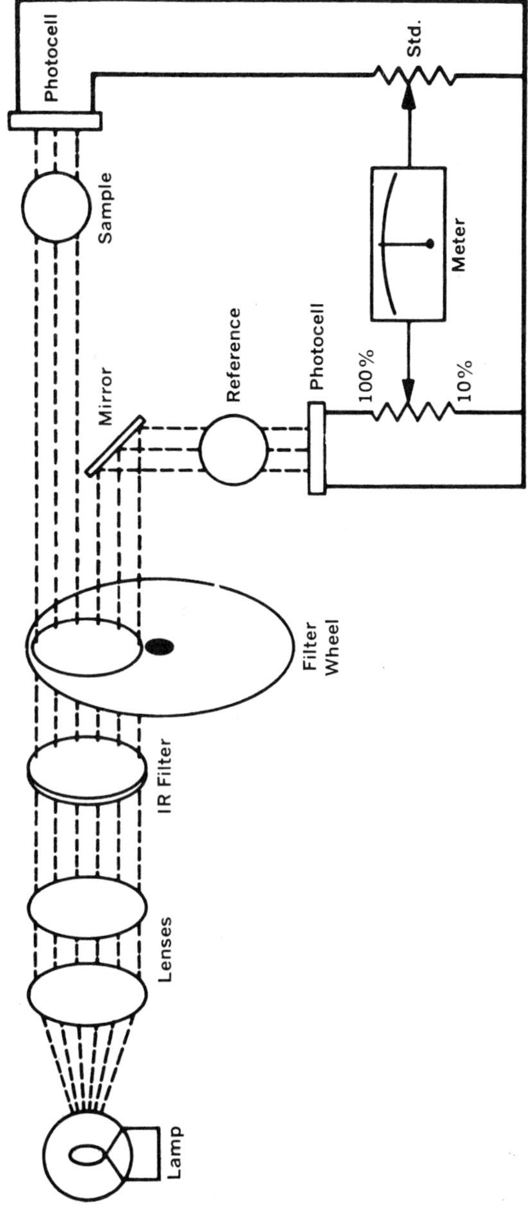

Fig. 4. Schematic diagram of Fisher Electrophotometer. (Courtesy Fisher Scientific Co.)

Table 3. Selected list of commercial spectrophotometers and filter photometers

Instrument mfg. and model		Design features
	Spectrophotometers	
Bausch and Lomb	Precision	Single beam, dbl. grating, 197–700 mμ
	Spectronic 505[a]	Dbl. beam, dbl. grating, recording 200–700 mμ
	Spectronic 600	Dbl. beam, dbl. grating, manual
	Spectronic 20	Single beam, single grating, manual
Beckman	DB	Dbl. beam, single prism, or grating, manual or recording, 205–770 mμ
	DU-2[a]	Single beam, single prism, manual 210–1000 mμ
	DK-2[a]	Dbl. beam, single prism, 185–3500 mμ recording
Cary	11	Dbl. beam, dbl. prism, recording 200–800 mμ
	14[a]	Dbl. beam, prism and grating, 186–2600 mμ recording
	15	Dbl. beam, bdl. prism, 185–800 mμ recording
	16	Dbl. beam, dbl. prism, 186–800, manual
Coleman	101	Single beam, single grating 220–900 mμ, manual
	Junior	Single beam, single grating 400–700 mμ manual
	EPS-3T	Dbl. beam, single prism, 170–2600 mμ
Perkin-Elmer	450[a]	Dbl. beam, dbl. prism, 165–2700 mμ recording
	202	Dbl. beam, single prism, 190–750 mμ recording
P-E Hitachi	139	Single beam, single grating, 195–800 mμ manual
	Filter Photometers	
Fisher, Electrophotometer II		Dual photovoltaic cells, filter wheel, source compensation
Klett, (Klett-Sumerson) Model 900-3		Dual photovoltaic cells, filter holder, source compensation
Photovolt, Lumetron 402		Dual photovoltaic cells, source compensation, convertible to fluorimeter

[a] Spectrophotometers for which a fluorescence measuring attachment is available.

Whether or not the cell compartment can accomodate longer absorption cells is another feature to be considered. In trace analysis it is often desirable to be able to use cells of longer optical path.

C. General Methodology

1. PREPARATION OF SAMPLE

Because the majority of samples of metallurgical interest are solids, the preparation of a colored solution suitable for photometric measurement constitutes an important step in the analytical procedure. In the dissolution process, care must be taken not to introduce traces of metals from the reagents being used. Most photometric methods for the determination of metals depend on either oxidation or chelation reactions for the development of suitable colored or absorptive systems. The acidity or pH is often critical for the maximum development of color. Other variables which may be relevant include amount and concentration of reagent, order of addition of reagents, temperature, time, and ionic strength. In storing dilute solutions of inorganic ions the possibility of changes in concentration due to adsorption and ion exchange processes with the glass surfaces must be considered as a significant factor in certain determinations. Minute traces of metal ions are also adsorbed and removed when paper has been used as a filtering media. In the selection of a photometric method and the development of a satisfactory absorptive system the following criteria should be considered: sensitivity, reproducibility, stability, specificity, and effect of other substances likely to be present.

2. ELIMINATION OF INTERFERENCES

Other substances present in the sample may interfere in the development of a suitable system for photometric measurement. These diverse substances may be self-absorptive or react with the reagent to produce absorptive species which absorb at the wavelength at which the desired constituent is being measured, produce a turbidity, or inhibit the reaction of the reagent with the desired constituent.

If there is no interaction between reagent and diverse substance, the external compensation method can be used in which the solution used in the reference cell has the same concentration of diverse substance as the solution being measured. Sometimes a solution of the sample is used as the reference, provided the reagents are nonabsorptive at the wavelength of measurement.

When interaction between diverse substances and reagent is observed, the internal compensation method is sometimes applicable. This method consists of adding a standard solution of the desired constituent to the sample and measuring the increase in absorbance with each increment of the standard solution. From the rate of increase in absorbance per unit increase in concentration the amount of desired constituent in the original sample can be determined. The validity of this method should be carefully checked by testing to ascertain that the slope of the resulting graph is virtually constant. (See Figure 5.)

When compensation methods are not applicable an interfering ion can sometimes be complexed by a specific ligand, e.g., EDTA, to form a stable, noninterfering complex. Infrequently a change in the oxidation state of an interfering ion minimizes the interference. Ringbom's book[6] should be consulted for theoretical and practical aspects of using the complexation or "masking" approach.

The ultimate answer to the problem of interferences is to perform an analytical separation to remove the interfering substance. Analytical separation methods involving liquid–liquid extraction, ion exchange, paper or thin layer chromatography, distillation, and electrodeposition are especially worthy of consideration.[84]

3. MEASUREMENT OF THE SYSTEM

After the preparation of a suitable ultraviolet or light absorptive system the next step in the general analytical procedure is the measurement of the absorptive capacity or absorbance, at a specific wavelength. The selection of the appropriate wavelength of measurement should be based on a

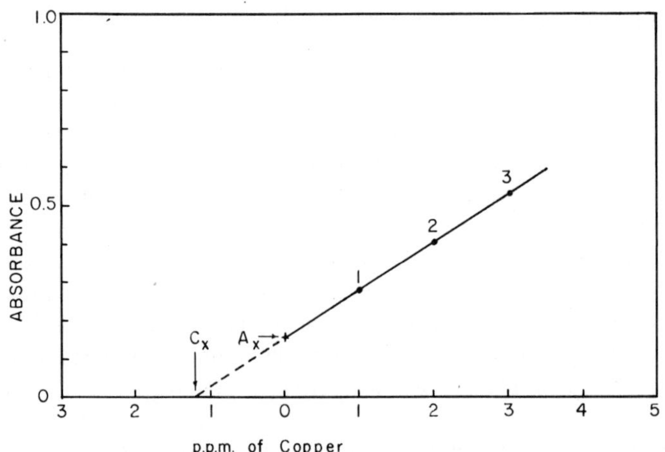

Fig. 5. Standard addition method.

thorough knowledge of the chemical system being measured and the instrument being used. In general, photometric measurements are made at the wavelength of maximum absorbance for the prepared solution. However, there are conditions under which measurement at the absorbance maximum is not recommended. For example, if the absorbance maximum is sensitive to slight variations in the pH of the solution or the amount of excess reagent, it may be preferable to make measurements at a shoulder on the spectrophotometric curve or at another absorbance band which although it may have a smaller absorptivity, is not sensitive to such solution variables. If the absorbance maximum is very narrow and a low-resolution instrument is being used the finite slit width effect becomes of concern and requires such careful control of slit width and setting of wavelength that it may be advisable to make photometric measurements at a wavelength region where the absorbance is still fairly high but changes more gradually. Conformity to Beer's law is often observed at several different wavelengths with the slope of the absorbance vs. concentration graph being proportional to absorptivity characteristic of the absorber at each wavelength. Often it may be more desirable to measure at a wavelength where the system is not as sensitive when a high-absorbance system is obtained than to resort either to dilution or to the use of the differential technique. Another consideration involves the avoidance of interference of another absorbing species also exhibiting absorptivity at the absorbance maximum by selecting a wavelength where the interferer does not absorb but the desired constituent has appreciable absorptivity.

As implied in the preceding discussion, attention should also be given to the absorbance range of the photometric measurements. For most manual spectrophotometers using either a deflection meter or a calibrated potentiometer with null point balance the scales are linear in respect to transmittance. The 20–60% transmittance (0.7–0.22 A) region is considered to correspond to the optimum concentration range inasmuch as the $\Delta T/\Delta C$ ratio is largest in this range. Although with exceptional precautions using a precision spectrophotometer photometric errors corresponding to 0.1–0.2% may be achieved, a more reasonable estimate for most spectrophotometric measurements would be 0.5%. Thus, if errors in preparing the sample are negligible, a relative error of less than 2% should be obtainable by confining measurements to the 20–60% transmittance region. When using recording spectrophotometers having linear absorbance scales the optimum concentration range is vastly different and depends on the optical and electronic systems of the instrument. Especially with double monochromator instruments for which the stray radiant energy is very low, it is possible to measure highly absorptive solutions having absorbance vlues as high as 2.0 (1% T)

with an error of less than 0.005 absorbance unit. In using recording spectrophotometers to measure sharp absorbance maxima it is advisable to use a slow scanning rate so that the slit servosystem and recorder can function to give maximum photometric accuracy.

4. SOURCES OF ERROR

Most errors in spectrophotometric analysis can be attributed either to (*1*) the nature of the chemical system prepared for measurement, (*2*) the operational characteristics of the instrument being used, and (*3*) faulty techniques.

The absorption system should be a homogeneous colored solution, free of turbidity. A few chromogenic agents have a tendency to plate out on the sides of the absorption cells. The stability of the solution must be sufficient to permit photometric measurements within a reasonable period of time. Certain metal chelates exhibit fluorescence which can result in the diminution of the absorbance value if the fluorescence emission overlaps the wavelength of photometric measurement.

Instrumental errors depend on the specific instrument being used and the manner it is being operated. The finite slit width effect refers to the fact that the minimum spectral slit width which can be achieved with a given spectrophotometer is related to the spectral emittance curve of the source, the spectral response curve of the phototube, and the resolving power of the monochromator. Only when the absorbance maximum (transmittance minimum) of the spectrophotometric curve is very narrow at the wavelength of measurement is an error likely to result. Figure 6 illustrates how the finite slit width effect can alter the corresponding photometric reading.

In a measuring solution exhibiting low absorbance an error can be caused by the multiple reflection path effect. This error is caused by the light beam being reflected repeatedly by the front and back, cell surfaces of the absorption cells and to a lesser extent by the exit slit lens and faces of the slit jaws. Hence, a fraction of the incident radiant energy traverses the absorption cell more than once and there is an exaltation in absorbance giving a positive deviation to Beer's law in the low-concentration range. Antireflection coating and the use of longer cells will minimize this error.[7]

In measuring solutions of high absorbance ($A > 1.5$) the stray radiation effect may introduce an error, especially in using single monochromator instruments.[8] A very small percentage of the total radiant power of the radiant energy emerging from the exit slit of a monochromator is actually due to heterochromatic (stray) radiant energy which is superimposed on the virtually monochromatic beam. This stray light is due to leakage of light into the spectrophotometer and the scattering and reflection of light at

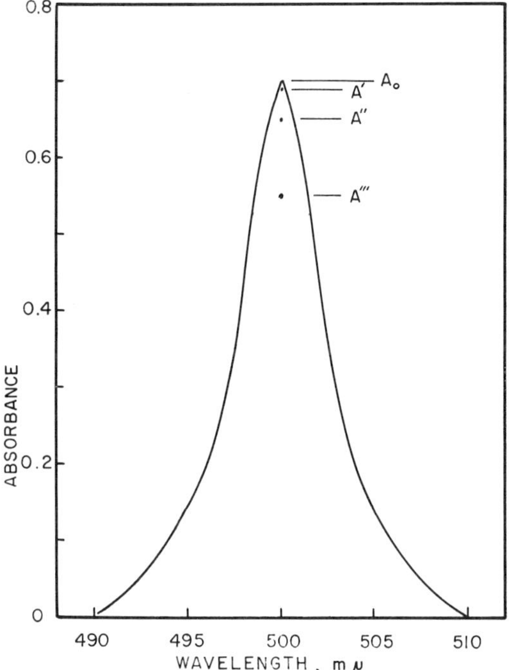

Fig. 6. Finite slit width effect.
A_0 = true absorbance (NBW = 5 mμ)
A^1 = observed absorbance (ESW = 1 mμ)
A^{11} = observed absorbance (ESW = 2 mμ)
A^{111} = observed absorbance (ESW) = 35 mμ)

surfaces of the optical components of the monochromator. Inasmuch as the absorbance readings are slightly lower when the stray radiation effect is operative, a negative deviation to Beer's law is observed.

The use of matched absorption cells is important in obtaining reliable spectrophotometric measurements. Not only should the lengths of the optical path for reference and sample cells be the same but the optical transmission properties of both cells should be identical. Alignment of absorption cells in the cell compartment in a reproducible manner is necessary.

Another instrumental source of error can be the nonlinearity of the photometer. In the case of barrier layer detectors, high resistance in the measuring circuit or fatigue effects due to too intense irradiation may result in a nonlinear response. In some photoemissive phototubes a departure from linearity is observed at high radiant power levels. Potentiometric slide wires and deflection meters may also have deficiencies in respect to linearity which results in corresponding errors in the absorbance or transmittance readings.

The accuracy of the wavelength scale of the spectrophotometer is often questionable and should be checked using a calibrated didymium glass filter or a mercury lamp as a reference standard.[9]

Errors in technique include: use of unclean or nonmatched absorption cells, failure to adjust instrument for 0% T dark current reading, measurement outside optimum concentration range, use of wide range of slit widths when measuring a narrow absorbance maximum, measurement on the steep segment of a spectrophotometric curve, failure to allow electronic components of instrument to achieve stability, misalignment of source so that extremely wide slit widths must be used, selection of improper filter for filter photometer, carelessness in the addition of reagents in the development of the colored system, dilution errors, and volumetric changes due to temperature.

5. PRESENTATION AND EVALUATION OF SPECTRAL DATA

Spectrophotometric data are commonly presented in graphical form by (*1*) plotting either absorbance or transmittance vs. wavelength to obtain the characteristic absorption spectrum of the absorber, and (*2*) plotting absorbance or log $^1/T$ values vs. concentration in order to obtain a calibration graph (Figs. 7 and 8). The absorption spectra are valuable in locating the most suitable wavelengths for photometric measurement. A linear relationship for an absorbance vs. concentration plot indicate conformity to Beer's law. As previously mentioned, there is an optimum concentration range for each spectrophotometric method and a specific instrument.

A Ringbom plot of per cent transmittance vs. logarithm of concentration gives a sigmoidal graph, the most linear segment corresponding to the optimum concentration range provided precise photometric measurements can be made under existent experimental conditions (Fig. 9).

In the presentation of spectral data it is recommended that the following information be included: length of absorption cell, concentration of solution, nature of solution in reference cell, wavelength of measurement, spectral or effective slit width, temperature of solution, and specification of instrument used for measurement. It is also helpful if the sensitivity of the method, in terms of molar absorptivity, and the optimum concentration range are indicated.

6. SPECIAL TECHNIQUES

The conventional method of measurement in spectrophotometric analysis requires that the photometer is adjusted to 0% transmittance when the de-

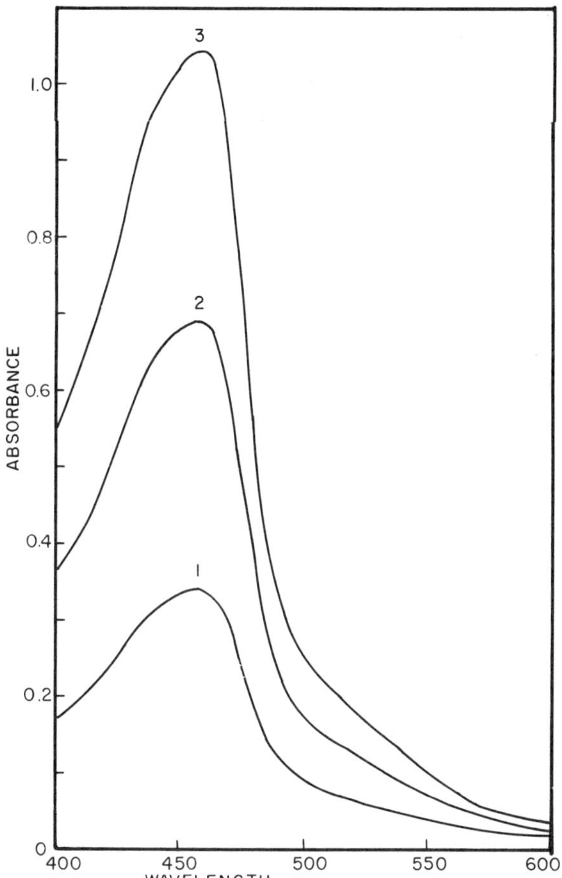

Fig. 7. Typical spectral transmittance curves.

tector is excluded from the optical beam. Then the photometer is adjusted to read 100% transmittance with the reference cell containing either the solvent or a reagent blank solution in the optical beam. The unknown solution in a matched absorption cell is placed in the optical beam and its transmittance measured. Thus, the absorbance being measured is due exclusively to the desired constituent. There are several other techniques which can be used advantageously in specific applications and these will be mentioned briefly.

a. Differential Spectrophotometry

The differential technique gives better precision in measuring solutions exhibiting either very high or very low absorbance or when the desired

Fig. 8. Plots testing for conformity to Beer's law.

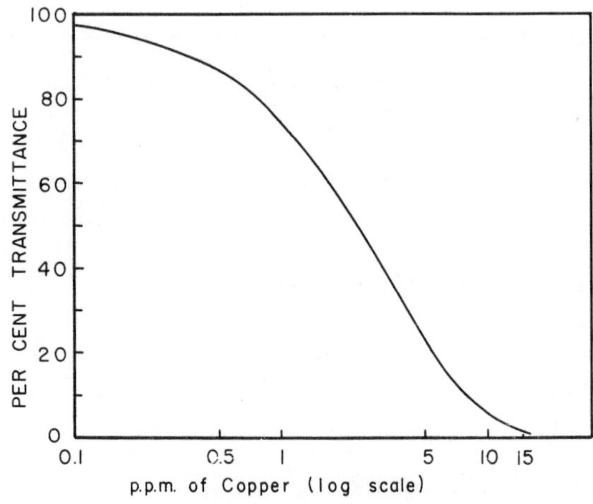

Fig. 9. Ringbom plot.

constituent reacts to decrease the absorbance of colored species, i.e., indirect method.[10–12]

b. Trace Technique

In applying the technique to very low absorbance systems, the 0% T or "dark current" setting is made using in the reference cell a standard reference solution slightly more concentrated than the sample, P_R, and 100% T setting is made with the solvent in the reference cell, P_S. Then, the transmittance of the unknown solution is measured, P_x. The absorbance of the desired constituent is obtained from the following expression:

$$A = \log(P_S - P_R)/(P_x - P_R)$$

This technique is not used very often.

c. Transmittance Ratio Technique

This technique in effect expands the photometric scale and gives increased precision in measuring high absorbance systems. The 0% T setting is made in the conventional manner but the 100% T setting is made using a standard solution in the reference cell which is slightly less concentrated than the unknown solution, T_R. When the transmittance of the unknown solution is measured, T_x, the ratio of the two solutions is obtained, T_x/T_R. The transmittance ratio technique has been used extensively in metallurgaical analysis in order to obtain relative standard deviations of less than 0.1–0.2%.[13]

d. Ultimate Precision Technique

This technique uses relatively concentrated and relatively dilute standard solutions in making the 0% T and 100% T settings. It is recommended that a calibration graph should be constructed when using differential techniques.

e. Indirect Technique

There are many indirect spectrophotometric methods based either on measuring the absorbance of a specimen equivalent to the amount of the desired constituent or by measuring the decrease in absorbance of a system due to a chemical reaction with the desired constituent. In the latter example, this often is referred to as a "bleaching effect." For example, barium chloranilate can be added to a sulfate solution and an equivalent amount of chloranilate liberated to give a colored solution whose absorbance is proportional to the sulfate precipitated. The bleaching effect of fluoride on the zirconium–eriochrome cyanine R complex serves as the basis of a spectrophotometric method for fluoride. In indirect spectrophotometry involving a bleaching effect, a solution having an absorbance higher than the unknown solutions is placed in the sample cell and the unknown solution is placed in the reference cell. The differential in absorbance is plotted against concentration to obtain a calibration graph.[14] The transmittance ratio method can be used also by using a partially bleached standard solution as the reference solution to measure the "unbleached" and unknown solutions.

f. Two-Component Technique

It is possible to determine two constituents in a sample by making photometric measurements at two wavelengths. In order for this technique to be feasible it is necessary that each constituent show adherence to Beer's law at the two wavelengths, that the absorbances attributable to each component be additive, and the ratio of their absorptivities at the two wavelengths be considerably different, preferably approaching minimum and maximum values at the two wavelengths. The following procedure outlines the spectrophotometric analysis of a binary system.[15]

1. Determine the spectrophotometric curve for each substance using the same reference solution.

2. Determine the wavelengths at which the relative absorbance for the two substances are at a minimum and a maximum.

3. Test each system for conformity to Beer's law at the two selected wavelengths.

4. Determine the absorptivity, a or ϵ, for each constituent at both wavelengths.

5. Prepare synthetic mixtures of the two substances and measure absorbances at the two wavelengths. Test for additivity of absorbances by plotting observed absorbance at each wavelength vs. calculated absorbances. If additive, a linear plot is obtained.

6. Solve the following simultaneous equations for c^{I} and c^{II}.

$$A_1 = a_1{}^{\mathrm{I}}bc^{\mathrm{I}} + a_1{}^{\mathrm{II}}bc^{\mathrm{II}}$$

$$A_2 = a_2{}^{\mathrm{I}}bc^{\mathrm{I}} + a_2{}^{\mathrm{II}}bc^{\mathrm{II}}$$

g. Photometric Titration

Another method of using a spectrophotometer or filter photometer in the determination of elements in metallurgical samples involves their use as a means of detecting the end point in precipitation, redox, or complexation titrations. Only when a satisfactory visual indicator is unavailable or a gradual color change occurs in the vicinity of the end point would this end point method become attractive. Typical examples are the titrimetric determinations of vanadium in steel using permanganate as titrant,[16] and magnesium in aluminum alloys using EDTA as titrant and chrome azurol S as indicator.[17] The end point is determined by plotting absorbance vs. ml of titrant. More detailed information is given in Headridge's monograph.[18]

D. Applications

Most metals and nonmetals can be determined spectrophotometrically. Whether or not a spectrophotometric method should be selected for a particular determination depends on many factors such as: concentration of desired constituent in sample, nature and amounts of other elements in sample, amount of sample, accuracy required, number of samples to be analyzed, availability of instruments for alternate methods of analysis, and the cost in terms of man-hours.

1. METALS

Over 170 different spectrophotometric methods for the determination of 46 metals, exclusive of the rare earth elements, have been compiled in tabular form.[19] Many of these methods are discussed critically in Sandell's treatise[20] and additional applications of these methods are cited in compendium of Snell and Snell.[21]

Most spectrophotometric methods for the determination of metals are based either on oxidation or chelation reactions. The oxidation of manganese(II) to manganese(VII) with periodate, and the determination of chromium as chromium(VI) are examples. Chelating agents which form metal chelates are extensively used in spectrophotometric analysis because of the high absorptivity of these compounds. Furthermore, metal chelates are usually soluble in organic solvents and the isolation of absorptive system by extraction from aqueous solution with an immiscible liquid is often an advantage in eliminating interferences. Thus, copper(I) forms a yellow metal chelate with 2,9-dimethyl-1,10-phenanthroline which is extractable in chloroform or hexanol. Cadmium(II) forms a stable colorless chelate with diethyldithiocarbamate which is extractable with chloroform and has an absorbance maximum at 262 mμ in the ultraviolet region.

2. NONMETALS

Spectrophotometric methods for most of the nonmetals and the application of these methods to the analysis of numerous metallurgical samples have been discussed.[22] Nonmetallic elements such as phosphorus and silicon form yellow heteropoly complexes which can be measured photometrically, or these complexes can be reduced to produce heteropoly blues which have unusually high absorptivity values.

3. SELECTED METHODS

In applying spectrophotometric methods to the analysis of specific metallurgical samples the effect of other elements present in the sample should be known or investigated. Representative spectrophotometric methods which have been used in the analysis of metallurgical materials are cited in Table 4. Recent developments in respect to new methods and applications are reviewed periodically in *Analytical Chemistry*.[68]

III. SPECTROFLUORIMETRIC ANALYSIS

A. Scope of Spectrofluorimetry[69–72]

Spectrofluorimetric methods are strictly applicable to the determination of trace quantities of substances, approximately in the 0.01–1 μg/ml concentration range with a relatively narrow operational range. In general, spectrofluorimetric methods are more sensitive and specific than spectrophotometric methods but are more susceptible to interference caused by other substances present in the sample. The experimental conditions, e.g., pH must be rigorously controlled. Although fluorescence and phosphorescence measurements have been used most extensively in the determination of organic, biological, and medicinal substances[73] many metal chelates and other inorganic substances are fluorescent. Analytical methods have been developed based not only on the measurement of fluorescence due to a specific inorganic substance, but also on the measurement of the decrease in fluorescence, or quenching, caused by a particular ion.[74,75]

B. Instrumentation

Fluorimeters, unlike spectrophotometers, must incorporate at least two wavelength-isolation systems, one to select the wavelength of the incident radiant energy for excitation and another system to isolate the emitted fluorescence or phosphorescence radiant energy. Hence, it is possible to obtain two types of spectra for a given luminescent system. The spectrum may be obtained by selecting the wavelength of maximum luminescence emission and continuously varying the wavelength of the incident radiant energy while measuring the radiant flux of the emitted fluorescence. This plot of luminescence radiant power vs. the wavelength of the excitation beam is the *excitation spectrum*. Another spectrum is obtained by irradiating with the wavelength of maximum excitation and continuously measuring radiant flux as a function of the wavelength of the emitted radiant energy. This plot of luminescence radiant power vs. the wavelength of the emitted radiant energy is the *emission spectrum*. Figure 10 illustrates an uncorrected emission and excitation spectra of an aqueous solution of uranyl nitrate in phosphoric acid. The exact nature of these two spectra will be highly dependent upon the design characteristics of the instrument being used.

1. ESSENTIAL COMPONENTS

Instruments used for fluorimetric measurements are comprised of the following components: (*1*) an intense source of radiant energy, (*2*) a wavelength isolation device for selecting the excitation wavelength, (*3*) a sample cell compartment, (*4*) a second wavelength isolation device, generally positioned with its optical path perpendicular to that of the excitation beam, thus permitting selection of the emission wavelength, and (*5*) a photosensitive detector with an associated readout system.

Table 4. Selected spectrophotometric methods

Constituent	Material	Reagent or method	Refs.
Aluminum	Carbon steel	8 Quinolinol	23
	Steel	Eriochrome cyanine R	24
	Zinc	Aluminon	25
Antimony	Germanium	Rhodamine B	26
	Lead	Iodide	27
Arsenic	Brass	Heteropoly blue	28
	Copper	Heteropoly blue	29
	Steel	Heteropoly blue	30
Beryllium	Steel	Eriochrome cyanine R	31
Boron	Alloys	Quinalizarin	32
Chromium	Alloys	EDTA	33
	Copper alloys	Diphenylcarbazide	34
Cobalt	Fe alloys	Nitros-R-salt	35
	Steel	2-Nitroso-1-naphthol	36
Copper	Nonferrous metals	Biscyclohexanone–Oxalyldihydrazone	37
	Steel	Pyrrolidine dithiocarbonate	38
	Steel	Neocuproine	39
	Steel	2,2′-Biquinoline	40
Iron	Nickel	2,2′-Bipyridine	41
	Tungsten	1,10-Phenanthroline	42
Lead	Steel	Dithizone	43
Magnesium	Nickel	8-Quinolinol	44
Manganese	Steel	Periodate	45
	Metals	PAN	46

(continued)

Table 4. *(continued)*

Constituent	Material	Reagent or method	Refs.
Molybdenum	Steel	Thiocyanate	47
Nickel	Tungsten	Dimethylglyoxime	48
Niobium	Zr alloys	Sulfochlorophenol	49
Phosphorus	Copper alloys	Heteropoly blue	50
	Steel	Heteropoly blue	51–54
	Steel	Molybdovanadophosphoric acid	55
Silicon	Copper	Heteropoly blue	56
Tin	Alloys	Rhodamine B	57
Titanium	Alloys	Disodium-1,2-dihydroxybenzene-3,5-disulfonate	58
	Steel	Hydrogen peroxide	59, 60
Tungsten	Steel	Thiocyanate	61
	Steel	8-Quinolinol	62
Vanadium	Aluminum	Vanadotungstophosphoric acid	63
	Al alloys	Peroxy-differential	64
Zinc	Nickel	Dithizone	65
Zirconium	Niobium	Catechol violet	66
	Steel	Chloranilic acid	67

A primary requirement of the radiation source of a fluorimeter is that it produces very intense radiant energy in the wavelength region necessary for excitation. High intensity is necessary to produce sufficient emission of fluorescent radiant energy so that it can be measured satisfactorily. Another desirable characteristic, although not always necessarily required, is the production of a continuum of uniform and constant intensity throughout the ultraviolet and visible region. A wide variety of sources has been used, but generally those most commonly encountered are the hydrogen, the mercury vapor, and the xenon arc lamps. Tungsten lamps produce a continuous spectrum of moderate intensity and have been used primarily for excitation in the visible region. Mercury vapor lamps exhibit a very intense output but with regions of extremely intense band and line emission. The high pressure xenon arc lamp is a popular source for fluorescence work since it produces a very intense continuum in the region of 200–800 mμ. For a more detailed description of various sources the reader is referred to the discussion by Lewin.[76]

Wavelength isolation devices consist of gratings, prisms, and filters. Gratings have been widely used in many commercial spectrofluorimeters. Occasionally, it is necessary to incorporate filters with grating systems to remove overlapping higher order spectra. Prisms, while not giving a linear dispersion, usually exhibit very good resolution, particularly at the lower wavelengths. The use of filters in fluorimeters has been quite extensive with several types being employed. Broad-band-pass filters have been used, however narrow-band-pass filters are to be preferred. A rather comprehensive listing of colored glass filters and their associated spectra may be found in the manufacturer's literature.[77,78] Due to their cost and relatively low transmittances, interference filters have gained only limited acceptance in filter fluorimetry.

The sample cell compartment is an extremely important part of the fluorimeter. It should be free from stray light and preferably provide for temperature control because fluorescence emission is often highly temperature dependent. The cell carrier should accommodate several cells and be adaptable for the use of cells of different shapes and sizes. The exact alignment of cells in a reproducible manner is necessary. Often a rotary cell holder is used which enables the operator to quickly position the blank, standard or unknown solutions for measurement. Fluorimeter cells should be constructed of fused quartz if excitation in the ultraviolet region is to be used. Rectangular cells having inside dimensions ranging from $10 \times 20 \times 50$ mm to $3 \times 3 \times 40$ mm and round cells with inside diameters ranging from about 15 to 3 mm are commonly used. These cells have a capacity of 0.2–7 ml. For certain applications, Pyrex cells can be used satisfactorily.

A variety of detector systems have been employed in fluorimeters. Some of the simple filter fluorimeters employ barrier layer or photoconductive cells as detectors. However, the disadvantages and limitations of the barrier layer cells and photoconductive cells has restricted their use in modern spectrofluorimetric instrumentation. Photoemissive type tubes are the most widely used detectors because of the following characteristics: (*1*) readily adaptable to modulation, (*2*) convenient amplification of photocurrent output, (*3*) excellent sensitivity, and (*4*) photocurrent output is essentially linear with respect to the radiant power. One particular type of photoemissive tube which has gained considerable popularity in the past decade is the photomultiplier. The principle limitation of photoemissive tubes is that they can rarely be operated at full sensitivity due to the rapid increase of electrical noise with higher sensitivities. Photomultiplier detectors are even more susceptible to this effect than are the single electrocathode photoemissive tubes.

2. SPECTROFLUORIMETERS

The development of new fluorimetric and phosphorimetric methods of analysis would be practically impossible without spectrofluorimeters. In order to develop a useful analytical method it is essential that both the excitation and emission spectra be known, since one must know what wavelength is to be used for excitation and also the wavelength at which the fluorescence emission will be observed. In many instances it is found that several excitation wavelengths, as well as fluorescence wavelengths, may be observed as seen in Figure 10. Generally, a wavelength is selected which results in the highest sensitivity.

The interpretation of excitation and emission spectra is frequently complicated due to the particular characteristics of the instrument being used. The fact that excitation sources commonly used exhibit nonlinear intensities with respect to wavelength often produces anomolous effects in the spectra obtained. Similar effects are observed which may be attributed to the nonlinear response of the detector with respect to wavelength. True spectra will be obtained only when these effects have been adequately compensated. Figure 10 is a typical example of uncorrected excitation and emission spectra. Ideally a corrected excitation spectrum should be coincident with the absorption spectrum of the system. It is difficult to determine the true relative intensities of the maxima observed without a knowledge of the spectral characteristics of the source and detector. Methods and standards for the correction of excitation and emission spectra have been reported.[79–81] There are several commercially available instruments which directly produce corrected spectra. Compensation of source and detector

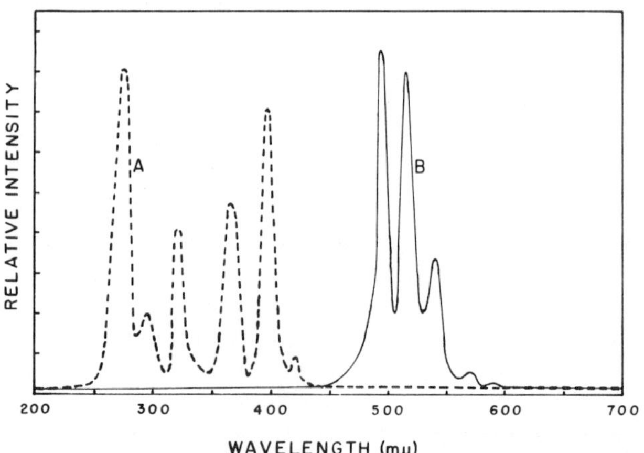

Fig. 10. Fluorescence excitation and emission spectra (uncorrected). Sample: 4 μg/ml uranyl nitrate in phosphoric acid. Instrument: Aminco-Bowman, Xenon Source 1P28 photomultiplier. Curve *A*. Excitation spectrum (emission wavelength of 435 mμ). Curve *B*. Emission spectrum (excitation wavelength of 270 mμ).

effects has been accomplished in a number of different ways. For detailed information concerning these factors the reader should consult the manufacture's literature.

Figure 11 illustrates a typical spectrofluorimeter. The radiant energy from the source is directed to the excitation monochromator which permits the selection of a particular excitation wavelength to irradiate the sample. The incident radiant energy which is not absorbed passes through the sample into a light trap. The fluorescence radiant energy is emitted in all directions, however only a small portion of this passes through the entrance

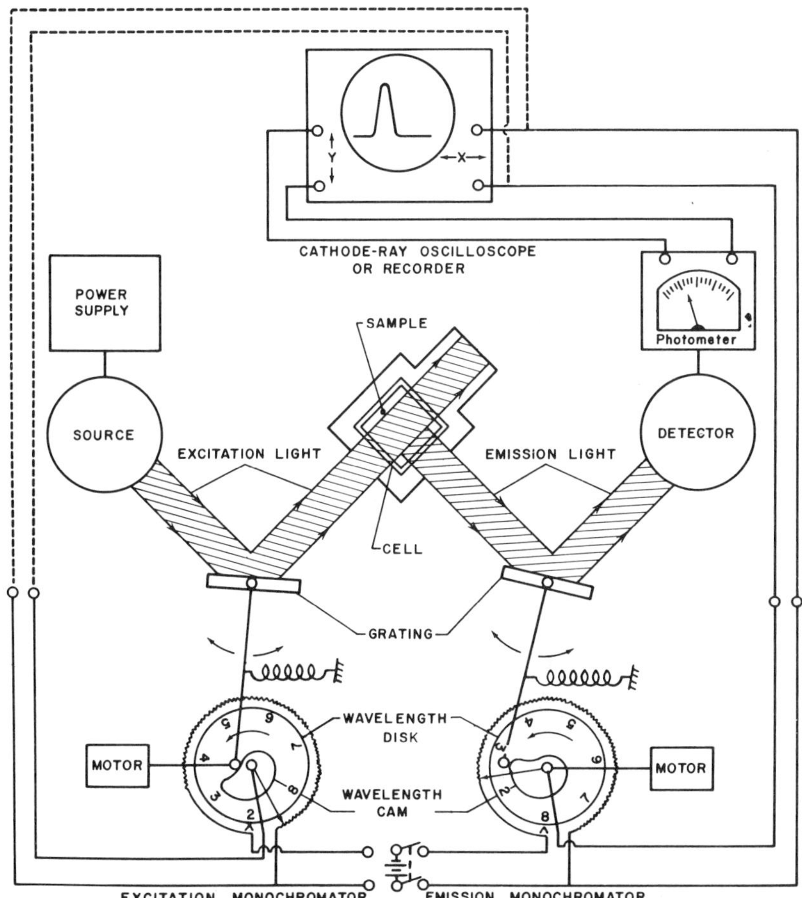

Fig. 11. Schematic diagram of Aminco Bowman spectrophotofluorimeter. (Courtesy of American Instrument Co.)

slit of the emission monochromator which is placed at right angles to the excitation beam. This second monochromator is used to select the wavelengths of fluorescence emission to reach the detector. This radiant energy produces an electrical signal which may be read from a panel meter, the *T* axis of a recorder, or the vertical axis of an oscilloscope.

The Baird-Atomic Fluorispec uses a dual grating system for both the excitation and emission monochromators in order to gain increased resolution and to minimize the effects of stray light. Generally speaking most of the commercially available spectrofluorimeters have similar wavelength ranges for both the excitation and emission monochromators covering the range from 200 to approximately 700 mμ. However, some of these instruments have the capability of extending their ranges into the near infrared region.

3. FILTER FLUORIMETERS

The design of filter fluorimeters generally is similar to that of the spectrofluorimeters, the principal exceptions being the substitution of filters in place of monochromators. A schematic diagram of a typical commercially available filter fluorimeter is shown in Figure 13. Radiant energy from the source passes through a range selector attenuator and the excitation filter which is designed so that only a relatively narrow band containing the

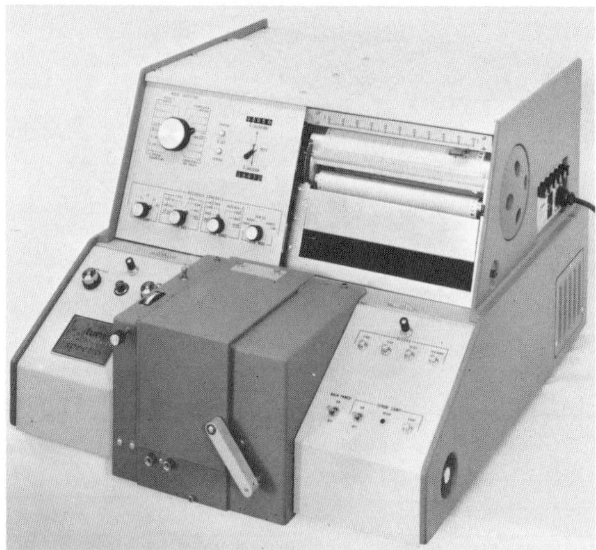

Fig. 12. The Turner Model 210 "Spectro." This instrument gives corrected excitation and emission spectra. (Courtesy of G. F. Turner Associates.)

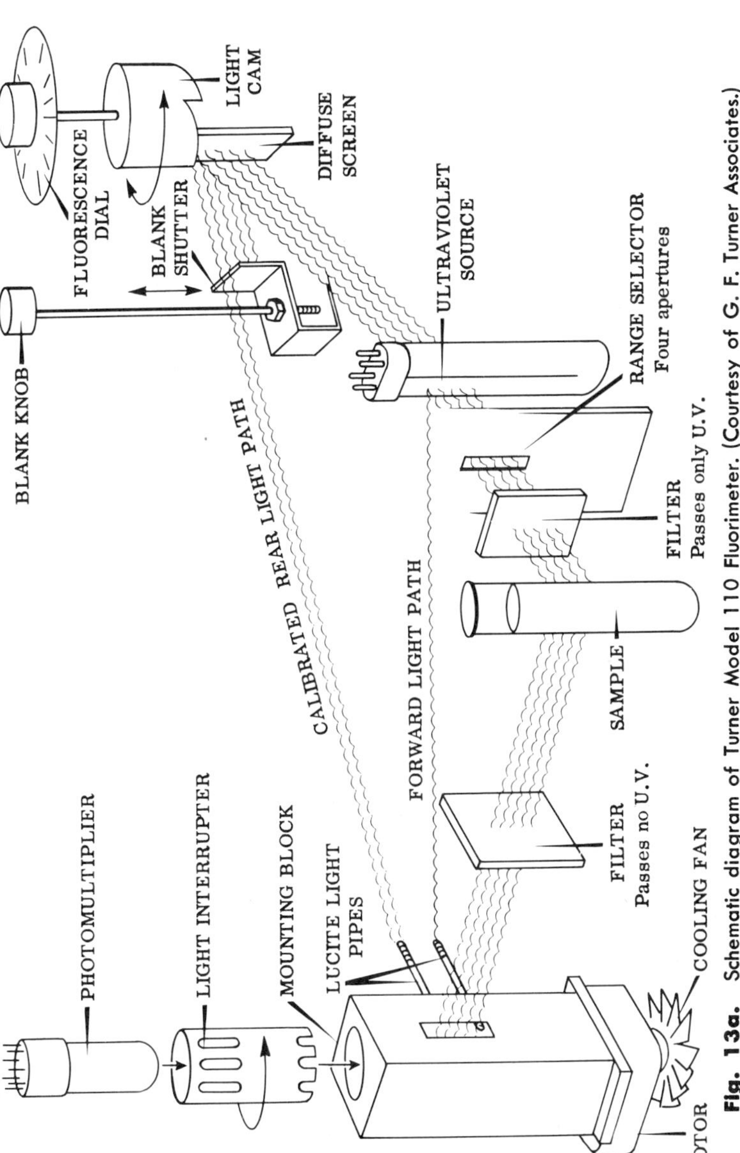

Fig. 13a. Schematic diagram of Turner Model 110 Fluorimeter. (Courtesy of G. F. Turner Associates.)

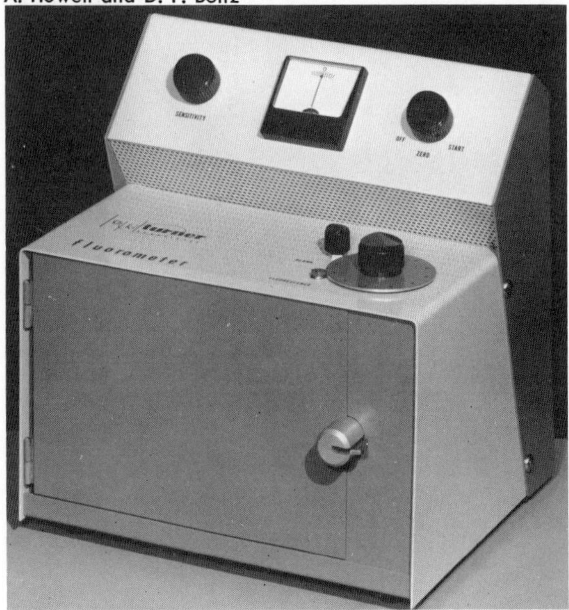

Fig. 13b. *(continued)*

desired excitation wavelength is transmitted. This is sometimes called the primary filter. The fluorescence emission induced when the sample is irradiated then passes through the secondary filter which is perpendicular to the excitation beam. The transmittance band of the emission filter should correspond to the emission wavelength band. The particular fluorimeter (Fig. 13) employs the optical null system. In this system a portion of the radiant energy from the source impinges on a diffuse mirror which is mechanically connected to a calibrated dial. The dial is rotated until an appropriate amount of reflected radiant energy falls upon a Lucite light pipe which transmits it to the photomultiplier. The photomultiplier receives alternating pulses of radiant energy arising from the sample beam and the reference beam due to the rotating light interrupter. This interrupter's rotation is synchronized to a relay-amplifier system which produces a differential signal. When the alternate pulses of radiant energy are not equivalent, an imbalance is observed on a null meter. The calibrated dial is then adjusted until a null is obtained. In addition to providing a means of measuring the fluorescence th is system also tends to compensate for any instability ofthe source. Other commercially available filter fluorimeters incorporate similar basic design features and are comparable in performance.

In general, it is usually possible to achieve a higher sensitivity in quantitative analysis with filter fluorimeters than with spectrofluorimeters. The main reason for this is that with low resolution the filter instruments are able to work with larger quantities of radiant energy. However, it is difficult to determine the excitation and emission wavelengths without having a large selection of filters and the expenditure of considerable time. Therefore, for routine fluorimetric methods for which the procedures and experimental parameters have been carefully developed, filter fluorimeters prove to be adequate and represent a relatively small financial investment.

4. PRACTICAL CONSIDERATIONS IN SELECTION OF INSTRUMENT

In choosing a specific instrument, one must determine exactly what is the nature and scope of their analytical problems and what restrictions may be imposed. Is the instrument to be used for routine determinations or is it to be used in the development of new fluorimetric methods? The accessible wavelength range and the capability of the instrument to accommodate pellets and plates as well as solutions and provision for temperature control should always be considered. If theoretical aspects of spectrofluorimetry are being studied, then energy compensated spectra and resolution characteristics are important in the selection of an instrument.

Consideration should also be given to the number and nature of accessories available for a specific instrument. Thus, temperature control devices, capillary cells, quartz dewar flasks for work at liquid nitrogen temperatures, polarizing accessories, and thin layer chromatography scanning accessories are often valuable attachments. If phosphorescence decay studies are anticipated, it is essential that some type of synchronous rotating slit system be available for the sample cell compartment as well as an oscilloscopic readout.

Lott[82] has evaluated fluorimeters and gave detailed descriptions of many of the commercially available instruments prior to 1964. Udenfriend[73] and Willard et al.[83] describe many of the commercial instruments available at the time of publication of their respective works. Tables 5 and 6 list a number of selected spectrofluorimeters and filter fluorimeters. Table 2 indicates those spectrophotometers having attachments for fluorescence measurements.

C. General Methodology

The techniques and methods employed in fluorescence and phosphorescence analysis frequently differ from one type of sample to another. The techniques to be used as well as the nature of the interferences in a vitamin assay of a particular biological sample would be significantly different from those involved in the analysis of zinc in a brass sample. Consequently, we shall limit our discussion of methodology to inorganic systems and particularly to metallurgical samples.

Table 5. Selected spectrofluorimeters[a]

Manufacturer and model	Design and operational features: Corrected spectra	Wavelength range ($m\mu$)	Additional features
American Instrument Co. (Aminco-Bowman SPF)	Available as accessory	200–800–1200 with accessory	Phosphorescence, transmittance, polarization, and solid sample accessories available
Baird Atomic Inc. (SF–1)	—	200–700	Dual grating system for both monochromators, solid samples
Farrand Optical Co., (Spectrofluorometer)	—	200–650	Transmittance accessory available
Perkin-Elmer Corp. (Model 236)	With standard model	200–800	Automatic slit control, phosphorescence, and transmittance
G. K. Turner Associates (Model 210 Spectro)	With standard model	200–650	Double beam optics, transmittance, and absorbance

[a] All of the models listed employ Xe arc sources, grating monochromators, and photomultiplier detectors.

1. PREPARATION OF SAMPLE

After obtaining a known amount of sample its solubilization into some appropriate solvent is one of the first important steps in the preparation of a sample for fluorescence analysis. This dissolution process is frequently accomplished by treatment with one or a combination of common laboratory acids. Consideration must be given to the possibility of ultimate quenching due to the presence of the anions from the acids. In selecting the dissolution process the final oxidation state of the various constituents in the sample must be considered. Occasionally certain samples require basic and/or acidic fusions before complete dissolution of the sample is obtained.

After dissolution, it is often necessary to perform a quantitative separation in order to remove any interfering substances. Colored ions which would absorb the fluorescent radiant energy are especially undesirable. A discussion of quantitative separations is not within the scope of this chapter. Other references should be consulted for detailed information on analytical separations.[84]

The next step in the general procedure is to develop the fluorescent system. In the determination of most inorganic substances, it is necessary to add a

Table 6. Selected filter fluorimeters

Manufacturer and model	Design and operational features: Source	Detector	Additional features
American Instrument Co. (Fluoro-microphotometer)	Choice of six selected sources	Photomultiplier	Direct reading meter or recorder. Sample changer and phosphorescence accessories available
Beckman Instruments, Inc. (Ratio Fluorimeter)	Low pressure diode type lamp	Photomultiplier	Double beam, direct reading meter or recording of ratio of sample to reference beams
Coleman Instruments, Inc. (Model 120)	Mercury vapor lamp	Blue sensitive phototube	Linear amplifier with direct reading meter
Farrand Optical Co.	Mercury vapor lamp	Photomultiplier	Decay studies accessory available
The Photovolt Corp. (Lumetron Fluorescence meter Model 402-EF)	Mercury vapor lamp	Barrier layer cells	Galvanometer null balance, source instability compensation
G. K. Turner Associates (Model 110)	—	Photomultiplier	Solids, null balance, excitation below 300 mμ. (Recording with Model 111)

reagent to produce luminescence. The reagents commonly used may be classified as: prior oxidants, prior reductants, pH buffers, chelating agents, or in the case of indirect fluorimetric methods, fluorescent reagents whose fluorescence will be significantly altered by the presence of the analyte. Chelating agents, e.g., 8-quinolinol, forms fluorescent systems with many metals.

The selection of a suitable solvent is important. For example, many metal chelates exhibit enhanced fluorescence in nonpolar solvents. Furthermore, the fluorescent species should be sufficiently soluble so that the fluorescence is measurable. Another important consideration relative to the choice of solvent and also any reagents used in the sample preparation procedure is the purity of the reagents and solvent. The extreme sensitivity of fluorescence and phosphorescence measurements demands that extreme caution be exercised to prevent the introduction of interferring trace impurities into the sample prior to its measurement. Consequently, it is a good policy to perform a blank determination, as well as several known standards, prior to preparing the sample for analysis. In this way troublesome contaminants in the reagents or solvent can be detected and appropriate remedial steps may be taken.

The adjustment of pH and dilution of the sample often play an extremely important role in the ultimate sensitivity, precision and accuracy of a fluorimetric determination. The stability of the fluorescent system with respect to time should be known. Those systems which exhibit a significant degree of time dependency should be measured at a definite time following preparation of the solution if reliable results are to be expected.

2. FACTORS AFFECTING FLUORESCENCE

Factors affecting fluorescence and phosphorescence are either chemical or instrumental in origin. Inasmuch as instrumental factors have already been discussed, chemical factors will be considered next. Bathochromic and hypsochromic shifts occur when the absorbing fluorescent species undergoes some change in the bonding of the chromophoric portion of the molecule. These shifts may be due to dissociation of the molecule as the result of temperature, photodecomposition, or pH effects. Shifts could also result from a higher degree of association due to solvent interaction, or by coordination with some species in the system.

Another factor affecting fluorescence is quenching. Perhaps one of the most commonly encountered forms of quenching is that arising from high concentration of the analyte or another absorbing specie. This phenomenon is referred to as concentration quenching. In dilute solutions concentration quenching is rather insignificant, however, with increasing concentration its effect becomes quite pronounced. This quenching effect is illustrated in Figure 14. Concentration quenching may be easily overcome by reducing the concentrations.

Quenching due to trace quantities of various substances is commonly encountered. The exact mechanisms causing quenching of this type assume a variety of interactions, many of which are not clearly understood. Trace quantities of dissolved oxygen in various organic solvents have been observed to inhibit fluorescence. However, it should be pointed out that occasionally trace quantities of impurities serve to enhance rather than quench fluorescence.

Fluorescence and particularly phosphorescence are often influenced by temperature changes. Some systems exhibit extreme sensitivity to temperature while others appear to be rather insensitive. Elevated temperatures tend to increase the kinetic energy of the molecules in the sample thus increasing considerably the numbers of collisions. These collisions permit alternate modes of deexcitation to occur which may result in a quenching effect. Increased temperatures may give bathochromic shifts in the excitation spectra.

The pH in aqueous media often exhibit pronounced effects upon fluorescence and phosphorescence. A large number of luminescent materials are themselves acids and bases. In some instances their dissociated forms are more susceptible to excitation than are their associated forms and often the converse is true. Consequently, the relative effect of pH on the enhancement and quenching of fluorescence is dependent upon the molecular structure of the fluorescent species.

Solvent effects may be attributed to the properties of the solvent such as acidity, dielectric constant, viscosity, absorptivity, solubility, coordination ability, etc. Solvent–solute interactions tend to be quite complex and much remains to be learned of their exact nature. Solvents for phosphorimetry frequently impose one additional requirement which is that the solvent

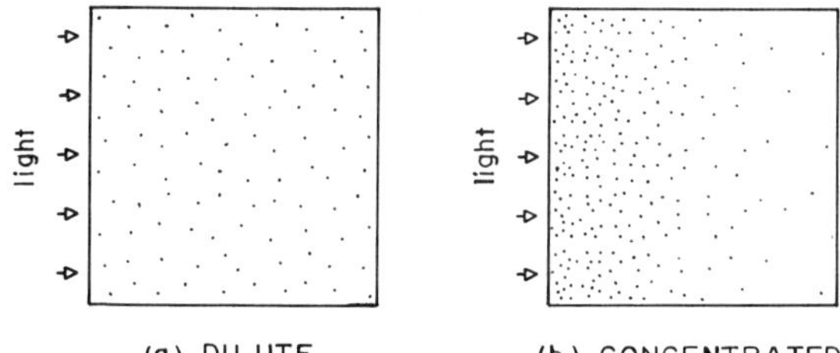

Fig. 14. Effect of concentration quenching. *a*. Dilute solution—uniform fluorescence. *b*. Concentrated solution—nonuniform fluorescence.

should be capable of forming a rigid clear glass at lower temperatures. Phosphorescent species suspended in such a media are relatively insulated from collisions thereby increasing the mean lifetime and quantum efficiency of the system. Table 7 lists a number of solvent systems commonly used in phosphorimetry. A more extensive table citing original references has been compiled by Keirs.[85]

Table 7. Selected solvents for phosphorimetry

Solvent	Temperature (°K)	Refs.
Acetone	77, 90	86
Acetic acid	77, 90	86
Boric acid	293, 233	86
Carbon tetrachloride	77, 90	86
Diethyl ether + isopentane + ethanol[a]	77, 90	87
Octane	77, 93	88
Phosphoric acid	193	89
Pyridine	77, 90	86
Sugar	294, 305	86
Sulfuric acid	77, 90	86
Triethanolamine	193	90
Water	77, 90	86

[a] A commonly used solvent, referred to as EPA, has a volume ratio of 5:5:2 of ether, isopentane, and ethanol, respectively.

3. MEASUREMENT OF THE SYSTEM

Reliable fluorescence measurements will be obtained only when the performance of the instrument to be used meets certain minimum specifications. These specifications and instructions for their evaluation are supplied by the manufacturer of the instrument. It is advisable to conduct periodic calibration tests in order to ascertain the proper intensity and stability of the source, optical focus, wavelength calibration, detector response, and signal to noise ratio of the readout system. A 1 μg/ml solution of quinine sulfate is one suitable standard solution commonly used to evaluate instrument performance.

After the sample solution has been prepared for measurement, the excitation and emission wavelength settings should be made very carefully. Also, a judicious choice of slit settings should be made in order to obtain the desired degree of resolution and sensitivity when certain instruments are being used. The sample cell must be free of any foreign contamination.

The cells should be cleaned and calibrated routinely. When calibration graphs are to be used to equate instrumental readings with concentrations it is necessary that all of the instrument settings and operational procedures used in obtaining the calibration data be used for the analysis of the sample.

4. SOURCES OF ERROR

Many of the sources of error have been discussed previously but perhaps can be briefly summarized. Care should always be exercised in reproducing both the excitation and emission wavelength settings from one measurement to another. Large changes in sensitivity may be observed if either or both of these settings are not adequately reproduced. Trace impurities in the solvent or reagents should always be maintained at the lowest possible level. Many solutions should be sufficiently buffered prior to measurement so that pH effects are kept to a minimum. The concentration range should be kept sufficiently low to avoid self-quenching. Slit settings, detectors, gain settings of the readout system and other adjustments for source and detectors should always be made in a reproducible manner.

5. PRESENTATION AND EVALUATION OF DATA

In any fluorimetric or phosphorimetric analysis it is imperative that notation of the source used, complete slit settings, wavelength settings, type of detectors, readout sensitivity, and gain settings be made in addition to the relative fluorimetric values. Data which are to be published should be corrected for source and detector characteristics or be specifically designated as uncorrected data. The presentation of fluorescent intensity values should reflect a correction for blank readings.

The most commonly used procedure to convert fluorimeter readings to the concentration of the analyte is performed by means of previously prepared calibration plots similar to that shown in Figure 15. As previously indicated, reliable results will be obtained only when the samples have been treated in an identical manner as were the knowns used in the preparation of the calibration graph.

Indirect fluorimetric methods of analysis also commonly employ calibration plots. However, when the desired substance produces a quenching effect, the calibration plot has a negative slope. The exact value of the slope will be determined by the mechanism by which the quenching takes place. Indirect methods utilizing fluorescence enhancement will producc calibration plots similar to those obtained in direct methods.

D. Applications

Although fluorimetric methods have been applied very extensively to the determination of organic substances in a variety of biological and com-

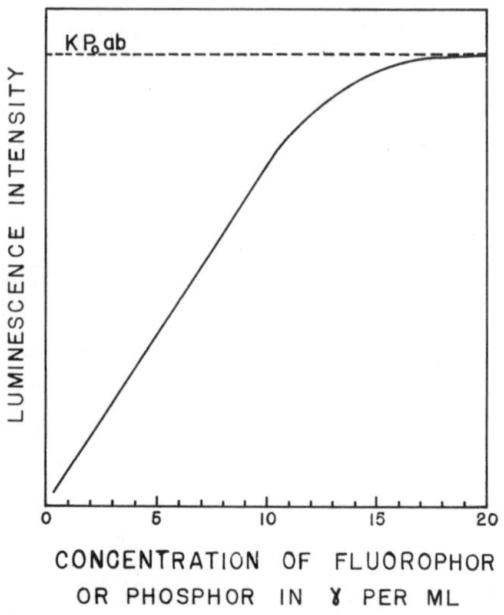

Fig. 15. Typical calibration plot.

mercial materials, this section deals primarily with those analytical methods involving the determination of inorganic substances. In general, a fluorimetric method would be selected only for the determination of trace quantities of an element so that the extreme sensitivity of this method can be utilized. Microanalytical methods involving spectrophotometry, atomic absorption spectrometry, polarography, etc. should be evaluated as alternate, competitive methods. The possible effect of other substances present in the sample, specificity, and analytical accuracy must also be considered.

1. METALS

Over 30 metals including several lanthanide elements can be determined fluorimetrically. Many of these methods are based on the formation of fluorescent metal chelates. For example, Diehl has described a sensitive method for the determination of magnesium in the presence of calcium by using *o,o′*-dihydroxyazobenzene as the chelating agent.[91,92] 3-Hydroxy-2-naphthoic acid gives an intense green fluorescence with aluminum and an intense blue fluorescence with beryllium.[93,94] Indirect fluorimetric methods are exemplified by the determination of cobalt by the quenching of the fluorescence due to the aluminum chelate of Super Chrome Blue-Black Extra,[95] and the determination of iron(III) by the quenching of the fluorescence of 4,4-diamino-2,2-disulfostilbene-*N*,*N*,*N′*,*N′*,-tetracetic acid.[96]

2. NONMETALS

Several nonmetals can be determined fluorimetrically. Perhaps the fluorimetric method for boron has been studied the most extensively with a number of fluorescent reagents having been proposed. Microgram quantities of boron in steel can be determined by measuring the fluorescence[97] of the borate–benzoin ester formed in a slightly basic 85% ethanolic solution. A methyl borate distillation is used to separate the boron from interfering substances. Cyanide in the 0.5–50 μg/ml range can be determined fluorimetrically.[98] Tellurium in arsenic has been determined by Butylrhodamine B[99] and selenium in semiconductors has been determined with 3,3′-diaminobenzidine.[100] Fluoride, oxygen, and sulfur are determined by indirect fluorimetric methods in which the quenching of fluorescence is proportional to concentration of the analyte.

3. SELECTED METHODS

Despite numerous publications on fluorimetric and phosphorimetric methods the actual number dealing with metallurgical materials is rather limited. In Table 8 a representative sampling of fluorimetric methods which have been used in the analysis of specific materials has been given. The

Table 8. Selected fluorimetric methods

Element	Reagent	Material	Refs.
Al	Pontachrome BBR	Steels and bronzes	101
B	Benzoin	Steels	102
Be	—	Bronzes	103
Cd	2(*o*-Hydroxyphenyl-benzoxazole	—	104
CN^-	Pd complex of 5-sulfo-8-quinolinol with Mg	—	105
Cu	Luminol	Zinc	106
F	Al complex of alizarin garnet R	—	107
Fe	4,4-Diamino-2,2-disulfostilbene-*N*,*N*,*N*′,*N*′-tetracetic acid	—	96
Mg	*o*,*o*′-Dihydroxyazobenzene	Minerals and serum	92
Pb	HCl		108
S	Fe(III) + *p*-phenylene diamine	Semiconductors	71
Sn	7-Amino-9-nitro-naphthalene–sulfonic acid	Copper-based alloys	109
Th	Diethylenetriaminepentacetic acid	Minerals and biological	110
U	NaF-flux	Ferrous alloys	111
Zn	Picolinealdehyde-2-quinolyl-hydrazone	Metals	112
Zr	Morin	Silicates	113

biennial review on fluorimetric analysis which appears in *Analytical Chemistry* should be consulted for more recent developments and practical applications.[114]

References

1. *Anal. Chem.*, **37,** 1814 (1965).
2. L. E. Orgel, *Quart. Rev.* (*London*), **8,** 422 (1954).
3. J. R. Dyer, *Applications of Absorption Spectroscopy of Organic Compounds*, Prentice-Hall, Englewood-Cliffs, N. J., 1965, Chap. 2, pp. 5–11.
4. R. P. Bauman, *Absorption Spectroscopy*, Wiley, New York, 1963, Chap, VI, pp. 249–282.
5. L. Meites, *Anal. Chim. Acta*, **27,** 131 (1962).
6. A. Ringbom, *Complexation in Analytical Chemistry*, Interscience, New York, 1963.
7. L. S. R. Goldringer, R. C. Hawes, G. H. Hare, A. O. Beckman, and M. E. Stickney, *Anal. Chem.*, **25,** 869 (1953).
8. W. Slavin, *Anal. Chem.*, **35,** 561 (1963).
9. M. G. Mellon, Ed., *Analytical Absorption Spectroscopy*, Wiley, New York, 1950, p. 244.
10. G. Srehla, *Talanta*, **13,** 641 (1966).
11. C. F. Hiskey et al., *Anal. Chem.*, **21,** 1440 (1949); **22,** 1464 (1950).
12. C. N. Reilley and C. M. Crawford, *Anal. Chem.*, **27,** 716 (1955).
13. R. Bastian, R. Weberling, and F. Palilla, *Anal. Chem.*, **21,** 972 (1949); **22,** 160 (1950).
14. J. A. Howell and D. F. Boltz, *Anal. Chem.*, **36,** 1799 (1966).
15. D. F. Boltz, *Selected Topics in Modern Instrumental Analysis*, Prentice-Hall, Englewood Cliffs, N. J., 1952, p. 149.
16. R. F. Goddu and D. N. Hume, *Anal. Chem.*, **22,** 1314 (1950).
17. T. Kanie, *Japan Analyst*, **6,** 711 (1957).
18. J. B. Headridge, *Photometric Titrations*, Pergamon, New York, 1961.
19. D. F. Boltz, in *Handbook of Analytical Chemistry*, L. Meites, Ed., McGraw-Hill, New York, 1963, Sec. 6, pp. 21-39.
20. C. B. Sandell, *Colorimetric Metal Analysis*, 3rd ed., Interscience, New York, 1959.
21. F. D. Snell and C. T. Snell, *Colorimetric Methods of Analysis*, Vol. 2, 3rd ed., D. Van. Nostrand, Princeton, N. J., 1949; Vol. 2A, 1959.
22. D. F. Boltz, Ed., *Colorimetric Determination of Nonmetals*, Interscience, New York, 1950.
23. R. M. Dagnall, T. S. West, and P. Young, *Analyst*, **90,** 13 (1965).
24. U. T. Hill, *Anal. Chem.*, **38,** 654 (1966).
25. E. F. Pellowe and F. R. F. Hardy, *Analyst*, **79,** 225 (1954).
26. C. L. Luke and M. E. Campbell, *Anal. Chem.*, **25,** 1588 (1953).
27. J. Bassett and J. C. H. Jones, *Analyst*, **91,** 176 (1966).
28. O. P. Case, *Anal. Chem.*, **20,** 902 (1948).
29. I. R. Scholes and W. R. Waterman, *Analyst*, **88,** 374 (1963).
30. H. A. Scapp and C. P. Evans, *Anal. Chem.*, **28,** 143 (1950).
31. U. T. Hill, *Anal. Chem.*, **30,** 521 (1958).
32. A. H. Jones, *Anal. Chem.*, **29,** 1101 (1957).
33. W. Nielsch and G. Boltz, *Metall*, **10,** 916 (1956).
34. H. Kitagawa and Y. Aimoto, *Japan Analyst*, **21,** 144 (1955).
35. B. E. McClellan and V. M. Benson, *Anal. Chem.*, **36,** 1985 (1964).

36. A. Claassen and A. Baamen, *Anal. Chim. Acta*, **12,** 547 (1955).
37. R. K. Rohde, *Anal. Chem.*, **38,** 911 (1966).
38. E. Kovacs, A. Guyer, and W. Luescher, *Z. Anal. Chem.*, **209,** 338 (1965).
39. A. R. Gahler, *Anal. Chem.*, **26,** 577 (1954).
40. W. T. Elwell, *Analyst*, **80,** 509 (1955).
41. S. Kertes, *Anal. Chim. Acta*, **15,** 73 (1956).
42. G. Norwitz, J. Cohen, and M. E. Everett, *Anal. Chem.*, **36,** 142 (1964).
43. J. A. Stobart, *Analyst*, **90,** 278 (1965).
44. C. L. Luke, *Anal. Chem.*, **28,** 1443 (1956).
45. M. D. Cooper, *Anal. Chem.*, **25,** 411 (1953).
46. E. M. Donaldson and W. R. Inman, *Talanta*, **13,** 489 (1966).
47. N. Lounamaa, *Anal. Chim. Acta*, **33,** 21 (1965).
48. G. Norwitz, J. Cohen, and M. E. Everett, *Anal. Chem.*, **37,** 417 (1965).
49. S. B. Savvin, V. A. Bortsova, and E. N. Malkina, *Z. Anal. Khim.*, **20,** 947 (1965).
50. H. C. Baghurst and V. J. Norman, *Anal. Chem.*, **29,** 778 (1957).
51. G. A. Bauer, *Anal. Chem.*, **37,** 154 (1965).
52. U. Bahnstedt and R. Budenz, *Z. Anal. Chem.*, **159,** 12 (1957).
53. H. L. Katz and K. L. Proctor, *Anal. Chem.*, **19,** 612 (1947).
54. N. Loynamaa and W. Fugmann, *Z. Anal. Chem.*, **199,** 352 (1964).
55. W. T. Elwell and H. N. Wilson, *Analyst*, **81,** 136 (1956).
56. J. M. Sturton, *Anal. Chim. Acta*, **32,** 394 (1965).
57. R. T. Arnesen and A. R. Selmer-Olson, *Anal. Chim. Acta*, **33,** 335 (1965).
58. B. T. Kenna and F. J. Conrad, *Anal. Chem.*, **35,** 1255 (1963).
59. W. T. Pickering, *Anal. Chim. Acta*, **12,** 572 (1955).
60. A. Weissler, *Ind. Eng. Chem., Anal. Ed.*, **17,** 775 (1945).
61. C. L. Luke, *Anal. Chem.*, **36,** 1327 (1964).
62. A. R. Eberle, *Anal. Chem.*, **35,** 669 (1963).
63. D. G. Biecher, D. E. Jordon, and W. D. Leslie, *Anal. Chem.*, **35,** 1685 (1963).
64. M. Q. Freeland and J. S. Fritz, *Anal. Chem.*, **27,** 1737 (1955).
65. W. L. Ott, H. R. MacMillan, and W. R. Hatch, *Anal. Chem.*, **36,** 363 (1964).
66. D. F. Wood and J. T. Jones, *Analyst*, **90,** 125 (1965).
67. R. B. Hahn and J. L. Johnson, *Anal. Chem.*, **29,** 902 (1957).
68. M. G. Mellon and D. F. Boltz, *Anal. Chem.*, **32,** 194 (1960); **34,** 232R (1962); **36,** 256R (1964); **38,** 317R (1966); **40,** 255R (1968).
69. D. M. Hercules, Ed., *Fluorescence and Phosphorescence Analysis*, Interscience, New York, 1966. See also D. M. Hercules, *Anal. Chem.*, **38,**(12), 29A (1966).
70. A. L. Conrad, in *Treatise on Analytical Chemistry*, Part I, Vol. V, I. M. Kolthoff and P. J. Elving, Eds., Interscience, New York, 1964, pp. 3057–3078.
71. C. A. Parker and W. T. Rees, in *Fluorometry and Spectrophotometry*, J. P. Cali, Ed., Pergamon, London, 1964, pp. 228–246.
72. C. A. Parker and W. T. Rees, *Analyst*, **87,** 83 (1962).
73. S. Udenfriend, *Fluorescence Assay in Biology and Medicine*, 3rd ed., Academic, New York, 1964.
74. C. E. White and A. Weissler in *Handbook of Analytical Chemistry*, L. Meites, Ed., McGraw-Hill, New York, 1963, Sec. 6, pp. 176–196.
75. C. E. White and A. Weissler, in *Standard Methods of Chemical Analysis*, Vol. IIIA: *Instrumental Analysis*, 6th ed., D. Van Nostrand, Princeton, N. J., 1966, Chap. 5, pp. 78–104.
76. S. Z. Lewin, *J. Chem. Educ.*, **42,** A165 (1965).

77. Bausch and Lomb, Inc., Rochester, N. Y., *Bausch and Lomb Multi-Films*, 1966.
78. Corning Glass Works, Corning, N. Y., *Glass Color Filters*, *CF*-3, 1965.
79. R. J. Argauer and C. E. White, *Anal. Chem.*, **36,** 368 (1964).
80. C. A. Parker and W. T. Rees, *Analyst*, **85,** 587 (1960).
81. C. E. White, M. Ho, and E. R. Weimer, *Anal. Chem.*, **32,** 438 (1960).
82. P. F. Lott, *J. Chem. Educ.*, **41,** A327, A421 (1964).
83. H. H. Willard, L. L. Merritt, and J. A. Dean, *Instrumental Methods of Analysis*, 4th ed., D. Van Nostrand, Princeton, N. J., 1965, Chap. 13, pp. 370–392.
84. E. W. Berg, *Physical and Chemical Methods of Separation*, McGraw-Hill, New York, 1963.
85. R. J. Keirs, *Aminco Instruction Manual No. 768F*, American Instrument Co., Inc., Silver Spring, Md., 1964, pp. 65–67.
86. V. V. Zelinskii, N. P. Emets, V. P. Kolobkov, and L. G. Pikulik, *Izvest. Akad. Nauk SSR, Ser. Fiz.*, **20,** 520 (1956).
87. G. N. Lewis and M. Kasha, *J. Am. Chem. Soc.*, **66,** 2100 (1944).
88. E. V. Shpol'ski and L. A. Klimiova, *Izvest. Akad. Nauk SSR, Ser. Fiz.*, **20,** 471 (1956).
89. G. N. Lewis, D. Lipkin, and T. T. Magel, *J. Am. Chem. Soc.*, **63,** 3005 (1941).
90. G. N. Lewis and D. Lipkin, *J. Am. Chem. Soc.*, **64,** 2801 (1942).
91. H. Diehl, *Calcein, Calmagite, and O,O'-Dihydroxyazobenzene. Titrimetric, Colorimetric, and Fluorometric Reagents for Calcium and Magnesium*, The G. Frederick Smith Chemical Co., Columbus, Ohio, 1964, Part VII, pp. 102–104.
92. H. Diehl, R. Olsen, G. I. Spielholtz, and R. Jensen, *Anal. Chem.*, **35,** 1144 (1963).
93. A. I. Cherkesov and T. S. Zhigalkina, *Tr. Astrakhansk. Tekh. Inst. Rybn. Prom. i Khoz*, **1962,** 25; *Zh. Khim.*, **19GDE** (1963); Abst. No. **20GO**.
94. G. F. Kirkbright, T. S. West, and C. Woodward, *Anal. Chem.*, **37,** 137 (1965).
95. S. B. Zamochnick and G. A. Rechnitz, *Z. Anal. Chem.*, **199,** 424 (1963). *Dokl. Akad. Nauk SSSR*, **153,** 97 (1963).
96. E. A. Bozhevol'nov, S. U. Kreingol'd, R. P. Lastovskii, and V. V. Sidorenko, *Dokl. Akad. Nauk SSSR*, 153, 97 (1963).
97. C. E. White, A. Weissler, and D. Busker, *Anal. Chem.*, **19,** 802 (1947).
98. G. G. Guilbault and D. N. Kramer, *Anal. Chem.*, **37,** 918 (1965).
99. V. M. Vladimirova, N. K. Davidovich, G. I. Kuchmistaya, and L. S. Razumova, *Zavodsk. Lab*, **29,** 141 (1963).
100. V. M. Vladimirova and G. I. Kuchmistaya, *Zavosdk. Lab.*, **30,** 528 (1964).
101. A. Weissler and C. E. White, *Anal. Chem.* **18,** 530 (1946).
102. E. Elliott and J. A. Radley, *Analyst*, **86,** 62 (1961).
103. A. I. Cherkesov and T. S. Zhigalkina, *Zavodsk. Lab*, **27,** 658 (1961).
104 N. Evcim and L. Reber, *Anal. Chem.*, **26,** 936 (1954).
105. J. S. Hanker, A. Gelberg, and B. Wittern, *Anal. Chem.*, **30,** 93 (1958).
106. A. K. Babko and L. I. Dubovenko, *Zadovsk. Lab*, **30,** 1325 (1964).
107. W. A. Powell and J. H. Saylor, *Anal. Chem.*, **25,** 960 (1953).
108. E. A. Bozhevol'nov and E. A. Solovev, *Zavodsk, Lab*, **30,** 412 (1964).
109. J. R. A. Anderson and S. L. Lowy, *Anal. Chim. Acta*, **15,** 246 (1956).
110. C. W. Sill and C. P. Willis, *Anal. Chem.*, **36,** 622 (1964).
111. J. Korkish and I. Hazan, *Anal. Chem.*, **36,** 2464 (1964).
112. R. E. Jenson and R. T. Pflaum, *Anal. Chem.*, **38,** 1268 (1966).
113. R. A. Geiger and E. B. Sandell, *Anal. Chim. Acta*, **16,** 346 (1957).
114. C. E. White and A. Weissler, *Anal. Chem.*, **32,** 47 (1960); **34,** 81R (1962); **36,** 116R (1964); **38,** 155R (1966); **40,** 116R (1968).

Chapter 8

X-RAY SPECTROSCOPIC METHODS

E. A. HAKKILA, University of California, Los Alamos Scientific Laboratory, Los Alamos, New Mexico

I. Introduction 276

II. Theory 277
  A. Origin of X-Rays 277
    1. Continuous Spectrum 277
    2. Characteristic Spectrum 278
    3. Fluorescence Yield 280
  B. Diffraction of X-Rays 280
  C. Scattering of X-Rays 281
  D. Absorption of X-Rays 281
  E. Intensity of X-Ray Lines 284

III. Instrumentation for X-Ray Spectroscopy 286
  A. X-Ray Sources 286
    1. Electron Excitation 286
    2. X-Ray Excitation 286
    3. High Energy Particles 287
  B. Diffraction Crystals 288
  C. Spectrometer Construction 290
  D. X-Ray Detectors 292
    1. Film 292
    2. Geiger Counter 292
    3. Proportional Counter 292
    4. Scintillation Counter 293
    5. Solid State Detectors 293
  E. Pulse Height Analysis 294

IV. Applications of X-Ray Emission and Fluorescence 295
  A. Selection of a Line 295
  B. Sample Preparation 296
    1. Solids 296
    2. Powders 296
    3. Solutions 298

C. Quantitative Analysis . . . 298
1. Statistics of X-Ray Measurements . . . 298
2. X-Ray Intensity Techniques . . . 299
3. Use of Internal Standards . . . 300
4. Arithmetic Correction Methods . . . 301
D. Trace Analysis . . . 304

V. Applications of Absorption Spectroscopy . . . 306
A. Absorptiometry with Polychromatic X-Rays . . . 306
B. Absorptiometry with Monochromatic X-Rays . . . 308
C. X-Ray Absorption Edge Analysis . . . 309
D. Analysis by Absorption Edge Fine Structure . . . 314

VI. Analysis of Thin Films . . . 316
A. Absorption Analysis of Thin Films . . . 316
B. Fluorescence Analysis of Thin Films . . . 319
References . . . 320
Supplementary References . . . 323

I. INTRODUCTION

The potential of X-ray spectrochemical analysis has been known almost since the discovery of X-rays. The earlier work was largely of a qualitative nature as fas as analytical applications were concerned, but the understanding of the periodic nature of X-rays culminated in the discovery of hafnium by Coster and von Hevesey in 1923.[1] Both the emission and absorption of X-rays by the elements were well understood by the 1930's, and the theoretical groundwork for quantitative analysis was formed. Probably the biggest deterrent to X-ray spectroscopy taking its place alongside emission spectroscopy and electroanalytical techniques as a routine instrumental method was the X-ray detection system. Photographic film was the only detection system available and the long exposure times compared to emission spectroscopy probably discouraged many chemists from venturing into the field.

The advent of the Geiger-Mueller counter in the 1940's opened new vistas to X-ray spectroscopy, and the development since the first commercial instruments of the 1940's to the sophisticated automated production models of today has indeed been rapid. This chapter will attempt to familiarize the reader with some of the basic principles of X-ray spectrochemical analysis.

II. THEORY

A. Origin of X-Rays

X-rays are produced by the interaction of high energy photons or particles with matter, and consequently generation of high energy particles may lead to X-ray production. This radiation may be produced by bombardment of the sample with electrons, protons, or gamma rays (emission spectroscopy), or by bombardment of the sample with secondary X-rays from an X-ray tube (fluorescence spectroscopy). Or the sample may be interposed between the X-ray source and the detector and the attenuation of the X-ray beam by the sample may be measured (absorption spectroscopy). These definitions will be used throughout this chapter.

X-rays are a continuation of the electromagnetic spectrum, but have much greater energy, and therefore shorter wavelengths, than visible light. Whereas the visible portion of the spectrum occupies the wavelength region from approximately 4000–8000 Å, the wavelength range for X-rays covers the region from approximately 0.1–200 Å.

1. CONTINUOUS SPECTRUM

X-rays are generally classified into two types, continuous and characteristic. The continuous portion results from the interaction of the bombarding stream of electrons or other high energy particles with the field surrounding the nucleus. It is the primary source of energy for producing secondary or fluorescent X-rays in X-ray fluorescence analysis, but is a detriment to good sensitivity for emission analysis. The maximum energy or shortest wavelength, λ_0, that can be produced by an X-ray tube is limited by the voltage applied to the tube, and can be calculated by the equation

$$\lambda_0 = hc/Ve \tag{1}$$

In this equation h is Planck's constant, c is the velocity of light, e is the charge on the electron, and V is the maximum potential applied to the tube. For λ_0 expressed in angstrom units and V in volts, this may be written $\lambda_0 =$ 12,400 V.

The intensity of the continuous spectrum increases smoothly but not linearly as the wavelength is decreased, reaches a maximum value at a wavelength of about 1.5 λ_0, and drops sharply to zero at λ_0. The intensity I, of the continuous spectrum is directly proportional to the atomic number Z of the target material and to the square of the voltage, and may be expressed as

$$I = KZV^2 \tag{2}$$

where K is a constant. Thus, the efficiency of converting electron energy to X-ray energy decreases as the atomic number of the target element decreases. This efficiency is low, of the order of 2%, even for targets composed of the heavier elements such as tungsten and platinum, with the remainder of the applied power being transformed to heat. Because of this large heat generation, metals having high melting points generally are used as targets, and continuous water cooling is required.

2. CHARACTERISTIC SPECTRUM

Characteristic or line X-ray spectra are produced by interaction of high energy particles or rays with electrons of the target atom. The nucleus of an atom is surrounded by a series of electron shells in which each electron has a discrete energy. Electrons nearest the nucleus are most tightly bound and require the greatest energy to be removed from the atom. The X-ray energy levels are designated as shown in Figure 1, with the K shell having one energy level, the L shell three levels, the M shell five levels, etc. An incident particle or photon may eject an electron from the atom only if the energy of

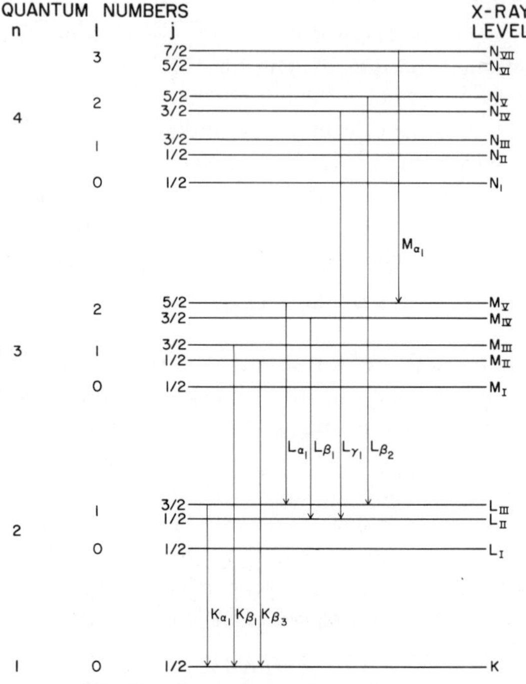

Fig. 1. X-ray energy level diagram.

the incident projectile is greater than the binding energy of the electron. If an electron is so ejected, the atom is left in an unstable state, and electrons of higher potential will attempt to fill the vacant layer. Certain electron transitions have a much greater probability of occurring than others, depending on quantum selection rules. The basic quantum selection rules for electron dipole transitions are, referring to Figure 1, that n should change, l should change by ± 1, and j should change by ± 1 or 0. Transitions to the K shell are most probable from the L_{III} level, giving rise to the K_{α_1} X-ray; transitions from the L_{I} to the K level have an extremely low probability and this X-ray is not observed in routine X-ray analysis.

With commercial X-ray spectographs the K_{α_1} and K_{α_2} X-rays cannot be resolved in the first order, and only one line is observed, designated the K_{α}. Likewise, the K_{β_1} and K_{β_3} appear as a single line, and the transitions $N_{\text{III}} \rightarrow K$ and $N_{\text{II}} \rightarrow K$ are designated the K_{β_2}. Thus, only three lines are observed in the first order K series in X-ray fluorescence spectroscopy. The L series is somewhat more complicated, but generally less than a dozen lines are observed.

In general, X-rays observed in the wavelength region below 2.5 Å arise from transitions of inner shell electrons so that wavelengths and intensities are unaffected by chemical combination. Thus, in contrast to emission spectroscopy, standards do not have to be in the same chemical state as unknowns, simplifying analytical techniques. X-ray spectra for light elements are even more simplified because fewer shells are filled in the atom. For example, the third period elements do not have electrons in the N level, and the K_{β_2} line is not observed. For these elements the $K_{\beta_{1,3}}$ X-ray arises from transitions involving valence electrons, and becomes quite sensitive to the chemical state of the atom. By the same reasoning the $K_{\beta_{1,3}}$ X-rays are not observed for period 2 elements (there are no M-shell electrons) and the K_{α} X-ray is sensitive to chemical combination effects. This same reasoning can be applied to determine which L and M series X-rays will be absent and which should be particularly sensitive to chemical effects.

The above considerations apply to X-ray fluorescence spectroscopy, in which case the sample is excited by an X-ray source. For electron excitation a somewhat more complicated spectrum is observed. The incident electron beam may cause multiple ionizations of the atom, leading to "sattellite" lines. The intensity of satellites increases inversely as the third power of the atomic number[2] and may be ignored for the heavier elements. However, for elements such as aluminum and magnesium they are more intense than the K_{β} band, and must be considered.

3. FLUORESCENCE YIELD

The probability that an X-ray will be produced when an electron is ejected from, for example, the K shell is referred to as the fluorescence yield for the K shell. At first thought one might expect that for ejection of a K shell electron the sum of the K series X-rays should account for all of the energy of the ejected electrons. However, some of the energy may be dissipated in multiple ionizations or Auger electrons. Measurements indicate that the fluorescence yield decreases from a value of approximately 0.8 for the heavier elements to approximately 0.3 for iron, cobalt, and nickel, and less than 0.05 for elements of atomic number less than 14 (silicon). Thus, for the lighter elements only a small fraction of the energy dissipated in the sample is obtained as measurable X-rays.

B. Diffraction of X-Rays

X-rays are subject to the same physical phenomena, including diffraction, as other electromagnetic waves, and a polychromatic X-ray beam may be separated into its components by diffraction with a suitable grating. The relation between wavelength and angle of diffraction, θ, is given by the Bragg equation:

$$n\lambda = 2d \sin \theta \tag{3}$$

In this equation n is the order of the diffracted X-ray, and is an integer, and d is the grating spacing. First order diffraction of X-rays in the region shorter than 10 Å requires grating spacings of the order of a few angstroms, and these cannot be prepared by ruling machines. However, the natural spacings of most inorganic crystals fall within these dimensions, and a single crystal, cleaved and polished to the desired plane, is suitable for diffraction of X-rays.

The more perfect a crystal is, the better is the resolution that can be attained. However, this resolution is achieved at some sacrifice in intensity, caused by primary extinction in the crystal. The intensity of the X-ray line is attributed to reinforcement of the X-rays diffracted from the topmost layers of the crystal by X-rays diffracted from lower planes, but if the crystal is sufficiently perfect to diffract the majority of the X-rays at the crystal surface reinforcement cannot occur. Therefore, some imperfection is introduced into the crystal during manufacture, either by mechanical abrasion or plastic or elastic deformation, to minimize primary extinction. This deformation must be carefully controlled so that excessive line broadening is not obtained. This would result in poor intensity as well as poor wavelength resolution. Crystals used in X-ray spectroscopy must be handled with care—the surface should never be handled with the fingers or allowed to become moist. Crystals should be stored in a desiccator when not in use, particularly in humid climates.

C. Scattering of X-Rays

X-rays from the tube target do not always interact with the sample to produce characteristic radiation. A large fraction of the primary beam is scattered by the sample, with this scattering being of two types: coherent or unmodified scattering and incoherent or Compton scattering. The unmodified rays arise from elastic collisions of the primary photons with the electrons in the sample, and are detected at the wavelength predicted by the Bragg equation. The Compton scattering arises from collisions of the primary X-ray quanta with electrons in the sample and transfer of a portion of the photon energy to the electron. The decrease in energy of the primary X-ray is dependent only on the angle ψ between the incident and scattered ray, and the change in wavelength is given by the equation:

$$\Delta\lambda = (h/mc) \operatorname{ver} \psi = a \operatorname{ver} \psi \tag{4}$$

where h is Planck's constant, m is the mass of the electron, c is the velocity of light, and $a = h/mc$.

The scattered radiation is a significant factor in X-ray fluorescence spectroscopy because it is the primary source of background intensity. The fraction of primary intensity, both characteristic and continuous, that is scattered by the sample, and hence transmitted to the detector, is dependent on the atomic number of the sample. Lighter elements produce more scattering and hence higher background intensities than heavier elements. The intensity of coherently scattered X-rays is proportional to the square of the atomic number of the scattering element, whereas the incoherently scattered fraction has an intensity proportional to the atomic number.

Although Compton scattering is a phenomenon generally associated with radiation arising from the X-ray tube, it can be observed from samples. Concentrated solutions of molybdenum, niobium, and zirconium each exhibit a weak line with 2θ approximately 1.1° (lithium fluoride crystal) to the long wavelength side of the K_α line. This can be attributed to Compton scattering at an angle of approximately 90° to the incident X-rays. These Compton lines are not readily observed from the solid metals which do not have light elements present to scatter the beam.

D. Absorption of X-Rays

Electromagnetic radiation, including X-rays, is absorbed by matter. The attenuation of an X-ray beam by matter is described by the equation

$$I = I_0 \exp\left[-(\mu/\rho)CL\right] \tag{5}$$

where I is the transmitted intensity, μ/ρ is the mass absorption coefficient of the absorbing material, C is the concentration, expressed in g/cm^3, and L is the path length in cm of the absorbing substance. The mass absorption coefficient is the ratio of the linear absorption coefficient, μ, to the density of the material, ρ, and includes two parts—the scattering component σ, and the photoelectric or true absorption coefficient, τ/ρ. The scattering component varies with wavelength and atomic number in a complex manner and becomes significant only for high energy X-rays and low atomic number elements. For low energy X-rays, σ for all elements approaches a value of approximately 0.12. The true absorption coefficient is a measure of the ability of the incident photon to eject an electron from the target atom, and varies with wavelength and atomic number of the absorbing element. Various empirical equations have been proposed in an attempt to relate true mass absorption coefficient, wavelength, and atomic number. They generally are of the form

$$\tau/\rho = k'Z^m\lambda^n \tag{6}$$

where k' is a constant which changes at the critical absorption wavelengths and m and n are constants for a particular element. This equation can be simplified to

$$\tau/\rho = k\lambda^n \tag{7}$$

where k is a constant for a particular element and n remains essentially constant for the element over a limited wavelength region. Values for n generally are in the range of 2.3–3.0.[3] In those cases for which the scattering component is relatively small, τ/ρ and μ/ρ approach the same value and may be used interchangeably. The relationship between mass absorption coefficient and wavelength is shown most readily by plotting the logarithm of wavelength as a function of the logarithm of mass absorption coefficient. This plot is approximately a straight line with the mass absorption coefficient increasing toward longer wavelength (Fig. 2). When a critical absorption limit is reached, the mass absorption coefficient drops abruptly. The critical absorption limit or absorption edge corresponds to the energy required to expel an electron from an inner shell. The ratio of the mass absorption coefficient at the short-wavelength side of the edge to the mass absorption coefficient at the long-wavelength side is referred to as the "jump" in mass absorption coefficient.

The wavelength at which an absorption edge occurs for a particular series decreases as atomic number increases.

The change in mass absorption coefficient as a function of wavelength for K and L_{III} edges is plotted in Figure 3, and is an important criterion in application of absorption edge analysis, as will be discussed later.

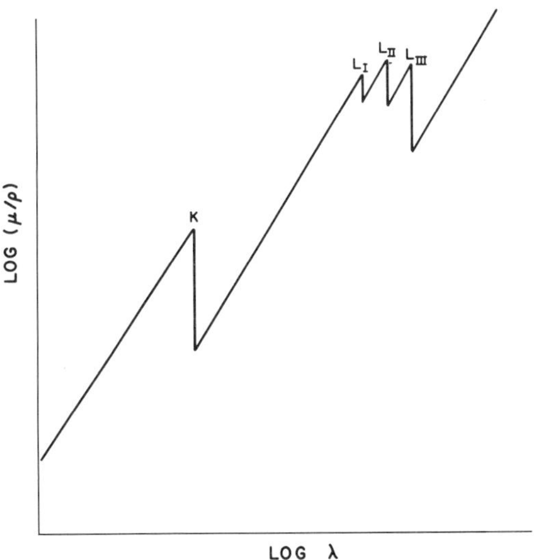

Fig. 2. Mass absorption coefficient as a function of wavelength.

With the coarse collimation generally used for X-ray spectroscopy, the absorption edge appears as a rather abrupt change in mass absorption coefficient. Actually, the absorption edge is more complicated than this simplified picture would indicate. The wavelength of the absorption edge represents the energy required to remove an electron from the shell in question to some outer energy band of the atom. The energy of this band is a function of the molecular structure in which the atom occurs, is different for gases, liquids, and solids, and is strongly affected by chemical combination. At the long-wavelength side of the edge, the first inflection point in the mass absorption coefficient curve corresponds to the nearest energy level where the electron may come to rest. The absorption edge generally has a short-wavelength maximum with a considerably higher mass absorption coefficient than indicated from the straight-line log–log plot of Figure 2. The absorption edge fine structure may extend several hundred electron volts to the short-wavelength side of the edge.

The property of elements to absorb X-rays has several important analytical chemistry applications: (*1*) the absorption of polychromatic X-rays is a nonselective measure of concentrations of all elements in a sample; (*2*) the absorption of monochromatic X-rays provides a more accurate determination of a particular element because mass absorption coefficients for monochromatic X-rays are known more accurately than for polychromatic X-rays; (*3*) the absorption edge procedure, in which measurements are made

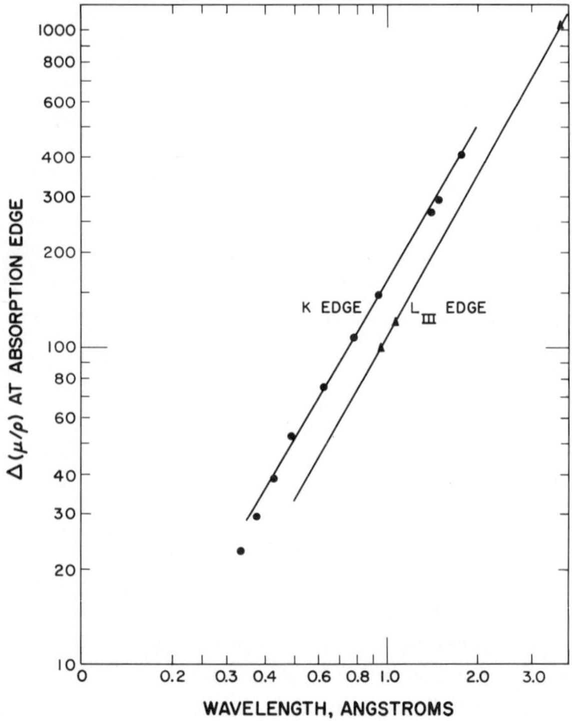

Fig. 3. Change in mass absorption coefficient at K and L_{III} edges as a function of wavelength.

at wavelengths on each side of an absorption edge, is a highly selective technique for determining the concentration of an element, and (*4*) the absorption edge fine structure can be used as an aid in elucidating chemical structure. These techniques will be described in more detail in a subsequent section of this chapter.

E. Intensity of X-Ray Lines

It was mentioned that the relative intensities of X-ray lines are a function of the probability of occurrence of electron transitions according to quantum principles. In the wavelength region below 10 Å the X-rays observed may arise from transitions to K, L, M, and for the actinide elements even to N levels. For practical purposes relative intensities of various lines in a series can be related, but relating intensities for different series is somewhat more difficult. In the K series the K_β line generally has an intensity 20–40% of the K_α line intensity for elements of atomic number greater than approximately

20, with the intensity of the K_β line increasing with increasing atomic number. In the L series the $L_{\beta 1}$, $L_{\beta 2}$, and $L_{\gamma 1}$ are approximately 50, 20, and 10% respectively, as intense as the $L_{\alpha 1}$, with other lines being weaker. Approximate intensities of various lines in a series, relative to the most intense line in that series, are summarized in Table 1.[4]

Table 1. Approximate intensities of X-ray lines relative to most intense line in the series

Line	Relative intensity
$K_{\alpha 1}$	100
$K_{\alpha 2}$	50
$K_{\beta 1}$	20–40
$K_{\beta 2}$	0.1–3
$L_{\alpha 1}$	100
$L_{\beta 1}$	50
$L_{\beta 2}$	10–30
$L_{\gamma 1}$	5–20
$L_{\beta 3}$	1–7
$L_{\beta 4}$	3–5
L_1	3
Others	<2
M_α	100
M_β	50
M_γ	?

Actually, the intensities observed with a spectrograph depend on several other variables such as excitation potential, the diffraction crystal, detector, and absorption between the sample and detector, and all of these must be considered in selecting an X-ray line for analysis. For example, absorption in the sample and the air path becomes appreciable at wavelengths greater than 2 Å. Whereas the $L_{\alpha 1}$ line is the most intense for the heavy rare earths as one would expect, the $L_{\beta 1}$ is significantly stronger than the $L_{\alpha 1}$ for cerium and lanthanum when aqueous samples are examined with an air path spectrometer. The M_α line for heavy elements has approximately twice the intensity of the M_β X-ray. However, when uranium is examined with an argon-filled proportional or Geiger counter the M_β line is significantly more intense than the M_α because the K absorption edge for argon falls just to the long-wavelength side of the M_β X-ray, thereby increasing sensitivity at this wavelength.

III. INSTRUMENTATION FOR X-RAY SPECTROSCOPY

Basically, an X-ray spectrograph consists of a means of producing X-rays in the sample, an X-ray optical system that can be either dispersive or non-dispersive, and an X-ray detector. Numerous combinations are available in selecting a spectrograph, but the choice is limited when one considers the applications for which the device is required.

A. X-Ray Sources

1. ELECTRON EXCITATION

As noted previously, X-rays arise from the collision of high energy rays or particles with the orbital electrons of the atom. In the early work of Siegbahn[5] and von Hevesy[6] on X-ray spectroscopy the sample was made the target of the X-ray tube and X-rays were produced by bombarding the sample with high energy electrons. This mode of excitation had two major drawbacks: sample interchange was time-consuming because the sample chamber had to be pumped between samples, and only solid conducting samples could be examined. Electron excitation is, however, more efficient than secondary X-ray excitation and has had an increased interest in the past few years, largely due to the advent of electron probe analysis. In addition to electron probes, X-ray spectrographs employing electron excitation have been custom built in the past few years[7,8] and are now commercially available,[9] primarily for long-wavelength analysis.

2. X-RAY EXCITATION

X-ray excitation is the primary mode of producing X-rays for spectrographic analysis. In this arrangement samples can be changed rapidly, can be analyzed in air, helium, hydrogen, or vacuum, depending on the requirements of the analysis, and high intensities of fluorescent X-rays can be obtained in the wavelength region below 10 Å. X-ray tube targets of either molybdenum or tungsten have been the workhorses of fluorescence analysis; however, in the past few years other target materials have been introduced in an effort to increase sensitivity for light elements. For analysis at wavelengths in the air path region (<2.5 Å) the X-ray tube continuum is usually the primary means of exciting X-rays, with the X-ray tube operated at a potential approximately three times the binding energy of the electron shell selected for analysis. However for determination of elements which have an absorption edge just to the long wavelength side of a characteristic X-ray from the tube target an increase in intensity of two- or three-fold compared to intensity expected from the X-ray continuum is observed. This is used to advantage for example by using a chromium target tube for determination of titanium and calcium, a silver target tube ($L_{\alpha 1}$ line at 4.154 Å) for determination of chlorine, or an aluminum-target tube for determination of magnesium or sodium. In selecting an X-ray tube target material, it is well to keep in mind the characteristic radiation contribution to intensity, particularly for trace analysis or for analysis at wavelengths longer than approximately 2.5 Å.

3. HIGH ENERGY PARTICLES

The production of characteristic X-rays by high energy particle bombardment has been realized since 1913.[10] However, applications for analytical purposes have not been attempted until comparatively recently, primarily because X-ray tubes were more readily available. Within the past ten years radioactive sources have received increased attention: β and γ emitters suitable as excitation sources are readily available as by-products of reactors; the sources are highly stable and do not have the high power requirements of X-ray tubes; the sources are cheaper and more compact than X-ray tubes; and sources are readily operated remotely, either at mills, mines, or on the moon. Recent work at Argonne[11] has been aimed at producing compact X-ray spectrographs for space applications such as lunar surface analysis. Using a tritium-zirconium source, a spectrograph weighing less than 10 lb and occupying a volume less than 1 ft^3 was described.

A number of α-, β-, and γ-emitting isotopes can be used as radioactive sources, and the various types of excitation have been compared by Cook et al.[12] Alpha-particle sources such as ^{210}Po produce an X-ray spectrum almost free of background; however, X-ray excitation efficiencies at energies greater than 14 kV are poor compared to beta ray excitation. The method should prove of value for excitation in the vacuum X-ray region.

Beta ray excitation is a much more efficient means of X-ray production, with excitation efficiencies as high as 15–20%.[13] The efficiency of excitation increases with increasing beta particle energy; however, bremsstrahlung background also increases. Beta particle sources can be used either by excitation directly with the beta rays, or by using the bremsstrahlung background. The former provides higher intensities of line radiation but with some bremsstrahlung background, particularly if long-wavelength X-rays are excited; the latter provide lower background but with some sacrifice in intensity. An efficient beta particle source can be prepared by mixing the beta emitter with an element which emits an X-ray just to the short-wavelength side of an absorption edge for the element being determined. A high intensity source for any application can be prepared in this manner.

K-capture X-rays are an efficient means of X-ray production and provide a low background. The applications of these sources are limited to elements having absorption edges at a slightly longer wavelength than the X-ray emitted from the element. Some Compton scattering will be observed from light elements, as with any X-ray source.

Gamma rays as X-ray sources have not found much favor to date, possibly because of the high background observed. Beta and X-ray sources can be readily shielded from the detector so that direct background production is not observed; this is not the case with gamma rays which may require several centimeters of lead to completely shield the detector from the source.

Some radioactive sources of value in X-ray excitation are summarized in Table 2.

Table 2. Radioactive sources of value for X-ray production

Source	Type radiation	Comments	Half-life
Po^{210}	alpha	Excites X-rays to 14 kV	138 days
Sr^{90}	beta		28 years
Kr^{85}	beta	Gamma contaminated	10.4 years
Pm^{147}	beta	Gamma contaminated	2.5 years
P^{32}	beta	Gamma free; short half-life	14.3 days
Tl^{204}	beta		
Y^{90}	beta	Gamma contaminated	64.2 hours
H^{3}	beta	Free of gamma	12.26 years
Ni^{63}	beta	Free of gamma	92 years
Co^{57}	gamma 125 kV, X-ray 6.4 kV		267 days
W^{181}	gamma 67 kV	Gamma contaminated	130 days
Cs^{131}	X-ray 35 kV	Free of gamma	9.7 days
Fe^{55}	X-ray 5.9 kV	Free of gamma	2.7 years
Am^{241}	X-ray 60 kV	Alpha contaminated	470 days

Excitation with protons has been investigated as a means of X-ray production. This technique provides better peak to background intensities at wavelengths longer than approximately 10 Å, but is not of interest at shorter wavelengths. It will be discussed in a subsequent chapter of this text.

B. Diffraction Crystals

In the wavelength region below 10 Å the primary means of dispersing an X-ray beam into its various components are crystals. In selecting a crystal for X-ray spectrographic analysis, the resolving power, reflection efficiency, and crystal fluorescence should be considered. Some of the more commonly used crystals used in the wavelength region below 10 Å are summarized in Table 3. These are not the only crystals that can be used, but include those which are available commercially and which are most commonly used.

Table 3. Analyzing crystals for the wavelength region below 10 Å

Crystal	$2d$ spacing	Comments
Topaz	2.712	Best crystal below 0.5 Å
LiF	4.0276	Good intensity and resolution in air path region
NaCl	5.639	Good in air path region; best crystal for Cl, S
Calcite	6.071	Excellent resolution; poor intensity
Quartz (1011)	6.686	
PET	8.742	Good in 5–8.5 Å region
EDDT	8.808	Good in 5–8.5 Å region; lower background than ADP.
ADP	10.65	Good for Al, Mg; high background in 4–6 Å region
Gypsum	15.12	Good for Al, Mg; may dehydrate in vacuum
Mica	19.92	Can be bent during use; good reflectivity for odd orders of diffraction.

In considering crystal resolution, the Bragg equation (equation 3) shows that for optimum resolution the smallest d-spacing crystal available in the wavelength region of interest will provide the greatest angular separation of X-rays. If resolution is the primary criterion in selecting a crystal, the crystal of shortest d-spacing should be selected.

The reflection efficiency of crystals varies not only for different types of crystals, but also for different crystals of the same type. As previously noted, this variation results from imperfections that are purposely introduced into the crystal during manufacture. Birks and Seal[14] have shown that significant increase in diffracted intensity can be obtained by careful preparation of the crystal surface.

The fluorescence of the crystal is another factor that can significantly affect analytical sensitivity. X-rays to the short-wavelength side of an absorption edge for an element in the crystal will be strongly absorbed by the crystal, and in turn characteristic radiation will be emitted by the crystal. An ADP crystal fluoresces strongly in the wavelength region of approximately 5.8–7 Å (the K absorption edge for phosphorus occurs at 5.78 Å). Ultimate sensitivity in X-ray spectroscopy is related to the peak-to-background ratio, and in trace analysis a three- to fourfold increase in background intensity can more than erase a twofold increase in peak intensity. Either an EDDT or PET crystal generally is preferred in the wavelength region from 4 to 7 Å because of potassium fluorescence from a KAP crystal and phosphorous fluorescence from the ADP crystal.

With any crystal one is limited from using low diffraction angles by excessive scatter of X-rays from the crystal; the low angle limit for flat crystal optics is in the region 10–20° 2θ. The high angle limit of a spectrometer is generally in the region 150° 2θ; at higher angles the detector begins to interfere with the X-ray source.

C. Spectrometer Construction

Crystal spectrometers are constructed on two geometrical arrangements, utilizing either flat or curved crystal optics. The flat crystal type uses either slits or parallel plate collimators to obtain the required resolution. The arrangement is nonfocusing (Fig. 4*a*) and some intensity is lost at the exit collimator because the total diffracting surface of the crystal cannot be seen by the detector. Nevertheless, the flat crystal spectrometer provides adequate intensity in the wavelength region below 10 Å except possibly where sensitivity in the submicrogram region is required.

The curved crystal spectrometer is a focussing system (Fig. 4*b*), based on the Rowland circle common in emission spectroscopy. The source slit, diffracting crystal, and detector are all located on a circle having a radius r, with the crystal bent to a radius $2r$. All X-rays of wavelength λ passing through the source slit are focused to the detector at an angle defined by the Bragg equation. The mechanics of the curved crystal spectrometer are more complicated than the flat-crystal type, but the curved optics are preferred where maximum intensity is required.

Various designs can be used in construction of curved crystal spectrometers. Crystals can be ground flat, then bent to radius $2r$ (Johann type). In this arrangement only the portion of the crystal tangent to the Rowland circle produces true focusing, and some broadening of the X-ray line results. In the Johansson arrangement the crystal is bent to radius $2r$, then ground to radius r so that all of the crystal surface is on the Rowland circle and better focusing is obtained. Both the Johann and Johansson spectrometers are reflection instruments. In the Cauchois[15] arrangement X-rays are diffracted by transmission through the crystal, limiting the use of the spectrometer to short wavelengths.

For mechanically changing the diffracted wavelength Birks and Brooks[16] moved the crystal on the Rowland circle to an angle θ and the detector to 2θ. This system is relatively simple to construct, but suffers from the disadvantage that the spectrometer views different areas of the sample at

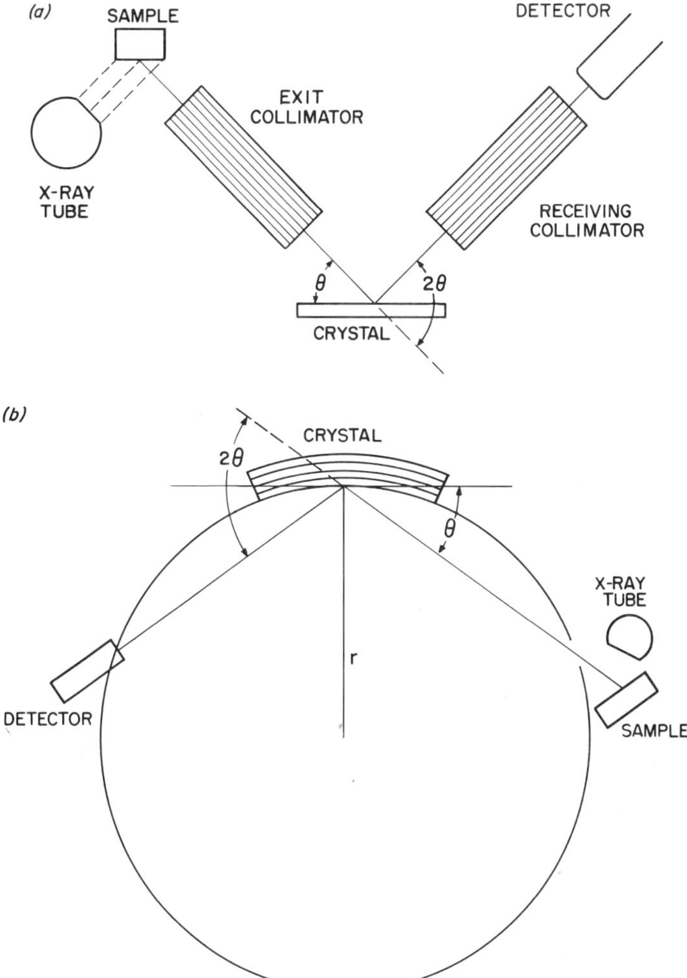

Fig. 4. (*a*) Geometry of flat crystal X-ray optics. (*b*) Geometry of curved X-ray optics.

different wavelengths (the sample is slightly behind the slit, and the angle through the slit to the crystal changes). The ARL spectrograph[17] moves the crystal in a straight line away from the sample as well as moving it on the Rowland circle so that the crystal is always in a line with the same position on the sample.

A problem in construction of curved crystal spectrometers is selection of the proper radius of curvature for the Rowland circle. If a small radius of curvature is used, poor resolution is obtained, particularly in the low-wavelength range. If the radius of curvature is too large the spectrometer can become bulky and awkward to position in a vacuum chamber, and the wavelength range that can be conveniently covered is limited. To circumvent this problem, Elion and Olgilvie[18] have constructed a variably bent mica spectrometer whereby the diameter of the Rowland circle decreases as wavelength is increased. This arrangement permits one spectrometer to cover a wide range of wavelengths without having to change crystals.

D. X-Ray Detectors

1. FILM

The observation by Roentgen in 1895 that photographic film is fogged by X-rays led to the obvious conclusion that film could be used for X-ray detection, and most of the work in X-ray spectroscopy to the early 1940's was performed with film. The main disadvantage of film is that data are obtained slowly, and this probably resulted in the reluctance by earlier analytical chemists to accept the X-ray spectrograph as an analytical tool. The advent of electronic detection systems such as Geiger, scintillation, and proportional detectors in the 1940's brought a new dimension to X-ray spectroscopy.

2. GEIGER COUNTER

The Geiger-Mueller tube, the earliest of the electronic detection systems, utilizes the principle that an X-ray absorbed by a high pressure inert gas between two terminals of widely different potential will produce a self-sustaining electric discharge in the gas. As the electrons are accelerated toward the positive electrode they produce additional ionizations so that a high amplification of incident X-rays is obtained, and simple electronic circuitry can be used for measurement of the X-ray intensity. The large munber of ionizations which are produced by each photon during the discharge results in a long "dead time," or time required for the tube to recover before another pulse can be detected. Also, all X-ray photons absorbed by the gas, regardless of energy, produce pulses of equal amplitude, making pulse height discrimination inapplicable with this detector. Thus, the Geiger counter measures the number of X-rays absorbed, but not their energies.

3. PROPORTIONAL COUNTER

The proportional detector is basically of the same design as the Geiger counter but is operated at a voltage below that required for a self-sustaining discharge. The lower voltage causes the discharge to be confined to the plane in which the quantum is absorbed. A much smaller pulse is produced than in the Geiger counter, but the remainder of the tube is free to detect additional pulses. The counter dead time is much shorter, of the order of 0.2 μsec, and much higher count rates can be measured without deviations from linearity. The small signal from the proportional counter makes additional amplification necessary in order to produce a signal sufficiently large to activate the counting circuit. The number of ionizations produced by an X-ray quantum in the proportional counter is proportional to the quantum energy, a distinct advantage over the Geiger counter. Thus by selecting only those pulses in a desired energy range (pulse height discrimination), undesired radiation such as second-order X-rays or electronic noise of the circuit can be eliminated. The counting efficiency of the proportional counter is of the same order as that of the Geiger counter, both being limited by the efficiency of the inert gas to absorb the X-rays entering the detector. The noise level of the proportional counter is extremely low, even for long X-ray wavelengths. X-rays having wavelengths longer than approximately 5 Å are detected by using a thin porous window in place of the normal mica window and slowly flowing the detector gas through the tube.

4. SCINTILLATION COUNTER

The scintillation counter consists of a multiplier phototube to which is attached a fluorescent crystal, usually sodium iodide activated with thallium iodide. X-rays absorbed by the crystal produce photoelectrons and Auger electrons in quantities proportional to the incident X-ray energy. These in turn produce visible fluorescence which is amplified and converted to electrical energy by the tube. The possibility for pulse height discrimination exists although energy resolution is not as good as that obtained in a proportional counter. The main advantage of the scintillation counter is its high quantum counting efficiency which is essentially 100% through the wavelength region between 0.3 and 3 Å. However, at longer wavelengths (lower energies) the amplitude of the pulses produced by the scintillation detector becomes so small that it is difficult to distinguish the small pulses from the noise pulses of the counting circuit.

5. SOLID STATE DETECTORS

Solid state detectors have gained increased interest as X-ray detectors over the past few years. Either germanium or silicon semiconductors can absorb almost 100% of radiation having energy less than 100 kV.[19] These detectors have been coupled with a radioactive source[19] and a liquid nitrogen cooled preamplifier into a nondispersive X-ray spectrograph which provides an energy resolution of 1.1 kV. This is sufficiently good to resolve the

K_α lines of adjacent heavy elements. This type of system lacks the sensitivity of conventional detectors, but it must be remembered that they are still in early stages of development and may well provide the X-ray spectrographic system of the future.

The useful wavelength ranges of some of the various detectors are summarized in Figure 5. The wavelength ranges given are only approximate and are a function of such parameters as sensitivity required, window material and thickness, detector geometry, and gas pressure.

E. Pulse Height Analysis

It has been noted in the discussion of detectors that some types of detection systems are energy sensitive, others are not. If the detector output can be made to change with the incident energy it is detecting, it is conceivable that one may do away with the dispersing device. Even with a dispersing spectrograph energy discrimination can be of considerable value in sorting out higher order Bragg reflections by the crystal—for example, eliminating second order zirconium radiation in measuring the $L_{\alpha 1}$ line for hafnium. Of the electronic detectors, the Geiger detector is not energy sensitive. The scintillation detector has an energy resolution of the order of 50%, or a 10 kV X-ray can be resolved from 5 or 15 kV X-rays. For proportional detectors the energy resolution is in the range of 8–25%, depending upon the detector construction and the gas filling. Even for the best proportional detectors adjacent elements cannot be resolved although elements two atomic numbers apart in the periodic table can be differentiated. With solid state detectors such as lithium-drifted germanium it is possible to discriminate adjacent heavy elements such as gold and platinum,[19] although for lighter elements only elements two or more atomic numbers apart could be discriminated.

Clearly, energy discrimination can be an important consideration in the construction of a spectrometer. If the dispersive system can be eliminated,

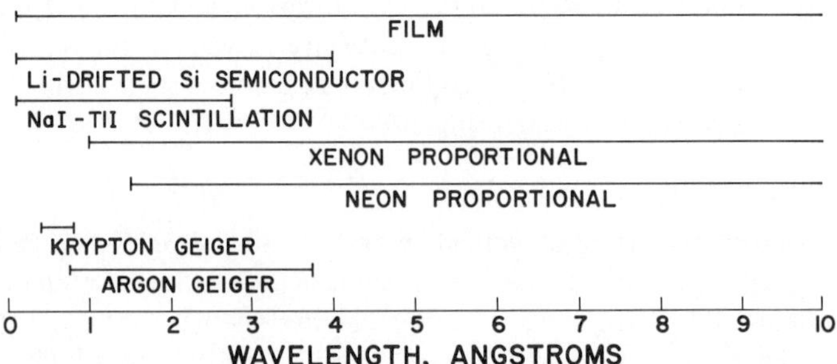

Fig. 5. Useful wavelength region for various X-ray detectors.

the detector can be placed much closer to the sample and a larger portion of the X-rays excited in the sample can be transmitted to the detector. This could be of considerable value for example in trace and thin film anaylses.

IV. APPLICATIONS OF X-RAY EMISSION AND FLUORESCENCE

A. Selection of a Line

For quantitative X-ray spectrochemical analysis the intensities of X-rays for the element being determined are measured and these intensities are related to the quantity present. The first consideration in X-ray fluorescence analysis is the selection of a suitable line. Published tables of X-ray wavelengths should be consulted in this selection.[20–25] The X-ray selected is generally a function of the equipment available as well as of the type of sample to be analyzed. Using an air path spectrometer and a 50 kV power supply, one is limited to a K-series line for elements of atomic number 22 (titanium) to 53 (iodine), and L-series line for elements of atomic numbers 58 (cerium) or greater, and either a K- or L-series line for elements between iodine and cerium. If a 100 kV supply and X-ray tube are available, K-series lines can be excited for the rare earth elements. A vacuum spectrograph with an ADP crystal extends the applicable wavelength region to 10 Å, and permits analysis of elements having atomic numbers greater than 12 (magnesium) by K spectra, atomic numbers greater than 32 (germanium) with L spectra, and elements of atomic number greater than 64 (gadolinium) with M-series X-rays.

The long-wavelength limit for X-ray fluorescence analysis is determined by absorption effects, either in the path traversed by the X-rays in the sample, between the sample and detector, or in the detector window. Absorption by the air path becomes serious for wavelenghts much greater than 2.5 Å, and either a helium or vacuum path is required. With a properly designed helium path this system can be used at wavelengths up to 10 Å with only slight loss in sensitivity compared to a vacuum path spectrometer.[26]

The relative intensities of various X-ray lines has been discussed previously. From Table 1 one may conclude that one is generally restricted to using the K_α, K_β, $L_{\alpha 1}$, $L_{\beta 1}$, $L_{\beta 2}$, $L_{\gamma 1}$, M_α, or M_β X-rays for analysis.

Where there is a choice of K or L, or L or M series X-rays, several factors must be considered. Molybdenum can be measured in air path using the K_α X-ray at 0.710 Å, or in a vacuum spectrograph using the $L_{\alpha 1}$ X-ray at 5.406 Å. At the wavelength of the K_α X-ray a high continuum from the X-ray tube is present, and cannot be eliminated by electronic discrimination. Background intensities in excess of 100 counts/sec are not uncommon in this wavelength region, and although the absolute intensity of the $L_{\alpha 1}$ line is less than that of the K_α, greater peak-to-background ratios, and hence better sensitivity, may be attained using the longer wavelength X-ray.

B. Sample Preparation

1. SOLIDS

Solid samples would appear at first glance to be the easiest to analyze by X-ray spectrographic methods because all that is required is a flat surface for analysis. This is actually the case for many routine samples such as are encountered in steel mills or other plant analyses, but difficulties in standard preparation preclude application to nonroutine analysis.

The most important steps in solid sample preparation are obtaining a flat surface, and preparing a homogeneous sample. For products such as metal ingots or even rocks the flat surface is obtainable by normal metallographic grinding and polishing operations. For wavelengths shorter than approximately 3 Å, grinding through 240-grit paper on a belt sander often is adequate for producing a suitable surface, but for longer wavelengths polishing with 3 or even 1/4 μ diamond or alumina is required. The requirement for a finer polish for analysis with longer wavelengths results from the lower penetrating power of the X-rays into the sample so that surface irregularities become more critical. At the K_α line for aluminum the X-rays observed by the detector may originate from a layer no greater than 5 μ in depth, and 3 μ scratches can affect X-ray intensities.

Samples of irregular shape may be melted in a button furnace[27] to obtain a suitable surface. Obtaining homogenous samples, however, is another problem, and may often be the deciding factor in resorting to solution techniques for analysis. Rocks are an obvious case of heterogeneous samples where standard preparation may be a serious problem for analysis as solids, but these same problems may be encountered in metals. Michaelis and Kilday[28] observed high and erratic results for determination of lead in leaded steels when only grinding techniques were used for sample preparation. This was attributed to smearing of the lead, and the problem was eliminated by metallographic polishing with 1/4 μ diamond abrasive.

2. POWDERS

For analysis of rocks and other brittle materials the first step in attempting to homogenize the specimen is grinding. The fact that a sample has been ground to a fine powder does not assure accuracy or precision in analysis. The X-ray is much more sensitive to particle size than the human eye, as many investigators have discovered. In an excellent theoretical and experimental study of particle size effects, Claisse and Samson[29] derived equations relating intensity of the analytical line as a function of matrix, exciting radiation, and particle size. Although the theoretical treatment is complex, involving several assumptions and approximations, they were able to qualitatively predict the effect of grinding on intensity of various X-rays. For example, intensity of strontium K_α excited by molybdenum K_α increases linearly as a function of concentration for particle sizes greater than 1000 μ. Grinding to a 7 μ particle size results in a sharp increase in intensity at low strontium concentration, with a leveling off of intensity at strontium concentrations greater than approximately 40%. For a particular wavelength of an analytical line a "critical size" was shown to exist which is a function of the exciting radiation. At particle sizes greater or less than the critical size particle diameter is not critical, but in the critical zone intensity is strongly affected by grinding. Hand grinding in a mortar is usually not adequate for sample preparation, and contrary to common practice, grinding to the finest size that can be obtained will not always attain a suitable sample.

Just as precipitation in solid samples becomes more serious at longer wavelengths, particle size effects in powders are generally more pronounced at longer wavelengths. Gunn[30] has plotted intensities of the K_α X-rays for molybdenum, copper, germanium, and iron in various matrices as a function of particle size. At 50 μ he observed intensity decreases of 40, 30, 10, and 0% relative to 1 μ size particles for K_α lines of iron (1.938 Å), copper (1.541 Å), germanium (1.255 Å), and molybdenum (0.710 Å), respectively. Matrices were generally light elements, and particle size effect would probably be more pronounced in more highly absorbing matrices. An excellent theroretical treatment of particle size effect as a function of wavelength has been given in a series of articles by Bernstein.[31–33]

The particle size effect in powders could be eliminated if one could grind all samples and standards to a uniform particle size. However, the rate of grinding is a function of the hardness of the material, and standards prepared from a silicate, for example, could not be applied to determination of silicon in quartz. One means of effectively producing a powder of uniform particle size is dissolution of the sample, followed by precipitation. This method has been applied to the determination of uranium dioxide in bismuth by dissolution of the sample in nitric acid, followed by precipitation of both bismuth and uranium with ammonia.[34]

Another problem in the analysis of powders is the variation in line intensity that may result from packing effects. This can be eliminated by using an internal standard, or adding a binder such as starch and pressing into a briquet with a metallurgical specimen press.

3. SOLUTIONS

Solutions are the only sample form which are not susceptible to particle size or heterogeneity effects in X-ray spectrographic analysis. Solutions may occur as liquids or as fused solids. The liquid form is easier to prepare, and is readily applicable with internal standards. The main disadvantages of liquids are that standards are difficult to keep for extended periods of time, and the method is difficult to apply with vacuum spectrographs. Solids are easier to use in vacuum systems, and standards and samples may be kept for years if desired without adverse affects. Various fluxes have been proposed for fusions, and some of these are summarized in Table 4.

Table 4. Fluxes for fusion X-ray spectrographic analyses

Flux	Type of sample	Fusion temp., ° C	Ref.
Borax, 10 g	Iron ore, 1 g	to melt	35
Borax, 9.5 g	Manganese nodules, 0.5 g	400°	36
Borax, 15 g	Rocks, 5 g	Meker burner	37
Borax, 5 g	Iron ore, 1 g	to melt	38
Borax, 10 g, BaO, 2 g	Ferrous alloy, 300 mg	to melt	39
$K_2S_2O_7$, 10 g, then borax, 2 g	Sulfide ores	Meker burner	35
$K_2S_2O_7$, 9 g, and NaF, 1 g	Silicous samples	350–700°	40
$K_2S_2O_7$, 10 g	Zinc ore; copper alloy 0.2 g	300–700°	41
$Li_2B_4O_7$,1 g	Cement, 1 g	1370°	42
$Li_2B_4O_7$, 1 g, and La_2O_3, 0.125 g	Rocks, 0.125 g	1100°	43
Na_2BO_3, 30 g	SiO_2 and Al_2O_3-base catalysts	1000°	44
Na_2CO_3, 4 g	Wolframite, 1 g	1000°	45
Sulfur, then borax, 1 g	Cadmium-tin-zinc alloys	to melt	46

C. Quantitative Analysis

1. STATISTICS OF X-RAY MEASUREMENTS

The measurement of X-ray intensity follows a Gaussian normal error curve. The standard counting error or standard deviation, σ, is equal to the square root of the number of events measured, or the relative standard deviation is expressed as:

$$\sigma, \% = 100\sqrt{N}/N \quad (8)$$

This equation defines the best precision that one may expect in any measurement, and obviously one may theoretically increase the precision of an analysis by accumulating a larger number of counts. Practical considerations generally set some upper limit to the number of counts one should accumulate. Aside from the time factor, one should take into account such parameters as sample heating and bubbling from prolonged exposure under the X-ray beam, and electronic instability in the instruments.

Most commercial X-ray instruments permit either fixed count or fixed time measurements to be performed. With the fixed count mode, counting errors are equal for line and background if the same number of counts are accumulated. In measurements involving a background correction, the relative error in the background can be considerably greater than the error in the line measurement if large peak-to-background ratios are involved. Using the fixed time mode the relative errors in peak and background measurements may be considerably different, but absolute errors are more nearly equalized. The statistics of fixed time vs. fixed count measurements have been discussed by Birks and Brown[46a] and by Gaylor.[46b]

2. X-RAY INTENSITY TECHNIQUES

The direct measurement of the intensity of an X-ray line from the element being determined is the most rapid technique for qantitative X-ray spectrometric analysis, but is the most susceptible to interferences. Interference may be caused by variations in sample composition, by overlapping X-ray lines originating either in the sample or in the X-ray tube, by enhancement, by absorption, or by fluctuations in instrument operating condition. The simplest method to compensate for these variations is to compare X-ray intensity from the sample to intensity from standards of known compositions prepared in the same manner as the samples. Corrections for background intensity may, but need not always, be applied. Any change in matrix or solvent may affect the intensity, and this absorption effect becomes more serious at longer wavelengths and in solvents of high mass absorption coefficient. In the determination of chromium in either $3N$ hydrochloric or $3N$ nitric acid, a variation in hydrochloric acid content of 0.04 ml/10 ml produced a 1% variation in chromium intensity, whereas 0.4 ml variations in nitric acid content produced the same error.[47] Absorption interference also results when elements of widely different mass absorption coefficients are substituted for each other in the sample.

Enhancement interference results if a major constituent in the sample has an intense X-ray just to the short-wavelength side of the absorption edge responsible for the analytical line. Enhancement is not so serious as overlap or absorption, and generally only K_{α}, K_{β}, $L_{\alpha 1}$, and $L_{\beta 1}$ X-rays need be considered.

In spite of its sensitivity to instrumental and sample variations, the line intensity technique is capable of good precision and accuracy for routine analysis if sample composition is predictable. Precision of 0.5% or better can be attained in many applications.

The line intensity technique can be made less sensitive to matrix composition by heavy element dilution of the sample.[35] By addition of a known quantity of barium oxide to borax fusion samples, not only is the fusion of the ore facilitated, but the high mass absorption coefficient of the barium minimized variations in sample absorbtivity. In the same manner Chodos et al. greatly improved the analysis of aluminum and magnesium in rocks by addition of 40% of tungstic acid to the pulverized sample.[48]

3. USE OF INTERNAL STANDARDS

The use of internal standards is a well-known means in emission spectroscopy to minimize instrumental and sample variations, and is equally applicable in X-ray spectroscopy. In the internal standard method the intensity of the X-ray line of interest is compared to another X-ray intensity which is in some manner affected by instrumental variables and matrix composition. This second X-ray may originate from another element present in the sample in constant concentration or from an element added to the sample, or it may be radiation originating in the X-ray tube and scattered by the sample.

To minimize matrix effects the internal standard should be affected to approximately the same degree by absorption and enhancement as the analytical line. If an element is used as the internal standard, the X-ray line preferably should be of the same series as the analytical line, should be of the same order of magnitude in intensity, and should be reasonably close in wavelength to the analytical line.

As already noted, variations in sample absorption will affect the line intensity method, severely limiting the precision and accuracy. Using the internal standard method, both lines will be affected to an equal degree by variations in sample absorptivity. The only exception occurs when a matrix element has an absorption edge between the analytical line and the internal standard X-ray; in this case the shorter wavelength X-ray will be attenuated by the matrix to a greater degree than the longer wavelength.

Enhancement effects also are minimized by the internal standard because if the X-rays are reasonably close in wavelength, both X-rays will be enhanced approximately to the same degree. However, if a major constituent of the sample has an intense X-ray between the absorption edges giving rise to the analytical and internal standard lines, only the X-ray originating from the longer wavelength absorption edge will be enhanced.

In selecting an X-ray line for an internal standard, a line originating from the same edge as the analytical line should be selected. The intensity of any line is proportional to the absorption jump at the edge. The K edge jump is greater than the L_{III} jump at the same wavelength so that if a K_{α} line for rubidium is used as an internal standard for the $L_{\alpha 1}$ line for uranium, the rubidium X-ray would be enhanced to a greater degree than the uranium X-ray. Likewise, the L_{III} jump is greater than the L_{II} so that an $L_{\alpha 1}$ line is enhanced to a greater degree than an $L_{\beta 1}$ line in the same wavelength region.

As noted, either line radiation or background from the X-ray tube may be used as an internal standard. This method is sensitive to instrumental variations, but is not so sensitive to matrix effects as an X-ray line originating in the sample. Kemp and Andermann[49] used the scattered background for determination of several elements in alloys with iron. The Compton-scattered $L_{\alpha 1}$ X-ray intensity from a tungsten-target tube can be used as an internal standard for the determination of tungsten.[50] The $L_{\alpha 1}$ X-ray intensity will be sensitive to tungsten content of the sample as well as to tube voltage and current whereas the Compton-scattered component will be affected primarily by tube operating parameters. Actually, some error may be introduced if the solvent varies over a wide composition range because the intensity of the Compton component of the scattered radiation is proportional to Z, the atomic number of the scattering substance, whereas the intensity of the coherently scattered portion of the tube radiation is proportional to Z^2.

Cullen[51] has used the coherently scattered $L_{\beta 1}$ line for tungsten as an internal standard for the determination of copper in solutions from refining operations. The X-ray results generally agree to 1% with values obtained by chemical methods.

The scattered white radiation from the tube is a useful internal standard for determination of low concentrations, but is generally not practical for major constituents because measured ratios will be too high. The background is again affected by instrumental variable and to some degree by the sample composition.

4. ARITHMETIC CORRECTION METHODS

Variations in matrix composition may be compensated by applying arithmetic corrections to X-ray intensities. The correction factors may be determined theoretically or empirically, with the latter generally being more reliable. Sherman[52] considered the general case of theoretical corrections to compensate for absorption and enhancement effects. For absorption corrections the mass absorption coefficients for all components in the sample are considered using X-ray absorption theory. The excited intensity of the X-ray line is somewhat more difficult to calculate. Such factors as the excitation efficiency of a polychromatic X-ray beam as well as excitation by *K*- and *L*-series X-rays from the sample and X-ray tube must be considered. The equations derived by Sherman are somewhat cumbersome for calculations but are usable with a computer. The uncertainty in experimentally determined values for some of the physical constants such as mass absorption coefficients and excitation efficiencies somewhat limit the use of this technique.

To eliminate the uncertainties resulting from sample composition and physical constants, Sherman[53] derived equations for calculations of concentrations using intensities measured from the sample before and after dilution. Both inert and active dilution techniques were discussed, with inert dilution involving addition of an element which did not contribute to the X-ray intensity, and active dilution involving addition of the element being determined. Precision of better than 1% can be obtained using this technique. This procedure suffers from the limitation that an essentially monochromatic source should be used for excitation.

The use of empirical correction factors relies on the calculation of the equation for the calibration curve, and the effect of matrix elements on the slope or intercept of the curve. For a straight-line calibration curve the concentration, C, is related to the line intensity, I, by the equation:

$$C = aI + b \tag{9}$$

where a is the slope of the curve and b is the intercept. From known standards, variations of a and b as a function of any other element concentration can be calculated. This method was used by Mitchell[54] to determine tantalum, niobium, iron, and titanium in oxide mixtures with a precision of 1% at the 50% concentration level or 10% in the range 0.1–1%. For nonlinear calibration curves a series of linear curves was used over a short concentration range.

The determination of zirconium and molybdenum in mixed carbides with uranium is complicated by serious overlap and absorption effects. Zirconium can be determined with a precision of approximately 1% using niobium as an internal standard and applying arithmetic corrections to the intensity ratio to compensate for the effect of uranium on the intercept.[55] Molybdenum is determined in the same sample using the K_β line and the K_β line for niobium as an internal standard. A correction factor is applied to the intensity ratio to compensate for effects of uranium on the slope of the calibration curve.

A semitheoretical approach to correction factors for either linear or nonlinear calibration curves for multicomponent systems has recently been proposed by Andermann.[56] The measured intensity, I_m, is related to true intensity, I_t, by the exponential equation:

$$I_t = I_m \exp\left[F(aC_3 + bC_4 + ---)\right] \tag{10}$$

where F is an experimentally determined factor which is constant for a given instrument and binary system of elements 1 and 2, and the factors $a, b, \ldots$ represent the different effects of elements 3, 4, etc. relative to element 2, on intensity from element 1. The factors a, b, . . . are related to mass absorption coefficients of the elements. The value of the semitheoretical approach is indicated by an average deviation of 0.28% for determination of 70% of silica in various glass samples having wide variations in composition of other constituents. Using the technique for determination of 25% of silica in cement, an accuracy of 0.1% was claimed.

Nonlinear calibration curves are somewhat more difficult to handle than linear curves using hand calculations, but are readily calculated with computers. The calibration curve can usually be expressed by an equation of the form[57]

$$(1-k)/k = a(1-c)/c \tag{11}$$

where k is the measured ratio of intensity from a sample or standard to intensity from a 100% standard, a is a measure of the deviation of the curve from a straight line, and c is the weight fraction of element being determined. Figure 6 shows typical curves where strong fluorescence effects or absorption effects are encountered. For binary systems the parameter a can be calculated from three standards having weight fractions of elements A and B of 0 and 1.00; 0.50 and 0.50, and 1.00 and 0.00. The equation is applicable to solids or liquids, with the weight fraction of liquids neglecting the solvent. For ternary or higher systems values for a can be determined for binary mixtures, and applied to the sample by the equation:

$$(1-k_A)k_A = (a_B^A c_B + a_C^A c_C)(1-c_A)/(c_B + c_C)c_A \tag{12}$$

where k_A is the measured intensity ratio for element A, c_A, c_B, and c_C are weight fractions of elements A, B, and C; and a_B^A and a_C^A are measured parameters describing the effects of elements B and C on the calibration curve for element A. Because all three concentrations are unknown initially, intensity

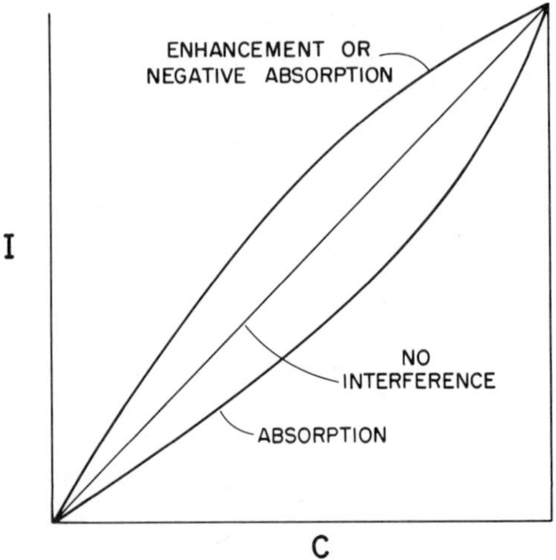

Fig. 6. Effect of fluorescence and absorption on X-ray intensity–concentration curves.

ratios k_B and k_C are first substituted for c_B and c_C and calculations are repeated until further changes are insignificant. As an example of this technique applied to solutions, ternary alloys of chromium, iron, and nickel were dissolved in nitric acid and analyzed. In the concentration range 1–99% the relative standard deviation ranges from 1.7% at the 1% level to 0.8% at concentrations greater than 75%.[47]

Another empirical approach to corrections for interelement effects was suggested by Lucas-Tooth and Pyne.[58] X-ray intensities for several constituents in steel samples were measured for a series of alloys of known composition. These intensities, along with concentrations of all elements, were fed into a computer to determine interelement effects. These effects were considered to be additive terms affecting the slope of the calibration curve by either negative or positive absorption (enhancement was considered equivalent to negative absorption). If a large number of known standards are available for determination of correction factors, this method will provide excellent results. For steel samples, ten or more elements either as major or minor constituents can be determined in less than 1 hr with a standard deviation of better than 1% for major constituents.

D. Trace Analysis

The principles of X-ray fluorescence applications to trace analysis are essentially the same as described for macroanalysis. However, as in any trace analysis technique, the care one observes in setting instrument parameters and in sample manipulations can directly affect the ultimate sensitivity and precision. It is not the purpose of this chapter to demonstrate the importance of laboratory cleanliness and possible contamination from laboratory ware, reagents, and water, but these points cannot be overemphasized.

In trace element analysis one is concerned with obtaining the highest possible signal for a given amount of sample, measured above the lowest possible background. The signal may be increased in several ways: separation and preconcentration of the desired element or elements, as well as judicious selection of instrumentation and instrument operating conditions. The background may also be decreased in many instances: filtration of the primary X-ray beam to reduce intensity from tube contamination, pulse height selection to reduce higher order radiation, and selection of high atomic number backing material to minimize scatter of primary tube radiation.

Trace analysis without separation is limited by the absorption of X-rays in the sample (either solid or solution) as well as by the dilution effect. For example, limits of detection of 40–50 ppm for determining molybdenum, niobium, and zirconium in 1-g tantalum samples dissolved in 5 ml of 36% hydrofluoric acid have been reported.[50] As an example of a heavy element in a light element matrix, a detection limit of 4 ppm of tin in aluminum or magnesium metal was measured by Jenkins.[59]

As examples of high- and low-absorption materials, Campbell and Thatcher,[59a] estimated detection limits for various elements in iron and beryllium matrices. For the iron matrix, detection limits ranged from 1 ppm for titanium to 170 ppm for silicon. As a comparison of sensitivity from the two matrices, detection limits for manganese and nickel were 0.5 ppm in beryllium, and 1.4 and 5.4 ppm, respectively, in iron.

For the best sensitivity in trace analysis separation of the elements to be determined is necessary. Conventional separation techniques such as ion exchange, solvent extraction, or precipitation may be employed. The separated elements may then be concentrated by evaporation on a substrate such as Mylar, filter paper, glass, or any other material which does not contain the elements to be determined. Alternatively, the elements may be adsorbed on ion-exchange membranes[60] or ion-exchange resin-loaded filter paper.[61] Campbell and co-workers[61] list detection limits for a number of elements concentrated on the resin-loaded filter papers, with a detection limit for nickel of 0.06 μg. A flat crystal spectograph was used in this work. Using a curved crystal focusing spectrograph, Luke[60] reported detection limits were improved by approximately a factor of 10. The difference in detection limit for the two procedures can be at least partly attributed to instrumentation.

With a focusing spectrometer a larger fraction of the X-rays from the sample is transmitted to the detector.

The detection limit in X-ray spectroscopy is also a function of counting statistics.[62] The line intensity is measured above some finite background which cannot be reduced to zero. However, for optimum sensitivity it should be reduced to as low a value as possible even at some sacrifice in line intensity. This can be illustrated as follows. The limit of detection is generally defined as that amount of element which yields an intensity above background which is three times the standard deviation of measuring the background. For measuring an element having 1 count/sec/ per μg. above a background of 10 counts/sec, a total of 1000 counts will yield a counting error of approximately 0.33 counts/sec, or a detection limit of 1 μg. An analysis time of 100 sec is required. For the same element measured above a background of 100/ counts/sec, a total of 100,000 counts is required to obtain the same limit of detection, or 1000 sec is required for the analysis. Theoretically, one might assume that any desired sensitivity can be attained merely by accumulating a sufficiently large number of counts. In practice many other factors enter into the intensity measurement, and the time selected for the analysis should be limited by such factors as sample reproducibility and instrumental fluctuations.

Trace analysis by X-ray fluorescence is now at a point where it is competitive with other analytical techniques, and probably deserves more consideration than has been given in the past.

V. APPLICATIONS OF ABSORPTION SPECTROSCOPY

The absorption properties of elements can be applied in several ways for determining the composition of a sample. Applications of X-ray absorptiometry to chemical analysis vary in complexity depending on the needs of the analysis.

A. Absorptiometry with Polychromatic X-Rays

The simplest technique for absorption analysis from the point of view of instrumentation is absorption of polychromatic radiation. In this technique the total radiation from the X-ray source is transmitted through the sample, and the attenuation of the beam is related to attenuation measured from samples of known composition. The only equipment required is an X-ray tube with a highly stable power supply and a detection and counting system. The fundamental absorption equation applies:

$$2.3 \log (I_0/I) = [(\mu/\rho)_A C_A + \sum (\mu/\rho)_i C_i]L \tag{13}$$

where $(\mu/\rho)_A$ and C_A are mass absorption coefficient and concentration, respectively, for the element determined, and $(\mu/\rho)_i$ and C_i are mass absorption coefficients and concentrations for each of the other components of the sample.

Because the total radiation output of the X-ray source is utilized, mass absorption coefficients are for the "effective wavelength" of the tube, which in turn is a function of tube voltage. The tube voltage is selected to maximize $(\mu/\rho)_A C_A - \sum (\mu/\rho)_i C_i$ so that small variations in sample matrix will not produce serious errors in the analysis. The effective wavelength of the source can be determined by measuring I_0/I for a pure material of known thickness and known mass absorption coefficients. From the comparison of the measured mass absorption coefficient with the known mass absorption coefficients, the effective wavelength of the X-ray source can be determined. The effective wavelength at a designated tube potential will vary somewhat for different elements.

One of the most serious limitations to absorptiometry with polychromatic X-radiation is instability in the X-ray source. Zemany et al.[63] have shown that the X-ray tube voltage must be maintained with a precision of 0.01% if absorption measurements are to be reproduced to 1% because tube intensity varies as the 24th power of voltage in some cases. To circumvent this instability, measurements of I_0 and I generally are obtained simultaneously, using such techniques as swinging the sample in and out of the beam[64,65] or using beam-splitting techniques.[63,66]

The applications of polychromatic X-ray absorption to quantitative analysis are many and varied. The technique has been used for determination of ratios of inorganic to organic elements in plant and animal tissue,[65] chlorine in plastics,[67] and lead in gasoline and sulfur in crude oil,[63] to mention only a few. In the latter determination the matrix is generally a mixture of hydrogen, carbon, and oxygen, and an effective wavelength of 0.5–0.6 Å is often used because mass absorption coefficients for hydrogen, carbon, and oxygen are approximately equal in this wavelength region (0.40, 0.305, and 0.508, respectively, for hydrogen, carbon, and oxygen at 0.5 Å). Thus, variations in hydrocarbon composition do not seriously affect the transmitted intensity.

The theory and instrumentation for absorptiometry of inorganic samples with polychromatic X-rays have been discussed by Lambert,[68] and he has applied the technique to the routine precision determination of plutonium[66] and uranium[69] in metals and alloys. For the determination of plutonium in solutions a standard deviation of 0.15% was reported using a differential "criss-cross" technique, certainly comparable to precision obtainable by X-ray fluorescence techniques.

The popularity of polychromatic absorptiometry stems from its adaptability to on-stream analysis. An absorptiometer in a gasoline line can be instrumented to measure and adjust lead content before wet chemical samples could be transmitted to a laboratory. Besides the numerous quantitative applications, qualitative applications of this technique encompass all forms of radiography, as well as study of high explosives, measurement of gas voids in steam, and controlling the liquid level in filling beer cans.

B. Absorptiometry with Monochromatic X-Rays

Absorptiometry with monochromatic radiation is a logical refinement of the polychromatic technique. The monochromatic technique requires somewhat more sophisticated instrumentation but is more selective than the polychromatic method. If an X-ray tube is used as a source, a spectrometer or filter is required to isolate the desired radiation from the remainder of the X-ray beam. This necessitates somewhat more costly equipment, and this, coupled with the decrease in intensity resulting from the use of a monochromator, is the main disadvantage compared to polychromatic absorptiometry. Because of the sensitivity decrease, intensity measurements are generally obtained by integration rather than instantaneous measurement. These disadvantages are offset by the increased selectivity of the method because the wavelength used can be selected just to the long-wavelength side of an absorption edge for the element determined. In addition, the power supply stability is not as serious a problem as in polychromatic absorptiometry.

As in polychromatic absorptiometry, equation 13 is used to relate measured intensities to concentration. However, mass absorption coefficients are more accurately known, and one has better control over the wavelength measured at the detector. For example, Hughes and Wilczewski[70] used the K_α X-ray for molybdenum from a molybdenum-target X-ray tube to determine sulfur in hydrocarbons. The carbon/hydrogen ratio was known, and empirically determined correction factors were applied to the measured transmitted intensity to correct for variations. The availability of radioactive sources led to a simplification of the instrumentation.[71] A 4 mCi ^{55}Fe source was substituted for the molybdenum X-ray tube, providing a monochromatic beam of 2.07 Å X-rays which were transmitted through a cell of 0.5 cm path length. For sulfur concentrations in the range 0.02–1.2% of sulfur, average differences between X-ray and chemical analyses of 0.026% absolute were reported.

The selection of a suitable monochromatic X-ray wavelength and a proper path length are of fundamental importance in X-ray absorptiometry. From a consideration of equation 13 it is apparent that for the best sensitivity in measuring C_a, the product $L(\mu/\rho)_A$ should be large. The wavelength therefore is selected so that mass absorption coefficients for the other elements are as nearly equal as practical, thus making $\sum(\mu/\rho)_i C_i$ in equation 13 a constant, and yet providing a large value for $(\mu/\rho)_A$. A practical limit in select ing values for $L(\mu/\rho)_A$, however, is the count rate observed at the detector. For a 1% counting error a minimum of 10^4 counts are required, and to obtain this number of counts in a reasonable length of time transmitted count rates should be in excess of 100 counts/sec.

Ogilvie[72] has discussed the possible application of monochromatic absorption to the study of diffusion couples. Thin sections are prepared with the polished surface perpendicular to the diffusion zone. For binary systems AB the intensity of an X-ray is measured through air and through equal thicknesses of A, B, and the diffusion zone, and the weight fraction, C_A, of A is given by the equation:

$$\frac{\ln (I/I_0)_{A+B}}{\ln (I/I_0)_A} = C_A + (1-C_A)\frac{(\mu/\rho)_B}{(\mu/\rho)_A}\frac{\rho_{A+B}}{\rho_A} \tag{14}$$

In this equation I_0 is intensity transmitted through air, C_A is weight fraction of A in the diffusion zone, $(\mu/\rho)_A$ and $(\mu/\rho)_B$ are mass absorption coefficients for pure A and B, respectively, and ρ_A and ρ_{BA} are densities of pure A and the binary alloy AB. The main difficulty in applying this technique is preparing equal thicknesses of A, B, and AB. In addition, the density of the unknown alloy AB is not always easy to determine.

The use of radioactive X-ray and gamma ray sources is well suited to monochromatic absorptiometry. Sources such as ^{55}Fe can provide a pure wavelength at a steady output. These have been used by Vose[73] to determine the ratio of organic to inorganic matter in dried bone. Sample thicknesses of 1 mm were used and average deviations of repeated measurements in the range 0.5–1.2% were reported.

The analysis of an n-component system by monochromatic X-ray absorption was reported by Jacobson and Lundberg.[74] Measurements were made at n wavelengths, selected so that mass absorption coefficients are considerably higher for a different element at each wavelength than for the remaining elements. In the proposed method a series of wedges are motor driven into the X-ray beam to balance a desired signal, and the wedge displacement is used as an indication of absorption. Data were not presented for analysis of samples, but the technique should lend itself to continuous on-stream analysis, or rapid analysis of a large number of similar samples.

C. X-Ray Absorption Edge Analysis

As has been noted, X-ray absorptiometry with polychromatic or monochromatic radiations require only simple instrumentation and can be applied quickly. They both suffer from the same basic drawback: they are nonspecific and therefore subject to serious errors if an element having a significantly different mass absorption coefficient than the remainder of the sample is unknowingly introduced. If absorption measurements are made at two wavelengths, selected such that the element being determined has an absorption edge between these wavelengths, the technique is selective for this element. This is commonly referred to as the absorption edge technique, and was first proposed by Glocker and Frohnmeyer.[75] They suggested the relation between transmitted intensities I_1 and I_2 at wavelengths λ_1 and λ_2 and concentration, C, as

$$I_2/I_1 = \exp\,[-\Delta(\mu/\rho)C] \tag{15}$$

where $\Delta(\mu/\rho)$ is the difference in mass absorption coefficients at the two wavelengths for the element being determined. Barium concentrations greater than 20 mg/ml were analyzed with a precision generally better than 4%, which is remarkably good considering that measurements were performed using photographic techniques. It is interesting to note that von Hevesy[76] commented that because of poor sensitivity the absorption edge technique was not applicable at wavelengths longer than 0.7 Å.

The absorption edge method did not find much favor among X-ray spectroscopists until the 1940's when both the instrumentation and theory were refined. The equation of Glocker and Frohnmeyer was actually an oversimplification of the theory, and probably accounted for the poor accuracy they obtained at low concentrations. If we consider the total absorption by the sample at the two wavelengths λ_1 and λ_2, the equations relating incident and transmitted intensities are:

$$I_{\lambda_1} = I^0{}_{\lambda_1} \exp\,\{-[(\mu/\rho)_{\mathrm{A}}^{\lambda_1} C_{\mathrm{A}} + (\mu/\rho)_{\mathrm{B}}^{\lambda_1} C_{\mathrm{B}}]L\} \tag{16}$$

and

$$I_{\lambda_2} = I^0{}_{\lambda_2} \exp\,\{-[(\mu/\rho)_{\mathrm{A}}^{\lambda_2} C_{\mathrm{A}} + (\mu/\rho)_{\mathrm{B}}^{\lambda_2} C_{\mathrm{B}}]L\} \tag{17}$$

where I^0 and I are the intensities transmitted through the sample cell filled with a reference solution such as water, and the intensity transmitted through a standard or the sample, $(\mu/\rho)_{\mathrm{A}}$ is the mass absorption coefficient of the element being determined at the designated wavelength, $(\mu/\rho)_{\mathrm{B}}$ is the mass absorption coefficient of all other components in the sample, C_{A} and C_{B} are concentrations of A and B, respectively, and L is the path length of the sample.

We may remember[3] that the mass absorption coefficient is related to wavelength by the expression:

$$(\mu/\rho) = k\lambda^n \tag{18}$$

where k is a constant through a wavelength region where no absorption edge occurs, and n is constant over a limited wavelength region but can vary from approximately 2.3 to 3 .By combining equations 16, 17, and 18, the equation:

$$C = k_1 \log(I_1^0/I_1) - k_2 \log(I_2^0/I_2) = k_1 \log R_1 - k_2 \log R_2 \tag{19}$$

is obtained. In this equation C is the concentration of element A in mg/ml, and k_1 and k_2 can be calculated by the equations:

$$k_1 = \frac{2303\lambda_2^n}{L[\lambda_2^n(\mu/\rho)_{\lambda_1} - \lambda_1^n(\mu/\rho)_{\lambda_2}]} \tag{20}$$

$$k_2 = \frac{2303\lambda_1^n}{L[\lambda_2^n(\mu/\rho)\lambda_1 - \lambda_1^n(\mu/\rho)_{\lambda_2}]} \tag{21}$$

Equation 19 has been derived by Engstrom,[77] Peed and Dunn,[78] and Dodd[79] and eliminates the effect of wide variations in absorptivity of the sample. Close examination of these equations shows that the denominator in both equations is the same. This value is a measure of the difference in mass absorption coefficients for the element determined at the two wavelengths, and is therefore an indication of the sensitivity of the method. If we divide equation 20 by equation 21 we obtain

$$k_1/k_2 = (\lambda_2/\lambda_1)^n \tag{22}$$

The value for n is essentially dependent only on the impurity elements in the sample (but not on their concentrations), and this value is therefore a measure of the matrix effect for absorption edge analysis. The matrix effect can be minimized by selecting λ_1 and λ_2 as close together as possible so that variations in n will not drastically affect k_1/k_2.

For measuring the concentration of element A in a complex mixture, one measures the parameter k_1/k_2 using equation 19 for a solution containing only impurity elements, so that $C = 0$. Then with a solution of known concentration, C, plus the impurity elements, values for k_1 and k_2 are measured and used for measuring C of the unknown solution.

The importance of determining k_1/k_2 in the presence of the impurity elements expected in the sample is shown by the data in Table 5. Some values for k_1/k_2 are shown using the K_α X-rays for copper and nickel (used for determination of cobalt)[80] and for the $L_{\alpha 1}$ X-rays for bismuth and thallium (used for determination of rhenium).[81]

As shown by the data, k_1/k_2 decreases with increasing atomic number, increases as an absorption edge is crossed, then decreases again to the next absorption edge. It may also be noted that once k_1/k_2 has been measured, equation 22 can be used to calculate the value for n in the wavelength region of the X-rays used.

Table 5. Measurement of matrix effect for K_α X-rays of copper and nickel, and L_α X-rays of bismuth and thallium

Element	Atomic number	k_1/k_2 (Cu:Ni)	k_1/k_2 (Bi:Tl)
Na	11	1.272	—
Al	13	—	1.159
Cl	17	1.245	1.179
Cr	24	1.230	—
Fe	26	1.244	1.147
Ni	28	1.243	—
Cu	29	1.262	1.156
Zn	30	1.237	1.174
Rb	37	—	1.172
Y	39	1.222	—
Zr	40	—	1.162
Cd	48	1.220	1.167
Ba	56	1.217	—
Ce	58	1.212	1.161
Ta	73	—	1.165
W	74	—	1.160
Hg	80	—	1.179
Pb	82	1.240	—
Th	90	1.227	1.154
U	92	1.210	1.157

Several facts can be observed from the data of Table 5. The wavelength separation of copper K_α (1.541 Å) and nickel K_α (*1*.659 Å) is greater than for bismuth $L_{\alpha 1}$ (1.144 Å) and thallium $L_{\alpha 1}$ (1.207 Å), resulting in larger variations in the matrix effect as expressed by k_1/k_2. This is predicted by equation 22. The values for k_1/k_2 correspond to values for n approximately in the region 2.3–3.0. The value for k_1/k_2 generally decreases with increasing atomic number until an absorption edge is crossed. For elements having an absorption edge near the X-ray lines used, the value for k_1/k_2 increases then decreases again with increasing atomic number until another set of absorption edges is reached.

Several methods have been proposed for obtaining the monochromatic wavelengths required for the absorption edge method. In the work of Glocker and Frohnmeyer, Peed and Dunn, and Barieau,[82] an X-ray tube was positioned to transmit the tube output through the sample and this was monochromatized with a crystal spectrometer, as shown in Figure 7*a*. This method has the advantage that any X-ray wavelength can be used, and the wavelength can be selected as close to the absorption edge as permitted by the resolving power of the spectrometer. This minimizes the number of interfering absorption edges, and minimizes the matrix effect as defined by equation 22. It has the disadvantage that equipment is somewhat cumbersome to use interchangeably for fluorescence and absorption analysis. Hughs and Hochgesang[83] used the line radiation from a thorium-target X-ray tube for the determination of lead. This method provides high intensity, but requires a separate tube for each element to be determined. Secondary targets have been proposed by several investigators. Engstrom[84] used pure elements in a rotating sample holder, so that any combination of elements could be used. Hakkila[85] and Hakkila and Waterbury[80,86] used binary alloys or oxide mixtures for the secondary target, as shown in Figure 7*b*. Bertin et al.[87] listed a number of secondary targets, along with interfering elements, applicable in the wavelength region 0.30–2.75 Å. Secondary targets offer the advantage

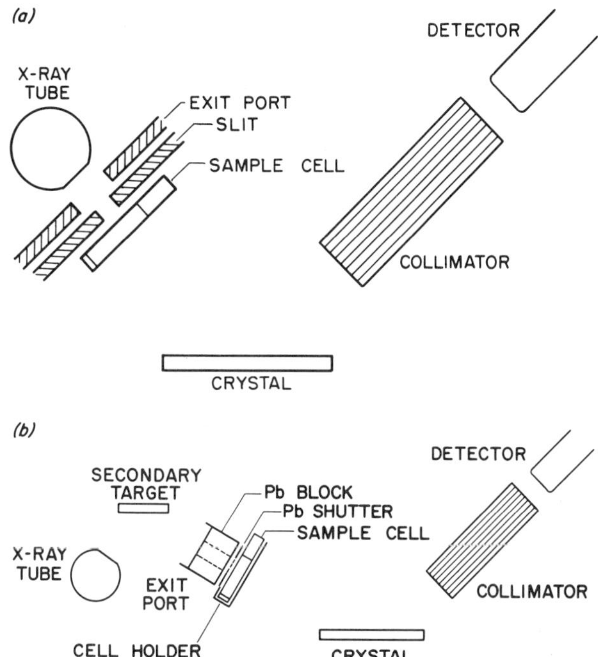

Fig. 7. Instrumental arrangement for absorption edge analysis. (*a*) Tube excitation. (*b*) Secondary target excitation.

that high intensities can be obtained, yet the spectrograph is rapidly interchanged from absorption to fluorescence analysis. The disadvantage of this technique is that wavelengths cannot always be obtained as close together as may be desired, so that more interference is encountered than when the tube white radiation is used. Also, if lines used are far apart the significance of the matrix effect is increased.

Cullen[88] used a novel form of the secondary target technique wherein the target elements were incorporated into the sample. In this method a conventional X-ray spectrograph was used without modifications, the X-ray intensities being measured as in fluorescence analysis. However, X-rays generated from elements in the sample or the X-ray tube may overlap the lines used for analysis—a problem not encountered in the transmission technique.

Cell construction is a critical factor in successful application of absorption edge analysis. Cells must have rigid windows so that cell path length does not change during analysis. This becomes particularly important at longer wavelengths where path lengths of the order of 1 or 2 mm are used. Windows also must be transparent to X-rays. Beryllium has been used as a material for windows, and is probably the best choice as far as X-ray transparency is concerned, but may be chemically attacked by acid solutions. Polystyrene is a convenient material in that it is easily machined and glued to form cells as shown in Figure 8, and is rigid enough that windows 1/32-in. thick or even thinner, can be used. Aluminium foil was used by Dodd,[79] and Dodd has also suggested the use of Mylar for analyses at long wavelengths. With the latter, cell windows should be replaced frequently because any tendency of the window to bulge will affect the cell path length.

The absorption edge technique has been applied to a variety of samples ranging from biological thin sections to complex alloys of uranium. The sensitivity of the absorption edge method is not as good as that available with conventional X-ray fluorescence methods. However, precision of 0.5% or better can be obtained for complex mixtures, with matrix effects generally much smaller than encountered in fluorescence analysis. For example, one blank and one known solution for determination of uranium will suffice for samples containing between 1 and 20 mg/ml of uranium in the presence of large variations in molybdenum, niobium, zirconium, cerium, lanthanum, and ruthenium concentrations.[85]

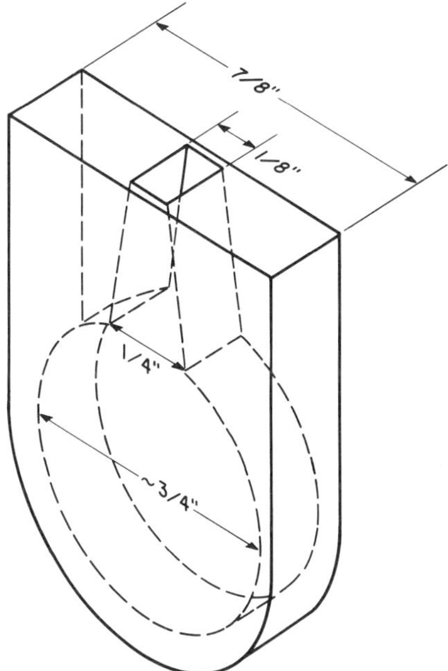

Fig. 8. Construction of polystyrene cells for absorption edge analysis.

D. Analysis by Absorption Edge Fine Structure

Absorption edge fine structure analysis is probably not considered by many analytical chemists as an analytical tool, but has the potential of providing useful analytical data, and should be included with other X-ray absorption techniques. The dependence of absorption edge fine structure on chemical combination has been studied by physicists in order to better understand the nature of chemical bonding. It may be remembered that the absorption edge represents the energy required to eject an electron out of the normal electron shell of the atom. The electron may be captured in some unoccupied outer energy shell, and the fine structure is attributed to the energy difference between the original level and the several final levels available to the electron. The fine structure is generally observed extending only a few hundred electron volts to the short-wavelength side of the edge, and thin samples and fine resolution spectrometers are required to observe the phenomenon. The resolution is obtainable with commercial diffractometers and spectrometers by using finer slits than for normal spectrographic studies. The thin samples can be obtained in some cases by petrographic techniques, by evaporation, or grinding. However, the obtaining of a suitable analytical specimen is often the limiting factor for applications.

Some of the few applications that have been reported may be mentioned. These studies are generally of a qualitative nature, but can be valuable in elucidating the nature of the sample. The wavelength of the *K* edge for sulfur decreases with increasing oxidation state, varying from 5.220 Å for sulfide in Cr_2S_3 to 4.9976 Å for sulfate in magnesium sulfate.[89] This shift would also be expected from elements other than sulfur.

Van Norstrand[90] has studied the absorption edge fine structure from catalysts used in oil cracking. X-ray diffraction studies of $NaMnO_4$ supported on silica gel and on activated carbon did not indicate crystalline components in either sample. However, comparison of the absorption edge fine structure of the catalyst with absorption edge fine structure from MnO_2 and $NaMnO_4$ indicated that the permanganate on silica gel remained as permanganate whereas the permanganate supported on the charcoal was reduced to MnO_2. Studies of cobalt catalysts supported on charcoal show a distinct difference in absorption edge fine structure of "active" and "inactive" types. The fine structure from the "inactive" catalysts corresponded to that from cobalt metal; that from the "active" catalysts to CoO.

Absorption fine structure also has been used to elucidate chemical bonding in glass.[91] Comparison of the germanium *K* edge fine structure in glass containing germanium dioxide with fine structure of tetragonal and hexagonal GeO_2 indicated the germanium in the glass is hexagonally bonded. Physics of absorption edge fine structure for both solids and liquids is not completely understood, but the technique can provide useful qualitative data in many cases.

VI. ANALYSIS OF THIN FILMS

Thin film analysis may be considered a special application of either X-ray fluorescence or absorption. In both fluorescence and absorption methods discussed to this point, one was concerned primarily with determining the composition of a sample. In thin film analysis one is concerned with the thickness of a coating over a substrate of different composition. Both the coating and substrate may be a pure metal, alloy, or compound.

A. Absorption Analysis of Thin Films

Probably the more obvious method for measuring film thickness with X-rays is the absorption technique, and this was investigated first.[92] For the measurement of tin plate thickness on iron, the characteristic X-rays of iron were excited with a copper-target X-ray tube and the intensity was correlated to tin plate thickness using conventional absorption theory.

The experimental arrangement may consist of either a dispersive or a nondispersive detection system. A nondispersive system offers the advantage of high intensity so that a low wattage power supply can be used and statistically significant measurements can be obtained in a short time. The nondispersive system, however, detects all X-ray wavelengths transmitted to the detector, and includes those produced by the coating and any white radiation from the X-ray source that may be scattered by the sample. Therefore, a filter, pulse height analyzer, or selective excitation voltage must be used. For example, Beeghly[92] used a manganese filter which would pass the K_α X-ray for iron but absorb higher energy rays such as the copper K_α from the tube. An excitation voltage less than 29 kV was used to excite the iron but not the tin K_α from the plating. A spectrometer would effectively transmit only the desired monochromatic rays from the substrate to the detector, but with some loss of intensity compared to the nondispersive system. The spectrometer is a more efficient means of energy discrimination and may be a requirement in some cases. In the example above for determining tin plate on steel, the potential required to excite the substrate is lower than that required for the coating, and energy discrimination can be effectively applied to minimize unwanted radiation. However, for measurement of light element coatings on heavy element substrates, energy discrimination obviously is not applicable and a filter system is inefficient. For example, Lublin[93] used the $L_{\alpha 1}$ line for uranium, monochromatized with a lithium fluoride crystal spectrometer, to measure zirconium, aluminum, or stainless steel cladding on uranium fuel elements. X-rays for all three coating elements are excited at the potential required to excite the characteristic radiation of the uranium substrate, and the intensity from the coating could be several times as intense as that of the substrate.

The geometrical arrangement of the detector with respect to the specimen can be critical in using the absorption technique for measuring coating thickness. The path length through the coating traversed by the X-rays from the substrate increases as the angle between detector and substrate is decreased. The intensity I transmitted through a coating thickness, t, is related to the intensity I_0 measured from pure substrate by the equation:

$$I_t = I_0 \exp\left[-(\mu_1) \csc \phi + (\mu_2) \csc \theta\right] t \qquad (23)$$

where ϕ is the incident angle of the primary beam, θ is the emergent angle of the beam from the substrate, and μ_1 and μ_2 are linear absorption coefficients of the coating at the effective wavelength of the primary beam and the emergent beam. Equation 23 shows that the absorption method becomes more sensitive to film thickness as the emergent angle θ is decreased. However, Cass and Kelly[94] have shown that the substrate roughness becomes more serious at low angles. This is demonstrated in Figure 9 where the apparent thickness of the plating is much more seriously affected at the low emergent angle θ_1 than at the higher angle θ_2. At an emergent angle of approximately 40° the error in measuring csc θ begins to approach a minimal value.

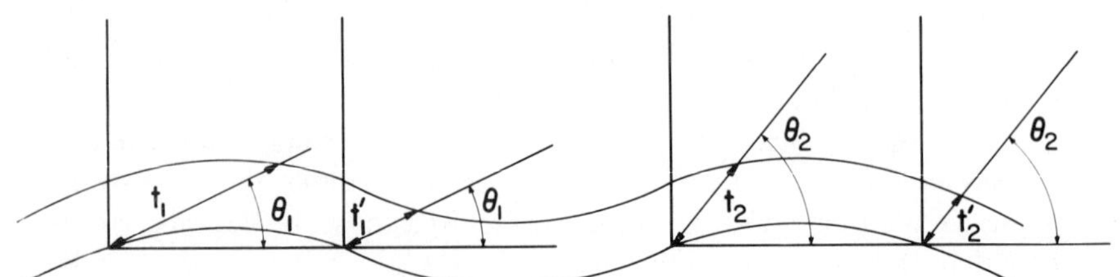

Fig. 9. Effect of substrate roughness on film thickness measurement by absorption.

From equation 22 it is seen that the sensitivity of the absorption method for film thickness is proportional to the mass absorption coefficient and density of the coating as well as to the incident and emergent angles of the X-rays. For incident and emergent angles of 45° (csc ϕ and csc θ each equal 1.4) we have

$$t = [2.3 \log(I_0/I)]/[1.4\rho(\mu_1 + \mu_2)] \tag{24}$$

For an average μ_1 and μ_2 of 100 and ρ of 10, it can be calculated that the method is applicable for measuring thicknesses roughly in the region 0.003 mils or greater.

Cameron and Rhodes[95] summarized thicknesses of various coatings on steel that have been measured using radioactive sources to excite the K_α X-ray for iron. Typical thickness ranges are tin plate on steel, $5\text{–}180 \times 10^{-6}$ cm; zinc on steel, $1.3\text{–}450 \times 10^{-5}$ cm; platinum or gold on titanium, $5\text{–}130 \times 10^{-6}$ cm. The precision of the measurements is generally of the order of 1–2%.

A comparison of uranium K- and L-series lines for measurement of aluminum, stainless steel, and zirconium cladding on uranium fuel elements was performed by Lowe, Sierer, and Ogilvie.[96] The $L_{\alpha 1}$ line was suitable for measurements of 20–40 mils of aluminum on uranium, but could not be used for zirconium or stainless steel in the thickness range investigated. Precision of aluminum thickness measurements was estimated to be 0.1 mils. The K series spectra, using a nondispersive system, were satisfactory for measuring zirconium cladding in the range approximately 4–40 mils with an accuracy of 3% at 30 mils, but mass absorption coefficients for aluminum and stainless steel were not high enough for precise measurements.

B. Fluorescence Analysis of Thin Films

The second method of determining film thickness relies on the measurement of the fluorescent intensity from the cladding material. This method was first reported by Koh and Caugherty[97] for measuring the film thickness of iron, nickel, and chromium. The intensity, I, of X-rays excited in the coating of thickness t can be expressed as

$$I = I_s\{1 - \exp[-((\mu/\rho)_1 \csc\phi + (\mu/\rho)_2 \csc\theta)\rho]\} \tag{25}$$

where I_s is the intensity from an infinitely thick sample of pure coating, ϕ and θ are angles of incident and emergent X-rays, and $(\mu/\rho)_1$ and $(\mu/\rho)_2$ are mass absorption coefficients for the exciting and emerging X-rays, respectively.

Cameron and Rhodes[94] have defined the maximum thickness that can be measured as

$$t_m = 3[(\mu/\rho)_1 + (\mu/\rho)_2] \tag{26}$$

or for the case of tin plating on steel, approximately 0.007 cm. The minimum thickness that can be measured is limited by the X-ray background that is observed from the sample, but is roughly 1% of the maximum thickness.

The comparative sensitivities of the absorption and fluorescent methods for film thickness are a function of the plating and base elements. Generally a plating element of high mass absorption coefficient will be more sensitive by the absorption method. A comparison of fluorescence and absorption methods for determining nickel plating on steel was reported by the National Bureau of Standards.[98] The fluorescence method using the nickel K_α X-ray provided an estimated minimum thickness of 0.0002 in., and the absorption method was applicable in the range 0.0001–0.001 in. It might be noted that the fluorescence method is limited in maximum thickness at the thickness where an infinitely thick sample is observed; the maximum range of the absorption method is generally higher, being limited by the point where X-rays no longer are obtained from the substrate. The theoretical limit is generally higher than the practical limit because of background radiation and the large numbers of counts that are required to differentiate background and line intensities.

Acknowledgments

The author acknowledges the encouragement of Dr. C. F. Metz, under whose supervision this work was performed, and of Dr. G. R. Waterbury for proofreading the manuscript and for valuable suggestions. The assistance of Mrs. Leona Kelly in preparing the manuscript is gratefully appreciated.

References

1. D. Coster and G. von Hevesey, *Nature*, **111**, 79 (1923).
2. A. E. Sandstrom, "Experimental Methods of X-Ray Spectroscopy," in *Encyclopedia of Physics*, Vol. 30, S. Flugge, Ed., Springer, Berlin, 1957, p. 238.
3. A. H. Compton and S. K. Allison, *X-Rays in Theory and Experiment*, 2nd ed., Van Nostrand, New York, 1935, p. 533.
4. op. cit. p. 637–646.
5. M. Siegbahn, *The Spectroscopy of X-Rays* (G. A. Lindsay, Translator), Oxford University Press, London, 1925.
6. G. von Hevesey, *Chemical Analysis by X-Rays and Its Applications*, McGraw-Hill, New York, 1932.
7. D. W. Fischer and W. L. Baun, *Advan. X-Ray Ana.*, Vol. 7, **7**, 489 (1964).
8. J. W. Thatcher and W. J. Campbell, *Advan. X-Ray Anal.*, **7**, 512 (1964).
9. R. A. Mattson, *Advan. X-Ray Anal.*, **8**, 333 (1965).
10. J. Chadwick, *Phil. Mag.* (*Ser. 6*), **25**, 193 (1913).
11. J. O. Karttunen, H. B. Evans, D. J. Henderson, R. C. Niemann, and P. J. Markovich, "A Portable Fluorescent X-Ray Instrument Utilizing Radioactive Sources," Paper No. 32, The 7th Conference on Analytical Chemistry in Nuclear Technology, Gatlinburg, Tenn., 1963.
12. G. B. Cook, C. E. Mellish, and J. A. Payne, "Applications of Fluorescent X-Ray Production by Electron Capture Isotopes" in *Proc. U. N. Intern. Conf. Peaceful Uses Ato. Energy, 2nd, Geneva, 1958*), **19**, p. 127.
13. N. Starfelt, J. Cederlund, and K. Liden, *Intern. J. Appl. Radiation Isotopes*, **2**(3), 4, 265 (1957).
14. L. S. Birks and R. T. Seal, *J. Appl. Phys.*, **28**, 541 (1957).
15. Y. Cauchois, *J. Phys. Radium*, **3**, 320 (1932).
16. L. S.Birks and E. Brooks, *Anal. Chem.*, **27**, 437 (1955).
17. J. W. Kemp, M. F. Hasler, J. L. Jones, and L. Zeitz, *Specırochim. Acta*, **7**, 141 (1955).
18. H. A. Elion and R. E. Ogilvie, *Rev. Sci. Instr.*, **33**, 753 (1962).

19. H. R. Bowman, E. K. Hyde, S. G. Thompson, and R. C. Jared, *Science*, **151**, 562 (1966).
20. J. A. Bearden, U. S. At. Energy Comm. Rept. NYO–10586 (1964).
21. J. A. Bearden and A. F. Burr, U. S. At. Energy Comm. Rept. NYO–2543–1 (1965).
22. ASTM Committee E–2, "X-Ray Emission Line Wavelength and Two-Theta Tables," American Society for Testing and Materials, Philadelphia, Pa., 1965.
23. "X-Ray Wavelengths for Spectrometer," General Electric Co., Milwaukee, Wisc., 1961.
24. M. C. Powers, "X-Ray Fluorescent Spectrometer Conversion Tables for Topaz, LiF, NaCl, EDDT, and ADT Crystals," Philips Electronics, Inc., Mount Vernon, N. Y., 1957.
25. E. W. White, G. V. Gibbs, G. G. Johnson, Jr., and G. R. Zechman, Jr., *X-Ray Wavelengths and Crystal Interchange Settings for Wavelength- Geared Curved Crystal Spectrometers*, Pennsylvania State Univ., University Park, Pa. 1964.
26. W. D. Ashby, V. E. Buhrke, and G. V. Patser, *Advan. X-Ray Anal.*, **7**, 623 (1964).
27. W. A. Fahlbusch, *Appl. Spectry.*, **17**, 72 (1963).
28. R. E. Michaelis and B. A. Kilday, *Advan. X-Ray Anal.*, **5**, 405 (1962).
29. F. Claisse and C. Samson, *Advan. X-Ray Anal.*, **5**, 335 (1962).
30. E. L. Gunn, "Film Depth and Particle Size in X-Ray Fluorescence Analysis," in *Encyclopedia of X-Rays and Gamma Rays*, G. L. Clark, Ed., Reinhold, New York, 1963, p. 373.
31. F. Bernstein, *Advan. X-Ray Anal.*, **5**, 486 (1962).
32. F. Bernstein, *Advan. X-Ray Anal.*, **6**, 436 (1963).
33. F. Bernstein, *Advan. X-Ray Anal.*, **8**, 555 (1965).
34. M. C. Lambert, "X-Ray Spectrographic Determination of Uranium and Plutonium in Aluminum and Other Reactor Fuel Materials," in *7th Ann. Conf. Ind. Appl. X-Ray Anal.*, W. M. Mueller, Ed., University of Denver, Denver, Colo., 1958, p. 193.
35. F. Claisse, Canadian Department of Mines and Technical Surveys Report PR #327 (1956); *Norelco Reporter*, **4**, 3 (1957).
36. G. M. Gordon, D. J. McNely, and J. L. Mero, *Advan. X-Ray Anal.*, **3**, 175 (1960).
37. A. K. Baird, R. S. MacColl, and D. B. McIntyre, *Advan. X-Ray Anal.*, **5**, 412 (1962).
38. B. R. Boyd and H. T. Dryer, "Analysis of Nonmetallics by X-Ray Fluorescence Techniques," in *Developments in Applied Spectroscopy*, Vol. 2, J. R. Ferraro and J. S. Ziomek, Eds., Plenum Press, New York, 1963, p. 335.
39. C. L. Luke, *Anal. Chem.*, **35**, 56 (1963).
40. T. J. Cullen, *Anal. Chem.*, **34**, 867 (1962).
41. T. J. Cullen, *Anal. Chem.*, **32**, 516 (1960).
42. G. Andermann, *Anal. Chem.*, **33**, 1689 (1961).
43. H. J. Rose, Jr., I. Adler, and F. J. Flanagan, *Appl. Spectry.*, **17**, 81 (1963).
44. J. E. Townsend, *Appl. Spectry.*, **17**, 37 (1963).
45. W. J. Campbell and J. W. Thatcher, "Determination of Calcium in Wolframite Concentrates by Fluorescent X-Ray Spectroscopy," *Proc., 7th Ann. Conf. Ind. Appl. X-Ray Anal.*, **7**, 313 (1959).
46. S. Sarian and H. W. Weart, *Anal. Chem.*, **35**, 115 (1963).
46a. L. S. Birks and D. M. Brown, *Anal. Chem.*, **34**, 240 (1960).

46b. D. W. Gaylor, *Anal. Chem.*, **34**, 1670 (1962).
47. E. A. Hakkila and G. R. Waterbury, *Anal. Chem.*, **37**, 1773 (1965).
48. A. A. Chodos, J. J. R. Branco, and C. G. Engel, "Rock Analysis by X-Ray Fluorescence Spectroscopy," in *Proc. 6th Ann. X-Ray Conf., Ind. Appl. X-Ray Anal.*, W. M. Mueller, Ed., University of Denver, Denver, Colo. 1957, p. 315.
49. J. W. Kemp and G. Andermann, *Spectrochim. Acta*, **8**, 118 (1956).
50. E. A. Hakkila and G. R. Waterbury, *Talanta*, **6**, 46 (1960).
51. T. J. Cullen, *Anal. Chem.*, **34**, 812 (1962).
52. J. Sherman, *Spectrochim. Acta*, **7**, 283 (1955).
53. J. Sherman, "A Theoretical Derivation of the Composition of Mixable Specimens from Fluorescent X-Ray Intensities," *Proc. 6th Ann. X-Ray Conf., Ind. Appl. X-Ray Anal.*, W. M. Mueller, Ed., University of Denver, Denver, Colo., 1957, p. 231.
54. B. J. Mitchell, *Anal. Chem.*, **33**, 917 (1961).
55. E. A. Hakkila, R. G. Hurley, and G. R. Waterbury, *Anal. Chem.*, **36**, 2094 (1964).
56. G. Andermann, *Anal. Chem.*, **38**, 82 (1966).
57. T. O. Ziebold and R. E. Ogilvie, *Anal. Chem.*, **36**, 322 (1964).
58. J. Lucas-Tooth and C. Pyne, *Advan. X-Ray Anal.*, **7**, 523 (1964).
59. R. Jenkins, *Limitations of Detection in Spectrochemical Analysis*, Hilger & Watts, Ltd., London, 1964.
59a. W. J. Campbell and J. W. Thatcher, *U. S. Bur. Mines, Rept. Invest.*, 5966 (1962); J. Adler and H. J. Rose, Jr., "X-Ray Emission Spectrography" in *Trace Analysis Physical Methods*, G. H. Morrison, Ed., Intesrcience, New York, 1965, p. 312.
60. C. L. Luke, *Anal. Chem.*, **36**, 318 (1964).
61. W. J. Campbell, E. F. Spano, and T. E. Green, *Anal. Chem.*, **38**, 987 (1966).
62. N. Spielberg and M. Bradenstein, *Appl. Spectry.*, **17**, 6 (1963).
63. P. D. Zemany, E. H. Winslow, G. S. Poellmitz, and H. A. Liebhafsky, *Anal. Chem.*. **21**, 493 (1949).
64. H. P. Hanson, W. E. Flynt, and J. E. Dowdey, *Rev. Sci. Instr.*, **29**, 1107 (1958).
65. G. P. Vose, *Anal. Chem.*, **30**, 1819 (1958).
66. M. C. Lambert, U. S. At. Energy Comm. Rept. HW–26399 (1952); reported by C. F. Metz and G. R. Waterbury, "The Transuranium Actinide Elements" in *Treatise on Analytical Chemistry*, Pt. I, Vol. 9, I. M. Kolthoff and P. J. Elving, Eds., Interscience, New York, 1962, p. 397.
67. H. A. Leibhafsky, *Anal. Chem.*, **19**, 861 (1947).
68. M. C. Lambert, U. S. At. Energy Comm. Rept. HW–SA–1972 (1960).
69. M. C. Lambert, U. S. At. Energy Comm. Rept. HW–24717 (1952).
70. H. K. Hughes and J. W.Wilczewski, *Proc. Am. Petroleum Inst.*, **30**, M (III), 11 (1950).
71. H. K. Hughes and J. W. Wilczewski, *Anal. Chem.*, **26**, 1889 (1954).
72. R. E. Ogilvie, "Analysis of Diffusion Couples by X-Ray Absorption," *Proc., 6th Ann. X-Ray Conf. Ind. Appl. X-Ray Anal.*, W. M. Mueller, Ed., University of Denver, Denver, Colo., 1957, p. 439.
73. G. P. Vose, *Microchemi. J.*, **4**, 537 (1960).
74. B. J. Jacobson and B. Lundberg, *Rev. Sci. Instr.*, **35**, 1316 (1964).
75. R. Glocker and W. Frohnmeyer, *Ann. Physik*, **76**, 369 (1925).
76. G. von Hevesey, *Chemical Analysis by X-Rays and Its Applications*, McGraw-Hill, New York, 1932, p. 132.
77. A. Engstrom, *Acta Radiol. Suppl.*, **63** (1946).
78. W. F. Peed and H. W. Dunn, U. S. At. Energy Comm. Rept. ORNL–1265 (1952).
79. C. G. Dodd, *Advan. X-Ray Anal.*, **3**, 11 (1960).

80. E. A. Hakkila and G. R. Waterbury, in *Advan. X-Ray Anal.*, **5**, 379 (1962).
81. E. A. Hakkila, R. G. Hurley, and G. R. Waterbury, U. S. At. Energy Comm. Rept. 3135 (1964).
82. R. E. Barieau, *Anal. Chem.*, **29**, 348 (1957).
83. H. K. Hughes and F. P. Hochgesang, *Anal. Chem.*, **22**, 248 (1950).
84. A. Engstrom, *Rev. Sci. Instr.*, **33**, 1012 (1961).
85. E. A. Hakkila, *Anal. Chem.*, **33**, 1012 (1961).
86. E. A. Hakkila and G. R. Waterbury, "Applications of X-Ray Absorption Edge Analysis," in *Developments in Applied Spectroscopy*, Vol. 2, J. R. Ferraro and J. S. Ziomek, Eds., Plenum Press, New York, 1963, p. 297.
87. E. P. Bertin, R. J. Longobucco, and R. J. Carver, *Anal. Chem.*, **36**, 641 (1964).
88. T. J. Cullen, *Anal. Chem.*, **37**, 711 (1965).
89. A. E. Lindh, "Rontgenspektroskopie," in *Handbuch der Experimentalphysik*, Vol. 24, No. 2, Akademische Verlagsgesellschaft, Leipzig, 1930, p. 291.
90. R. A. Van Nordstrand, *Advan. Catalysis*, **12**, 149 (1960).
91. W. F. Nelson, I. Siegel, and R. Wagner, *Phys. Rev.*, **127**, 2025 (1962).
92. H. F. Beeghly, *J. Electrochem. Soc.*, **97**, 152 (1950).
93. P. Lublin, *Norelco Reporter*, **6**, 57 (1959).
94. D. E. Cass and J. H. Kelly, *Norelco Reporter*, **10**, 49 (1963).
95. J. F. Cameron and J. R. Rhodes, "Coating Thickness Measurement and Chemical Analysis by Radioisotope X-Ray Spectrometry," in *Encyclopedia of X-Rays and Gamma Rays*, G. L. Clark, Ed., Reinhold, New York, 1963, p. 150.
96. B. J. Lowe, P. D. Sierer, Jr., and R. E. Ogilvie, "Cladding Thickness of Fuel Elements by X-Rays," in *Proc. 7th Ann. Conf. Ind. Appl. X-Ray Anal.*, W. M. Mueller, Ed., University of Denver, Denver, Colo., 1959, p. 275.
97. P. K. Koh and B. Caugherty, *J. Appl. Phys.*, **23**, 427 (1952).
98. B. F. Scribner, Ed., *NBS Tech. Note*, **272**, 45 (1965).

Supplementary References

I. Adler and H. J. Rose, Jr., "X-Ray Emission Spectrography" in *Trace Analysis, Physical Methods*, G. H. Morrison, Ed., Interscience, New York, 1965, p. 271.

L. S. Birks, *X-Ray Spectrochemical Analysis*, Interscience, New York, 1959.

M. A. Blokhin, *Methods of X-Ray Spectroscopic Research*, Pergamon Press, Oxford, 1965.

G. L. Clark, Ed., *Encyclopedia of X-Rays and Gamma Rays*, Reinhold, New York, 1963.

A. H. Compton and S. K. Allison, *X-Rays in Theory and Experiment*, 2nd ed., Van Nostrand, Princeton, N. J., 1954.

S. Flugge, Ed., *Encyclopedia of Physics*, Vol. 30, Springer, Berlin, 1957.

E. F. Kalble, Ed., *Handbook of X-Rays*, McGraw-Hill, New York, 1967.

F. Jaundrell-Thomson and W. J. Askworth, *X-Ray Physics and Equipment*, F. A. Davis Co., Philadelphia, 1965.

H. A. Liebhafsky, H. G. Pfeiffer, E. H. Winslow, and P. D. Zemany, *X-Ray Absorption and Emission in Analytical Chemistry*, Wiley, New York, 1960.

H. A. Liebhafsky, H. G. Pfeiffer, and E. H. Winslow in *Treatise on Analytical Chemistry*, Part I, Vol. 5, I. M. Kolthoff and P. J. Elving, Eds., Interscience, New York, 1964, Chap. 60.

Chapter 9

SOFT X-RAY SPECTROSCOPY IN METALS RESEARCH

J. E. HOLLIDAY, Edgar C. Bain Laboratory for Fundamental Research, United States Steel Corporation Research Center, Monroeville, Pennsylvania

I. Introduction 326
II. Spectrometer 327
III. Analyzers 328
A. Gratings 328
B. Crystals 336
C. Comparison of crystals and gratings 339
IV. The Detector 345
V. Excitation of Soft X-rays 349
A. Electron excitation 349
B. X-ray excitation 358
C. Penetration depth 363
VI. Emission Lines and Bands 364
A. Emission lines 365
B. Emission bands 370
C. Emission band from transition metals and alloys 373
VII. Changes in Shape and Wavelength of Emission Bands from Elements When Chemically Combined 383
A. Oxides 384
B. Nitrides 386
C. Carbides 388
D. Diborides 397
VIII. Electron Distribution and Bonding 398
A. Ionic bond 401
B. Metallic bond 403
C. Bond strength 405
IX. Quantitative Analysis 406
General References 414
References 414

I. INTRODUCTION

The field of soft X-ray spectroscopy has advanced rapidly in the past several years. The reason for this progress has been the interest in the use of soft X-rays as a tool in metals research, specifically to measure quantitatively the 2nd period elements and to study chemical bonding and electronic properties of metals, alloys, and compounds. The large changes in wavelength and intensity distribution of emission bands with alloying and chemical compounds, which have recently been observed, show that important information can be obtained regarding electronic properties of solids. It has been shown[1] that these changes are more easily correlated with electronegativity and bonding concepts than with theoretical band calculations.

The increased interest in soft X-ray spectroscopy has also resulted in advances in instrumentation. The most significant advance has been made in analyzers. Previously, crystals were not available with large enough *d* spacings to be used in the soft X-ray region. However, the improved state of the art has made it possible to obtain crystals with large enough *d* spacings so that the Be *K* band at 110 Å can be measured. The pioneer work of the author[2] in the use of blazed replica gratings has made it possible to obtain reliable gratings that consistently give high intensity and peak-to-background ratios (P/B). The blazed replica gratings have been particularly valuable in the 10–50 Å region where the conventional Siegbahn grating yielded erratic intensities and P/B ratios. In addition the large number of grooves per millimeter that are made available with the blazed replica grating allows improved resolution over the Siegbahn grating.

In any review of soft X-ray spectroscopy as a technique in metals research, the names of Skinner, Tomboulian, and Parratt must be mentioned. Their pioneer investigations in this field are the foundation for the present review. Skinner's classic article in the *Proceedings of the Royal Society*, Tamboulian's *Handbuch der Physik* article, and Parratt's *Review of Modern Physics* article provide good background material for the present article. Also Blochin's book *The Physics of X-rays* gives a good treatment of the whole field of X-ray physics. The reader is referred to the general references at the end of this chapter. The wavelength region covered in this article is from about 10–150 Å. Sections II through V are devoted to instrumentation. In these sections are discussed analyzers, detectors, electron, and X-ray excitation. Section VI gives a basic background of emission lines and bands, Sections VII and VIII cover changes in emission band fine structure with chemical changes and bonding. Section IX is devoted to quantitative analysis of the light elements.

II. SPECTROMETER

The mechanical details and alignment of the grating spectrometer have been covered by the author in other articles.[1,2] The mechanical details and alignment of a crystal spectrometer in the soft X-ray region are the same as those in the hard X-ray region of the spectrum. Sandstrom[3] has an excellent review article on crystal spectrometers in the *Handbuch der Physik*. Since the principle of a grating spectrometer is not as well known as that of the crystal spectrometer, and because its alignment is more critical than the alignment of a crystal spectrometer, the basic principle of the grating spectrometer will be briefly described.

The grazing incidence grating spectrometer is indicated schematically in Figure 1. The entrance slit S_1, the analyzer slit S_2 and the curved grating are on the Rowland circle, whose radius is one-half the radius of curvature of the grating. The X-rays from the target are directed by slit S_1 at glancing angle ϕ to the grating. They are then diffracted through angle θ and focused on the Rowland circle. The analyzer slit S_2 and counter are scanned across the focused X-rays by the lead screw B and the spectrometer arm A. Since diffraction from a grating is based on a surface interference principle, ϕ remains constant when S_2 is scanned across the diffracted X-rays. It is essential that the window of the proportional counter and S_2 remain collinear

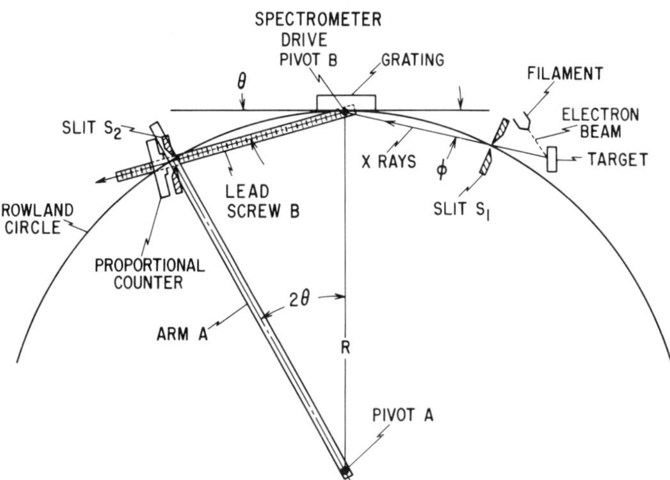

Fig. 1. Schematic of grazing incidence soft X-ray spectrometer showing placement of gratings and slits on Rowland circle, Lead screw *B* has 36 threads per inch and the spectrometer is driven from pivot *B*.

with the pole of the grating. This is accomplished by the lead screw B which is pivoted below the grating.

III. ANALYZERS

A. Gratings

Gratings have been used as analyzers in soft X-ray spectrometers since the early 1930's. The grating analyzer for soft X-rays involves two basic principles, diffraction and reflection. Figure 2*a* is a schematic drawing of the Seigbahn type diffraction grating showing the incident and diffracted X-rays. The relation between wavelength λ, glancing angle ϕ, grating constant σ, and the angle of diffraction θ is

$$n\lambda = \sigma(\cos\theta - \cos\phi) \tag{1}$$

where n is a negative integer denoting the spectral order of diffraction. This equation shows that the angle of diffraction does not depend on the shape of the reflecting surface. It is important to keep this in mind when discussing the blazed grating. Thus, for a given wavelength, ϕ and σ, the diffraction angle θ is always the same regardless of the type of periodic structure.

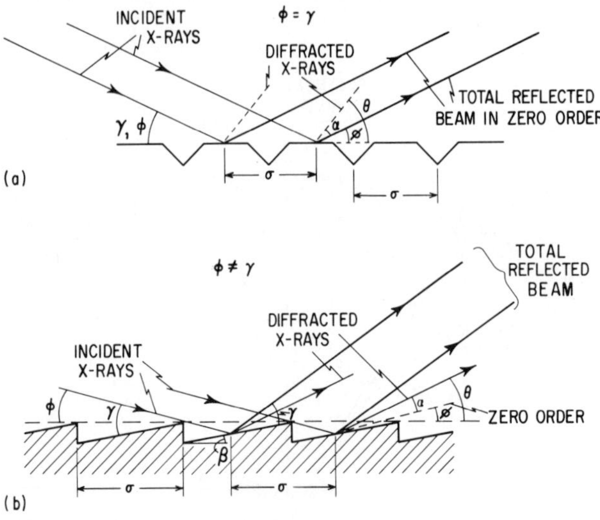

Fig. 2. Diffraction of X-rays showing relative position of totally reflected beam with respect to zero order: (*a*) diffraction of X-rays from Siegbahn type grating; (*b*) diffraction of X-rays from blazed grating.

In order to obtain X-ray intensity from a grating, it is necessary that the condition for reflection of X-rays from the grating surface be fulfilled. This condition can be understood from the law of refraction which is

$$\sin i/\sin r = \mu \tag{2}$$

Since μ is slightly less than 1 for X-rays, the angle of refraction, r, will always be greater than the angle of incidence, i. At an angle of incidence, i_c, whose sine equals μ, r equals 90° ($\mu/\sin r = \mu$). For angles of incidence larger than i_c, refraction ceases and the X-ray beam is totally reflected. It is expected from this analysis that the reflected intensity will jump from a very small value to a value near that of the incident beam. Such is not the case, however, and the critical angle i_c is rather indefinite.

Even though i_c is not sharp, it is possible to calculate the critical angle for X-rays from dispersion theory. The theory gives the following value for the refractive index

$$\mu = 1 + \frac{e^2}{2\pi m}\sum_s \frac{N_s}{V_s^2 - V^2} \tag{3}$$

where e and m are the charge and mass of the electron, N_s, the number of relatively free electrons per unit volume having the natural frequency V_s, and V is the frequency of the radiation. For X-rays $V \gg V_s$, therefore, the fractional term is negative, giving an index of refraction which is less than one. Thus equation 3 may be rewritten in the following form

$$\mu = 1 - (e^2N_s/2\pi mV^2) \tag{4}$$

In terms of wavelength the equation is

$$\mu = 1 - (e^2N_s\lambda^2/2\pi mc^2) \tag{5}$$

Putting this value of the refractive index into equation 2 and making $r = 90°$ (the condition for total reflection) the following relation is obtained:

$$\sin i_c = 1 - (e^2N_s\lambda^2/2\pi m_c{}^2) \tag{6}$$

In terms of the critical glancing angle,* ϕ_c ($\phi_c = 90° - i_c$), equation 6 is written

$$\cos \phi_c = [1 - (e^2N_s{}^2\lambda^2/2\pi m_c{}^2)]\sin r \tag{7}$$

Expanding $\cos \phi_c$ into an infinite series, and approximating, the angle ϕ_c is given by

$$\phi_c = (2\delta\lambda_c) \quad \text{where} \quad \delta = (e^2N^2/2\pi m_c{}^2) \tag{8}$$

* In previous publications by the author ϕ was labeled θ but it has been changed in order to reserve θ for the angle of diffraction so the notation will be the same as for crystal spectrometers.

where λ_c is the cutoff wavelength. Thus, the greater the free electron density, the larger the critical angle, ϕ_c, and the smaller the cutoff wavelength λ_c. Nicholson et al[4] have solved equation 8 for ϕ_c Using Al_2O_3, Au, and Pt, which have different N_s, as a function of λ_c. The results of their calculations are shown in Table 1. The critical angle, ϕ_c, is indefinite and thus λ_c is

Table 1. The critical glancing angle ϕ_c, as a function of λ_c calculated from equation 8. Reprinted by permission from J. B. Nicholson, G. Frank, Y. Mooney, and G. L. Griffin, in *Advances in X-ray Analysis*, G. R. Mallett, M. Fay, and W. M. Mueller, eds. Vol. 8, Plenum Press, New York, 1965, p. 306

	Al_2O_3	Au	Pt
23.7 Å (O *K*)	4.37°	8.77°	9.52°
31.6 Å (N *K*)	5.82°	11.69°	12.69°
44 Å (C *K*)	8.12°	12.28°	17.67°
65 Å (B *K*)	11.99°	24.06°	26.10°

also indefinite (eq. 8). This indefiniteness becomes greater with the increase in the wavelength at which the X-rays are cut off. The diffuse cutoff is seen in Figure 3 for Mo white radiation near the "cutoff wavelength." The X-ray intensity does not drop instantly to zero but is spread over several Angstroms.

The cutoff wavelength λ_c is reduced for materials that have a high N_s (Table 1). Holliday[2] has shown that the intensity of the Cu L_{III} band (13.3 Å) from a Pt-coated blazed replica grating increased four times over the value that was obtained from an Al-coated grating, and the P/B doubled. For the C K band (44.85 Å), however, the Pt grating gave about 10% less intensity than the Al grating.

The experiments performed in the last several years have shown the importance of groove shape in obtaining optimum results from a grating. The grating profile used until recently is shown in Figure 2*a*. This grating is lightly ruled with a sharp diamond point, and the diffracted X-ray intensity is contributed by the flat area between the grooves. The most favorable groove width is that in which the ruled and unruled portions are of equal width. Since this type of ruling for the soft X-ray region was pioneered by Siegbahn, it is referred to as the Siegbahn grating. Individual gratings of the Siegbahn type were found to have widely differing characteristics. For example, one grating would give good intensity and P/B while another

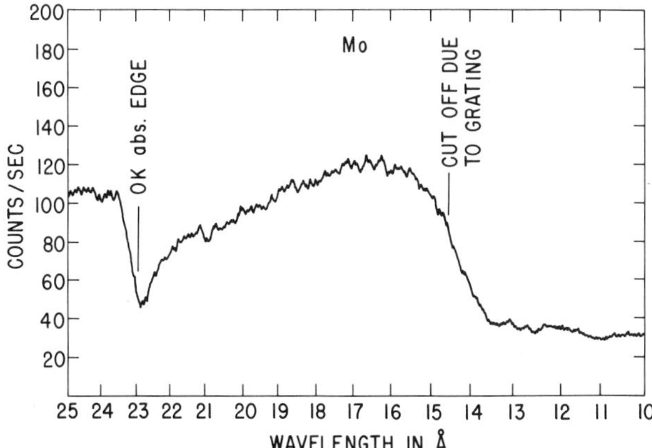

Fig. 3. White radiation from an Mo target showing cutoff of grating due to "critical wavelength" λ and the oxygen–absorption edge resulting from the oxygen in the counter window and the Al_2O_3 on the grating surface. The grating was aluminized with 600 grooves per/mm and $\beta = 1°$. The radiation below ~13A is due to scattered radiation from slits and grating. The target potential was 3.5 kV, beam current 1 mA, and slits 40 μ wide. Reprinted by permission, J. E. Holliday, *J. Appl. Phys.* **33**, 3260 (1962).

would not give any intensity beyond the zero order. One reason for this variation in characteristics is that during the ruling process, material will be piled up on the edge of the groove, resulting in random scattering of the X-rays.

It was believed by Richardson[5] of Bausch & Lomb and also by Holliday that an advantage could be gained by using a blazed grating. Original experiments by Holliday[6–8] confirmed this especially at the shorter wavelengths (10–44 Å). The profile of the blazed grating is shown in Figure 2*b*. There are two basic advantages of the blazed replica grating over that of the Siegbahn type. The first is that in the blazed replica grating there is no pileup of material at the groove edge. In the replication process the bottom of the groove in the master ruling, which is very sharp, becomes the top of the groove in the first generation replica. In ordering replica gratings, it is important to order the first generation replica (the grooves in the second generation replica correspond to the grooves in the master ruling). The second advantage of the blazed replica grating is the ability to diffract maximum intensity at discrete wavelengths. The fact that this can be done with a blazed grating is based on the principle mentioned earlier, namely, that the diffraction angle θ is a function of ϕ and σ and not on the angle δ that the X-rays make with the reflecting surface (Fig. 2*b*). Thus, the reflecting surface could be tilted with respect to the grating plane without changing the diffracting angle. There may be some who are under the misapprehension that ϕ in the grating equation (eq. 1) is the angle made with the reflecting surface rather than with the plane of the grating. It should be remembered that the geometrical considerations used in obtaining the grating equation are based on the fact that ϕ was measured with respect to the plane of the grating. Tilting the reflecting surface with respect to the plane of the grating will not have any effect on the condition for interference. Since diffracted X-ray intensity is based on the two separate principles of diffraction and reflection, the condition for reflection can be changed without affecting the diffraction condition. Thus, by tilting the reflecting surface ($\gamma \neq \phi$) the total reflected beam can be reflected at any discrete wavelength as is illustrated in Figure 2*b*. Since $\gamma = \phi$ in the Siegbahn grating, it can be seen from Figure 2*a* that the totally reflected beam is in the zero order which could result in a very inefficient grating. This fact could account for some Siegbahn-type gratings giving no diffracted intensity beyond the zero order. It is quite possible that the Siegbahn gratings that do give good diffracted intensity have some accidental blazing.

The experimental verification of the effect of blaze angle in obtaining a high reflecting efficiency for a given wavelength (selective effect) is shown in Figures 4 and 5. Figure 4 is the work of Holliday,[9] who varied γ by changing β and keeping φ constant. Figure 5 is the work of Nicholson and Hasler.[10] They kept β constant and varied γ by changing ϕ. The diffracting angle at which the maximum reflected intensity will occur is given approximately by $(\gamma \pm \beta)$ or $(\varphi \pm 2\beta)$. The sign of β is defined by the direction from which the incident X-rays strike the blaze. It is negative for the direction indicated in Figure 6*a* and positive for Figure 6*b*. Thus, the wavelength at which the maximum reflected intensity occurs can be varied by changing either γ or β. However, the values assigned to γ and β are limited by λ_c. According to the $(\gamma \pm \beta)$ relation the maximum reflected intensity for a 600 grooves/mm grating for $\varphi = 2°$, $\delta = 3°$, and $\beta = 1°$ will occur at 4° which is approximately the θ angle for the C *K* band (44.85 Å). This is confirmed by the experimental results shown in Figure 4 where the maximum diffracted intensity occurs between the C *K** and N *K* bands. For the above grating conditions, the diffraction angles for N *K* and C *K* bands are 4.4° and 4.33°, respectively for $\varphi = 2°$ and a 600 grooves/mm grating. The drop in intensity

* According to Nicholson and Hasler[10] the absolute efficiency of the 600 grooves per millimeter 1° blazed replica aluminized grating is 5.7%.

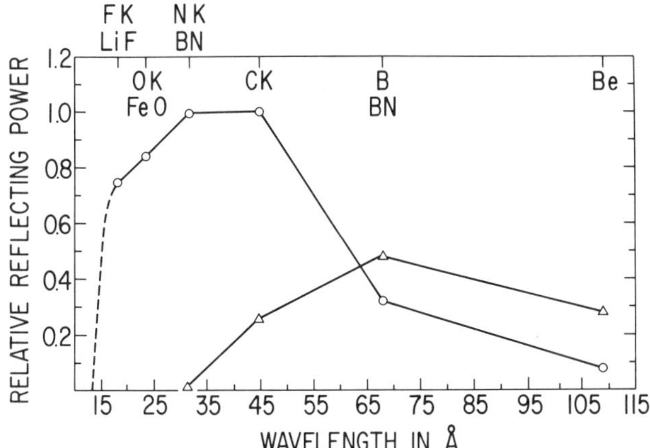

Fig. 4. Relative reflecting power referred to the C K band at 44.85 A for two aluminized gratings having 600 grooves/mm, concave radius of 1 m, and a $\beta = 1°(\bigcirc)$ and $7°55'$ ($\triangle$). For both gratings $\phi = 2°$, giving a γ of $3°(\bigcirc)$ and $9°55'(\triangle)$ .The radius of curvature is 1 meter for both gratings. Reprinted by permission, J. E. Holliday in *The Election Microprobe,* T. D. McKinley, K. F. J. Heinrich and D. W. Wittry Eds., Wiley, New York, 1966 p. 10.

indicated by the dashed line in Figure 4 is due to a combination of the above-mentioned selective effect and λ_c (cutoff wavelength). As would be expected for $\beta = 7.°55'$ blaze, the peak intensity occurs at a longer wavelength. Since Nicholson and Hasler,[10] varied δ by changing ϕ instead of β, their work (Fig. 5) does not show the effect of a greater intensity at a given ϕ as well as Holliday's results do (Fig. 4). When ϕ is varied, not only is ϕ changed but the diffracting angle θ is also changed. One can see from Figure 5 the effect of the blaze grating on maximum intensity at a given wavelength. For the BeK band the peak occurs at $\theta \cong 4°$ and then falls gradually for $\theta > 4°$. It would be expected that the maximum for Be K would occur at an angle much greater than 4°. A similar effect is observed for the 7° 55′ blazed grating in Figure 4. The maximum intensity occurs at N K (67.56 Å) while the $\gamma + \beta$ relation predicts that the peak should occur near 200Å. The reason for this discrepancy is that the $\gamma + \beta$ relation is an oversimplification of a complex phenomenon.

In a Siegbahn grating the direction from which the X-rays strike the reflecting surface of the grating is unimportant. This is not true for the blazed grating. The effect on the diffracted intensity for the two directions is shown schematically in Figures 6*a* and 6*b*. It was indicated above that the diffraction angle at which the maximum reflected intensity occurs is

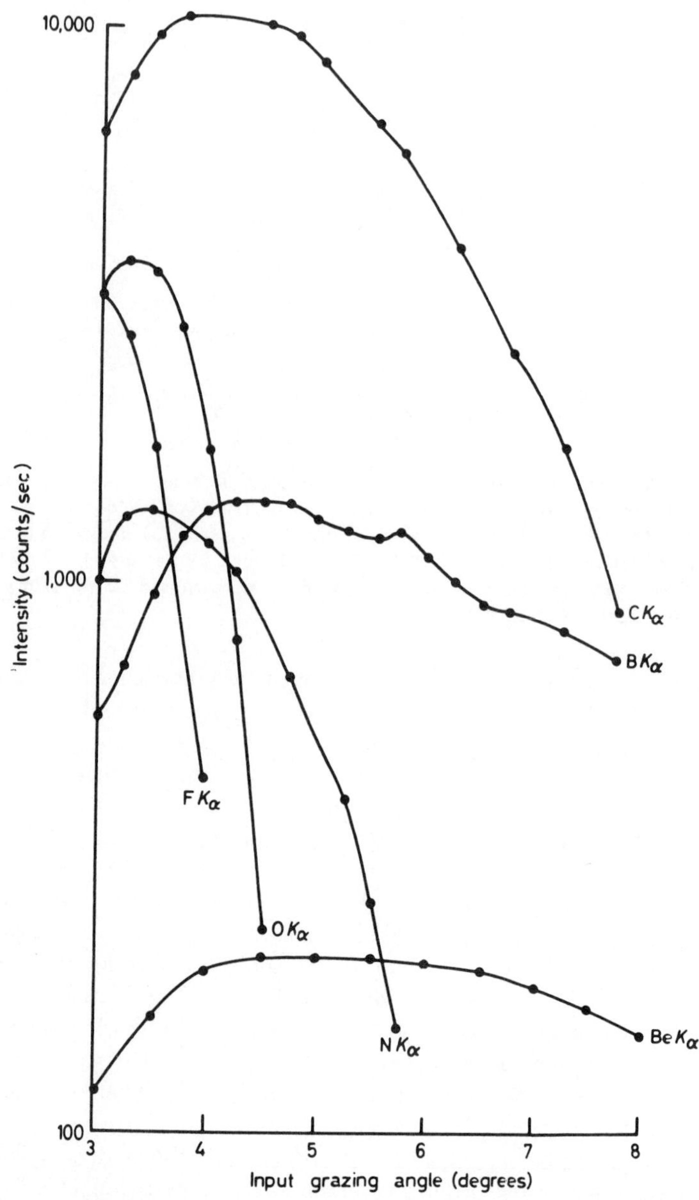

Fig. 5. Grating efficiency versus input grating angle γ for 600 grooves/mm, 1° blazed Al grating with a 1 meter radius of curvature. Reprinted by permission, J. B. Nicholson and M. F. Hasler, in *Advances in X-ray Analysis,* Vol. 9, Plenum Press, New York, 1966 p. 425.

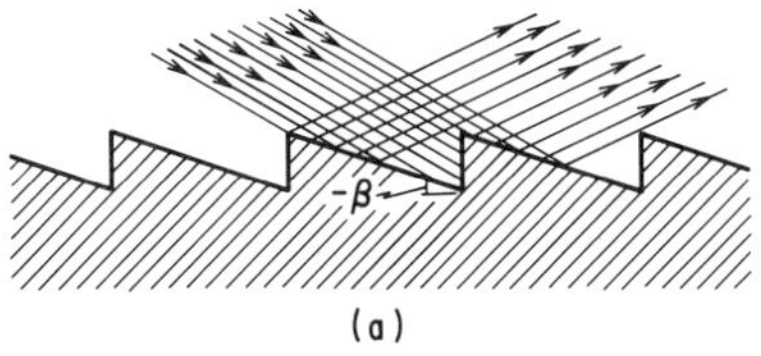

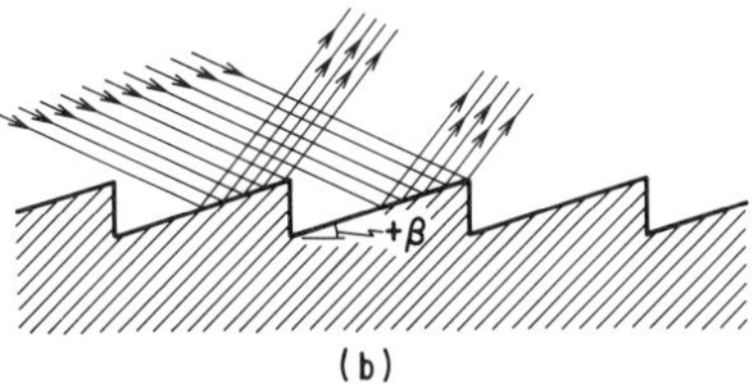

Fig. 6. (*a*) and (*b*) compare the two different directions at which the X-rays strike the blaze. For the direction indicated in (*a*), the blaze angle is $-\beta$ and for (*b*), it is $+\beta$. Reprinted by permission, J. E. Holliday in *Hnadbook of X-rays*, E. F. Kaelble Ed., McGraw-Hill, New York, 1967, p. 38-5.

$\gamma + \beta$. This relation is only for the direction indicated in Figure 6*b*. For the direction indicated in Figure 6*a* the maximum reflected intensity would occur at $\gamma - \beta$. Thus, for the direction indicated in Figure 6*a* the maximum reflected intensity occurs in the unusable positive orders. Also at low diffraction angles the majority of the intensity is lost in the step of the blaze. Holliday[2] has shown the effect of X-ray intensity for the two directions from which the X-rays strike the blaze for a 600, 1200, and 2160 grooves per millimeter, 1° blazed grating. For the 600 grooves/mm grating, the $+\beta$ direction gives about 12 times more intensity than the $-\beta$ direction. There is less difference between the two directions for the 2160 groove/mm grating. The $+\beta$ direction gives only twice the intensity of the $-\beta$ direction. This reduced difference indicates the blaze is becoming more rounded for the 2160 grooves/mm grating than for the lower number of grooves per millimeter gratings. If the step of the blaze is rounded rather than sharp (it is sharp in Fig. 2*b*), then the advantages of the blaze grating will be lost. The above results from the 2160 grooves/mm grating indicate that improvements in intensity for a large number of grooves/mm gratings can be obtained by making the blaze shape closer to that in Figure 2*b*.

The fact that improvements in intensity can be obtained from a grating with a large number of grooves/mm by improving the groove shape is shown in Table 2. In order to determine the contribution of the improved groove

Table 2. X-ray intensity and P/B from two 3600 grooves/mm and 1° blazed gratings[a] replicated from different masters

Grating	V L_{III} (24.47 Å)		Ti L_{III} (27.45 Å)		C K (44.85 Å)		B K (67.56 Å)	
1st Ruling	C/S	P/B	C/S	P/B	C/S	P/B	C/S	P/B
Al Surface	21	1.5	30	1.7	200	7	50	7.5
2nd Ruling Pt Surface	560	13	285	8	625	19	100	15

[a] Ruled by Bausch & Lomb Co.

surface, the effect of the Pt surface relative to Al will have to be considered. It was stated earlier that the Cu L_{III} band intensity was increased by a factor of 4 for a Pt surface, and it has been shown by Holliday[2] that the C K band intensity dropped slightly for a Pt surface as compared to an Al surface. Thus, from the results in Table 2, it can be stated that the improved groove shape has resulted in an increase in grating efficiency by at least a factor of 3.

B. Crystals

Recently a number of crystals have become available that have sufficiently large d spacings to be used in the soft X-ray region (10–150 Å). Crystals have certain advantages over gratings. A detailed comparison between gratings and crystals will be left to the next section. Since crystals accept a larger solid angle than gratings, it is expected that a greater intensity is possible from the crystal. Also, crystals are cheaper than gratings and the same spectrometer can be used for both the hard and the soft X-ray regions.

Two types of crystals are used in the soft X-ray region. They are the phthalic acid salts and soap film crystals. The phthalic acid salt crystals all have a $2d$ spacing of approximately 26 Å, and it would be possible to measure the O K band with these crystals. The following salts of phthalic acid have been found to give good results, K^+, Na^+, Rb^+, Cs^+, and NH_4^+. Potassium acid phthalate (KAP) has been found to give the greatest intensity and P/B of the above phthalic acid salts. An advantage of KAP over the soap film crystals is that it can be ground to the radius of the Rowland circle, which is not possible with the soap film.

The soap film crystals have a multilayer soap film structure. They are usually a metal stearate with a $2d$ spacing of approximately 110 Å. Recently Henke[11] and Ehlert and Mattson[12] successfully measured Be K at 110 Å with a lead lignocerate crystal which has a $2d$ spacing of 130 Å. Of the metal stearate crystals tested, lead stearate gives the best intensity and P/B.

Barium stearate, which was first used as a soft X-ray analyzer, has been replaced by the lead stearate crystal. The stearate crystal is made by repeated dippings of mica or a glass slide into a metal stearate solution. Each dipping produces one additional molecular layer. Enough layers should be deposited so that the crystal appears infinitely thick to the radiation being measured. Experiments have shown that structures larger than 100 layers do not give a noticeably improvement in intensity. Also, it becomes difficult to build good crystals with more than 100 layers. At present Ehlert and Mattson[12] are making lead stearate crystals with 60 layers. If too few layers are used, reflection can occur from the backing material which will result in spurious peaks in the emission spectra under investigation. Ong[13] has found that lead stearate crystals of less than 100 layers will result in reflection from the substrate when measuring shorter wavelength such as F K at 18.5 Å. Of course, lead stearate crystals of less than 100 layers would be satisfactory for longer wavelengths.

The resolving power obtainable from a stearate crystal is also related to the number of layers by the following relation:

$$\lambda/\Delta\lambda = N \tag{9}$$

where N is the number of effective layers. Henke[14] has found that for Al K (8.34 Å) the crystal guassian error has a width at half-maximum intensity ($W_{1/2}$) of 0.165 Å, which gives a resolving power of 50. The effect of varying the number of layers is demonstrated by Ehlert and Mattson[12] for Be K using a 10-, 20-, and 30-layer lead lignocerite crystal. Their results show a larger reduction in $W_{1/2}$ in going from a 10- to 20-layer soap film than from 20 to 30 layers. The intensity, however, increases from 2900 counts sec^{-1} W^{-1} for the 20-layer structure to 5100 counts sec^{-1} W^{-1} for the 30-layer structure. There is a shift in the peak wavelength in going from the 10- to 20-layer structure which shows that there was a change in the $2d$ spacing.

If curved crystal optics are used in the spectrometer it will be necessary to apply the multilayer soap film analyzer to a supporting material that can be bent, such as mica, and then cemented to a metal or glass blank that has a radius of curvature of twice the instrumental radius. Alternatively, the soap film analyzer can be deposited directly on a curved stainless steel or a glass blank without the necessity of the supporting material. Since pits and scratches reduce the resolution, the stainless steel must be highly polished. The soap film is not as fragile as it appears. For example, it is not damaged when put into water, but care must be taken not to scratch or abrade the surface. Although the apparatus for making the soap film analyzer is simple, the deposition of the film is an accomplished art. Henke[15] gives a detailed description of making these films. It would prove more satisfactory for the spectroscopist to purchase the soap film from an X-ray manufacturer rather than to make it himself.

As indicated above, the two most efficient crystals in the soft X-ray region (10–150 Å) are the lead stearate and KAP crystals. Since the KAP crystal can only be used for wavelengths below 26 Å, it is of interest to determine which crystal is most efficient in this wavelength region. A comparison of a KAP crystal and a lead stearate crystal as measured by Henke[15] appears in Table 3. It is seen that the KAP crystal gives a greater relative intensity

Table 3. Comparison of diffraction efficiency of a KAP crystal with that of a 100 layer lead stearate crystal. Reprinted by permission of Burton L. Henke, D. W. Fischer and W. L. Baun, *Advances in X-ray Analysis,* G. R. Mallett, M. Fay and W. M. Mueller, eds. Vol. 7, Plenum Press, New York, 1964, p. 471

Fluorescent source	λ, Å	LSD/KAP
Al *K*	8.3	1.8
Cu *L*	13.3	3.2
Ni *L*	14.6	4.0
Fe *L*	17.6	4.3
Mn *L*	19.4	4.5
Cr *L*	21.7	5.0
O *K*	23.6	7.7
V *L*	24.3	5.9

than the lead stearate crystal and this greater relative intensity improves with increasing wavelength. One reason for the poorer P/B of the stearate analyzer, at shorter wavelengths, is the long-wavelength specular reflection. The relation between the critical wavelength λ_c and critical angle ϕ_c at which X-rays began to reflect, for a grating was given by equation 8. The critical reflection of X-rays from a crystal surface is also governed by this equation. Unlike a grating, however, where all the totally reflected X-rays are either in the zero order or diffracted into higher orders, the total reflected X-rays from the crystal surface will be superimposed as background on the diffracted X-rays. This will increase the background at small 2θ (Fig. 1) angles, resulting in a decreased P/B at the shorter wavelengths. This situation is more severe for the lead stearate crystal than for the KAP crystal, because the stearate $2d$ spacing of 100 Å results in most of the spectra occurring at low 2θ angles. This is especially true where the spectrum of an element is measured from a material containing a lighter element than the one being measured. This is shown in Figure 7, where Henke[15] measured the O *K* bands from urea which also contained carbon and nitrogen. The O K band for a lead stearate crystal occurs at a 2θ of about 27°. Little O K radiation is totally reflected for $\theta = 13.5°$, but considerable C K radiation intensity is. This is shown by Ong[16] in Figure 8, where the C K radiation is measured as a function of 2θ .The C K specular reflection is the reason the background is so high under the O K band in Figure 8. Also, in Figure 7 the OK band from urea is shown using a KAP crystal. There is very little specular reflection of the C K radiation, using the KAP crystal because the O K band occurs at a 2θ of 125°. The specular reflection problem for the lead stearate crystal can be partly overcome by measuring the spectra at higher orders or by using pulse height analysis and filtering.

C. Comparison of Crystals and Gratings as Soft X-ray Analyzers

The last two sections have dealt with gratings and crystals as soft X-ray analyzers. The choice of analyzer will depend on which one or combination of the following parameters is the most important to the investigator, X-ray intensity, P/B, or resolution. It is generally thought that in a spectrometer where the target is behind the entrance slit the crystal will give a greater intensity than a grating for a given resolution. This is due to the fact that for a given resolution the slit widths are about 20 times larger in a crystal spectrometer than in a grating spectrometer. Also, the crystal will intercept

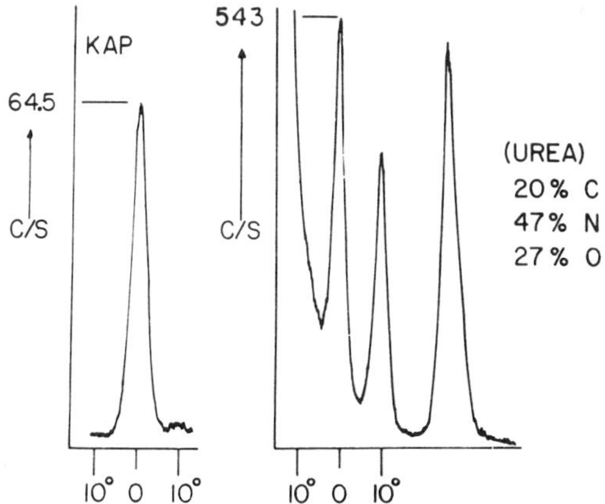

Fig. 7. Counting rate vs. 2θ scans of the O K (23.6 A), N K (31.6 A) and C K bands (44.85 A) with a lead stearate–deconoate crystal (right) and the same O K radiation as measured with a KAP crystal, illustrating the low-angle specular reflection background associated with the larger 2d-spacing crystal. Reprinted by permission, Burton L. Henke, in *Advances in X-ray Analysis,* G. R. Mallet, M. Fay, W. M. Mueller, eds. Vol. 7, Plenum Press, New York, 1964, p. 468.

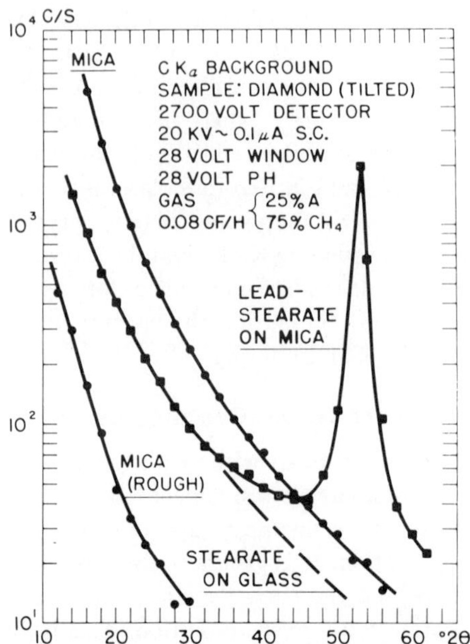

Fig. 8. Background for the C K band from diamond showing increase in background for decreasing 2θ angle due to specular reflection. Reprinted by permission, Poen Sing Ong, in *The Electron Microprobe,* T. D. McKinley, K. F. J. Heinrich, D. B. Wittry, eds. Wiley, New York, 1966, p. 50.

a larger solid angle of the X-ray intensity than the grating. Although considerable caution has to be used in comparing intensities obtained on different spectrometers, some idea of the differences in intensity between a crystal and grating can be obtained by comparing intensities obtained from spectrometers having approximately the same resolution. In Table 4 are listed peak intensities of the Be *K* and B *K* bands obtained by Ehlert and Mattson[12]using a lead stearate and lead lignocerate crystals and the intensities of the same spectra obtained by Holliday using a 600 grooves/mm, Al, 1° blazed replica grating. The lead stearate and lead lignocerate crystals give about 50 times more intensity per watt than the grating for B *K* and about 12 times more for Be *K*.

When a fine focus beam is in the place of the entrance slit, as in the microprobe, the grating has been found to give a greater intensity per watt than the lead stearate crystal. This conclusion is based on a study by Nicholson and Hasler,[10] whose results are shown in Table 5. For a 50 μ slit, the grating gives twice the X-ray intensity than that obtained from a crystal. Nicholson and Hasler show that the ratio of solid angle accepted by the crystal to that

Table 4. Comparison of peak X-ray intensities using soap film crystals and a 600 grooves/mm grating

	counts $sec^{-1}W^{-1}$	counts $sec^{-1}W^{-1}$
Analyzer	B K Band	Be K Band
Lead stearate	28,000	
Lead lignocerate		5,100
Grating with 600 grooves/mm 1° blazed Pt surface and 25μ slits	570	420

Table 5. Comparison of peak X-ray intensities and P/B ratios for a 600 grooves/mm 1° blazed grating and a lead stearate crystal using the electron microprobe as the X-ray source. Reprinted by permission of J. B. Nicholson and M. F. Hasler, *Advances in X-ray Analysis,* G. R. Mallett, M. Fay and W. M. Mueller, eds. Vol. 9, Plenum Press, New York, 1966, p. 426.

			Peak intensity (counts/sec)/line-to-background ratio		
			Grating		
Spectral band	Sample	λ(Å)	100-μ slit	50-μ slit	Crystal 0.020-in. slit
O *K*	Al_2O_3	23.62	3619/30	2123/38	47*/70
N *K*	BN	31.60	1600/13	1262/23	922/10
C *K*	Graphite	44.85	16,000/86	9277/110	4910/51
B *K*	BN	67.2	1495/19	778/22	598/22
Be *K*	Beryllium foil	114	230/45	115/49	

*Kap Crystal

accepted by the grating is 6.7. It was shown in the section on gratings that the efficiency of a 600 grooves/millimeter 1° blazed grating with a Pt surface was 5.7%. From this grating efficiency, the intensity measurements in Table 5 and the difference in solid angle acceptance, Nicholson and Hasler[10] calculated the efficiency of the lead stearate crystal to be 0.5% at 44.85 Å. As shown in Table 5 the P/B is better for the grating than for the lead stearate crystal. This is due in part to the specular reflection phenomenon of the lead stearate crystal mentioned in the section on crystals.

Several factors must be considered in comparing the resolving powers obtained with a crystal and a grating spectrometer. Statements such as "gratings have better resolving power than crystals" are meaningless unless the parameters affecting the resolving power are discussed at the same time. The factors affecting the resolving power of both a grating and a crystal spectrometer are the radius of the Rowland circle and the order of the spectrum being investigated. Those peculiar to a crystal are (*1*) the number of layers (stearate crystal) and (*2*) the quality of crystal. For a grating spectrometer they are (*1*) the number of grooves/mm and (*2*) slit widths. The resolving power $\lambda/\Delta\lambda$ for a grating spectrometer can be calculated from the following relation by Fisher[17]

$$\Delta\lambda = (\sigma/Rn)[s_1 + (s_2/5)\sin(\theta)] \tag{10}$$

where $\Delta\lambda$ is the resolution, σ the grating constant, s_1 and s_2 the slit widths, θ the angle of diffraction, R the radius of the Rowland circle, and n the order. In Section III-B it was stated that the $\lambda/\Delta\lambda$ for the first order from a lead stearate crystal was about 50. From equation 10 the parameters that give a $\lambda/\Delta\lambda$ of approximately 50 for a grating spectrometer at C K (44.85 Å) for the first-order band are 25 μ slits, and a 1-m radius of curvature 600 grooves/mm grating.

The grating spectrometer, however, has more parameters that can be varied to obtain higher resolving power than does the crystal spectrometer. For example, it is not possible to measure the C K band beyond the second order and the B K beyond the first order using a lead stearate crystal. In the case of the grating spectrometer, carbon has been measured out to the tenth order. Reducing the slit width on a crystal spectrometer does not appreciably affect the resolution, but there is an inverse linear relation between slit width and resolution in the case of the grating spectrometer (eq. 10). Another factor is the number of grooves/mm on the grating, $1/\sigma$. Bausch & Lomb is now making gratings that give good intensity and P/B with 3600 grooves/mm (Table 2). Although the resolution of the lead stearate crystal is a linear function of the number of layers, Ehlert[18] believes that there is little improvement in resolving power beyond 60 layers. In addition, the resolution of a grating spectrometer is very sensitive to the accuracy of the alignment of the grating and slits. The resolution of a crystal spectrometer is not as sensitive to alignment, and it is largely determined by the quality of the crystal. Using 25μ slits, the second order of a 1-m 3600 grooves/mm blazed replica grating and an improved method of alignment the author[1] was able to obtain as much detail for the Ni L_{III} band (14.56 Å) from Ni as that reported by Bonnelle[19a] using a gypsum crystal. Since gypsum has a $2d$ spacing of $\sim$15.18 Å, Bonnelle was measuring the Ni L_{III} band at large 2θ angles. An improvement in resolution cannot be expected for the crystal spectrometer at wavelengths larger than 14.56 Å. This is because other crystals with larger $2d$ spacings will have to be used at longer wavelengths which are inferior to gypsum resulting in poorer resolution. Bonnelle[20] stated that her spectrometer had a resolution of 0.3 eV or 0.01 Å at the Ni L_{III} band. This would indicate that Holliday's grating spectrometer resolution is also about 0.01 Å. However, from equation 10 the resolution in the 2nd order would be 0.05 Å, which indicates that a grating spectrometer has a resolution better than given by equation 10. Although it is difficult to state exactly how much greater a grating spectrometer's resolution could be over a lead stearate crystal spectrometer, the above results indicate a factor of at least six times greater. This has been shown for the B K bands from TiB_2 and ZrB_2 where the grating spectrometer showed considerably more detail than was observed with a lead stearate crystal spectrometer.

Another factor to consider in comparing crystal and grating spectrometers is dispersion. The relation for the dispersion in a crystal obtained by differentiating the equation for diffraction from a crystal is

$$d\theta/d\lambda = (n/2d \cos \theta) \tag{11}$$

and for a grating it is

$$d\alpha/d\lambda = (n/2\lambda\sigma)^{1/2} \tag{12}$$

where α is defined in Figure 2*b*. Since d is approximately 10^{-8} cm. for a crystal and σ is approximately 10^{-4} cm. for a grating, it can be seen that the dispersion of the grating is inferior to that of a crystal.

When measuring from target elements that are also contained in the crystal analyzer, variation in the shape and wavelength have been found for different crystals. This was clearly shown in a series of experiments by Mattson and Ehlert[21] on the O K band of CO_2 using lead stearate and KAP crystals. The results of these measurements are shown in Figure 9. Examination of the O K bands from CO_2 in Figure 9 shows that the alteration in shape is largely due to a change in the relative height of the peak at 23.3 Å. Holliday[22] has shown that the peak at 23.3 Å is due to the oxygen in the

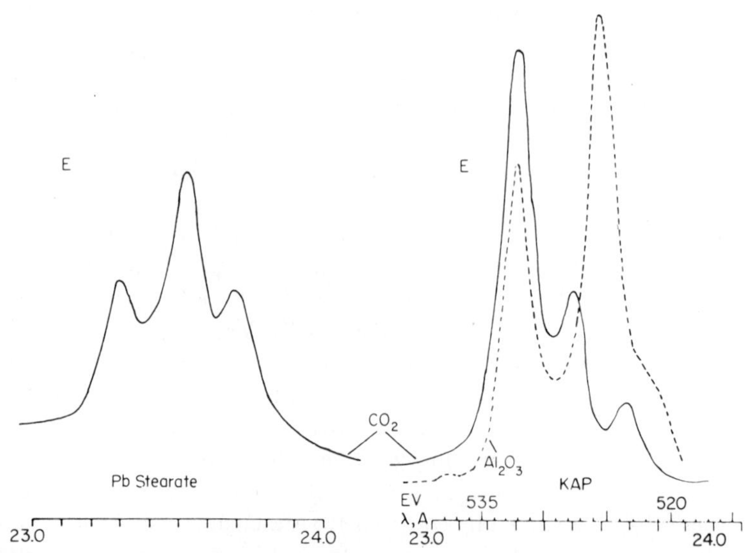

Fig. 9. The O *K* bands from CO_2 showing the affect of a KAP and lead stearate crystal on the intensity distribution of the O K bands. Reprinted by permission of R. A. Mattson and R. C. Ehlert, in *Advances in X-ray Analysis,* G. R. Mallet, M. Fay, W. M. Mueller, eds. Vol. 9, Plenum Press, New York, 1966, p. 481.

crystal and is not part of the O *K* band from the solid. The O *K* band from BeO in Figure 10 using a grating analyzer shows no peak at 23.3 Å, but the same band measured with a KAP crystal does show a strong peak at 23.3 Å.

As a result of the above comparison between grating and crystal spectrometers the following summary can be made:

(*a*) *Spectrometers that have the target behind an entrance slit:*

(*1*) The lead stearate and KAP crystal spectrometer give an intensity per watt which is approximately 10 times greater than that obtained with a grating spectrometer for comparable resolution.

(*2*) At wavelengths greater than 14 Å, greater resolving power can be obtained from the grating spectrometer. Better than 6 times more resolution can be obtained from the grating than for a lead stearate crystal spectrometer at 44.85 Å. At approximately 14 Å the greater resolving power and the large improvement in dispersion make the crystal a better analyzer than the grating.

(*3*) The P/B is about the same for crystal and grating spectrometers. At shorter wavelengths where specular reflection is a problem, the grating spectrometer gives much better P/B.

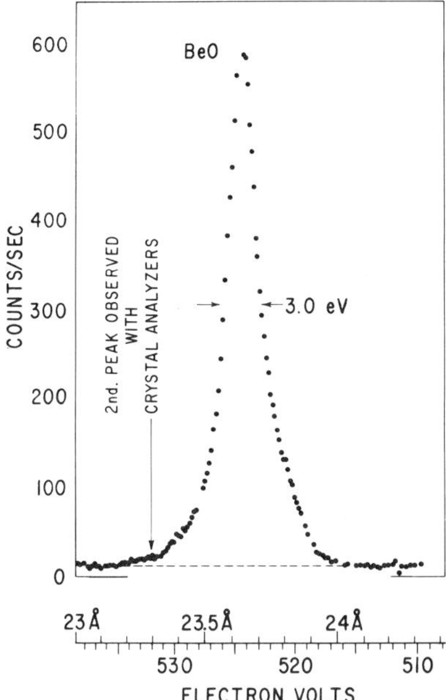

Fig. 10. The O K Band from BeO using a 1 meter, 2160 grooves/mm 1° blazed grating, with a Pt surface, and 3.5 kV excitation voltage. The arrow indicates the wavelength at which 2nd peak was obtained by Mattson and Ehlert[21] using a KAP crystal to measure the O K band from BeO. Reprinted by permission, J. E. Holliday *Norelco Rep.*, **14**, 84 (1967).

(*b*) *Spectrometers in which a fine focus beam is in the place of the entrance slit:*

(*1*) The grating spectrometer gives a greater intensity/watt in the 10–150 Å region than the crystal spectrometer.

(*2*) Since the grating spectrometer resolution is more dependent on the X-ray source size than the crystal spectrometer, a greater improvement in resolution will be obtained from the grating spectrometer by using a 1 or 2 μ beam in the place of the entrance slit than for the crystal spectrometer.

(*c*) *With either type of analyzer (grating or crystal), caution must be used in measuring spectra from elements that are in the crystal or on the surface of the grating.*

IV. THE DETECTOR

The three basic types of detectors used in the soft X-ray region are the gas counter, the photoelectric multiplier, and the scintillation counter. The basic operating mechanism of these detectors will not be discussed since there are numerous articles on these devices. Their characteristics as they are related to the detection of long-wavelength X-rays will be presented.

In Table 6 the intensity and background for a gas counter, a Bendix electron strip multiplier, and a scintillation counter are compared for Al K (8.34 Å) and Cu L_{III} (13.3 Å) bands. The results show that in this wavelength region, the gas counter is far superior in P/B and intensity to the other types of detectors. At wavelengths where the absorption of X-rays in the counter window becomes appreciable (~150 Å), the photoelectric multiplier is more efficient. At wavelengths where the gas counter is less efficient than the photoelectric multiplier it is still preferred. This is because the photosurface for converting X-rays to electrons in the photoelectric multiplier is unpredictable. A number of commercial CuBe photomultipliers were tested in this laboratory, and it was not possible to obtain a signal from any of them. Other investigators have reported similar difficulties. Apparently Fisher, Crisp and Williams,[23,24] and Catterall and Trotter[25] have been able to activate the CuBe surface, since they have reported good X-ray intensities and P/B. The activation of the photosurface is an art which is not well known or understood. The Bendix electron strip multiplier appears to give good results beyond 150 Å. Cuthill et al.[26] has obtained good intensities for the CuM_{III} band at 160 Å with the Bendix multiplier. Both Fischer and Baun,[27] and Lukirskii and Brytov[28] have found that CsI is the most efficient photocathode surface in the 1–20 Å region. Even with a CsI photocathode, the Bendix photomultiplier is considerably less efficient than the gas counter in the 10–150 Å region.

Table 6. Comparison of count rates for three different detectors. Reprinted by permission of David W. Fisher and William L. Baun, *Advances in X-ray Analysis,* G. R. Mallett, M. Fay and W. M. Mueller, eds., Vol. 7, Plenum Press, New York, 1964, p. 496.

	SPG-7 Flow counter		Bendix M-306		Windowless scintillation counter	
Line	Peak	Background	Peak	Background	Peak	Background
Al K_α	30,000	25	8,500	400	11,500	350
Cu L_α	16,000	150	900	300	440	300

Because the gas counter is the most reliable and efficient detector in the 10–150 Å region, the remainder of this section will be devoted to this detector. Since the thin windows used (2000–3000 Å) in the soft X-ray gas counter have some leakage, sealed counters are not reliable. The majority of gas counters used in the soft X-ray region are of the flow type. In addition, gas counters are used almost exclusively in the proportional region. Although the higher output voltage of the Geiger counter compared to the proportional counter simplifies the electronics, because of the higher signal to noise ratio and greater gas amplification, the gases required to operate the counter in the Geiger region are less efficient than those used in the proportional region. A gas counter operated as a Geiger counter and detecting C *K* radiation at 44.85 Å using 96% helium and 4% isobutane produces a considerably lower counting rate than for the counter operated in the proportional region using 90% argon and 10% methane (P-10).

The three basic parameters to consider when operating a proportional counter in the soft X-ray region are (*1*) type of gas, (*2*) gas pressure, and (*3*) the window. It was shown above that P-10 gas is more efficient than helium isobutane mixtures but it has been shown by Ong[29] that P-10 gas is not the best argon–methane mixture for pulse height analysis. This is because in the soft X-ray region the proportional counter does not operate as a true proportional counter. There is a large variation in pulses for a monoenergetic soft X-ray beam, as is shown by the oscillographs in Figure 11. The pulses are from the counter preamp when it is sitting on the O *K* band peak (524.8 eV) and the C *K* band peak (276.4 eV). It can be seen that there is as much as 50% variation in pulse height. This effect is due to statistical fluctuation in the number of ion pairs generated by X-ray quanta of a given energy. Ong[29] has shown that this fluctuation is minimum for 25% argon and 75% methane (P-75). Increasing or decreasing the argon from the 25% value increases the energy distribution. Lukirskii et al.[30] reported that methylal gas is best for proportional counters in the soft X-ray region. Other gases that have been used for the flow proportional counter are CO_2 and natural gas.

In order to prolong window life, counters are generally operated at about 10 torr in the soft X-ray region. In addition, operating the counter at this reduced pressure prevents the X-rays from being absorbed too near the counter windcw. If X-rays are absorbed near the counter window, there is an increased probability that the electrons and ions might recombine before reaching the wire which reduces counter efficiency. There are several advantages to operating the counter at atmospheric pressure. The energy spread of the pulses is reduced and the counter can be operated at a higher voltage, which improves the signal-to-noise ratio. The higher counter voltage has the disadvantage that spurious counts are more probable as the result of the higher probability of arcing within the counter. The counter voltage for P-10 gas at 10 cm Hg pressure is 900 V, while at atmospheric pressure it

(a)

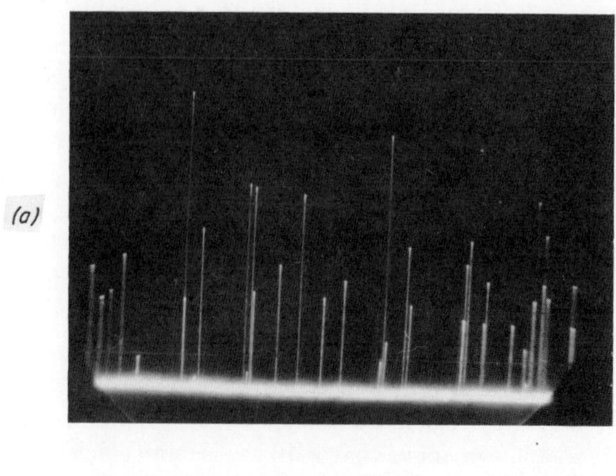

(b)

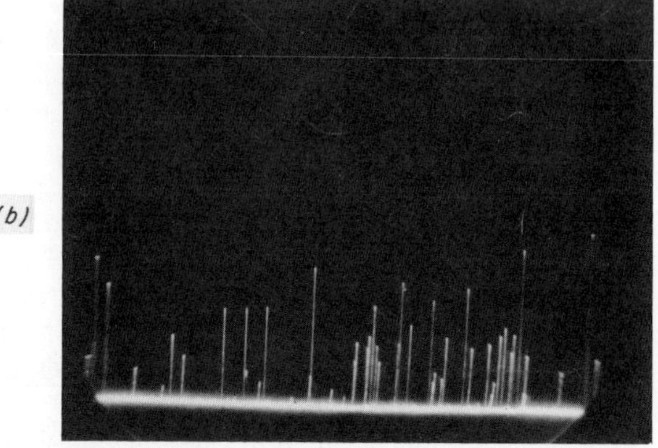

Fig. 11. Oscillograph of pulses from Proportional Counter Preamp using P-10 gas, showing variation in pulse height when 90% of X-ray photons have the same energy.

is 1800 V. The pulse output voltage for P-10 gas at 900 V is between 0.5 and 2 mV, while at 1800 V at atmospheric pressure it is 100 mV; for P-75 gas the counter output is 700 mV. When the output pulse is less than 1 mV, very special techniques in circuit amplification and noise elimination are required to keep the signal above the noise level of the recording electronics. In this laboratory the pulse height of the counter output is ~1 mV. A cascade circuit described by Wilson[31] is used as the first stage of pulse amplificaton.

The signal-to-noise ratio for the lowest voltage pulse going into the scaler is 2/1 with a noise level of 200 mV.

The third important factor in the gas counter is the window. The counter window has to be thin enough to keep absorption to a minimum in the 10–150 Å region, yet thick enough to prevent breakage. The thickness of the film is determined by measuring the mg/cm^2 and dividing by the density of the film. The mg/cm^2 of the cellulose nitrate film used at this laboratory varied between 0.020 and 0.028 mg/cm^2. For a bulk density of 0.77 g/cm^2, the cellulose nitrate film varied between 2600 and 3600 Å. The accuracy of the film thickness depends on how accurately the density of the film is known. The density of a thin film is usually less than the bulk density because of the large number of pin holes. Thus, the thickness of the film is slightly greater than indicated above. Lukirskii et al.,[32] Henke et al.,[33] and Holliday[2] have published values of absorption in cellulose nitrate for the 10–400 Å region. Cellulose nitrate and most other plastic films such as zapon are made by floating the plastic on the surface of water and picking the film up with a wire loop. It is more desirable to use several films whose total thickness is approximately 2500 Å rather than a single 2500 Å film because the multiple films will reduce the chances of leakage through pinholes. One method of mounting and attaching the window to the counter has been described by Holliday.[2] The cellulose nitrate film is found to have an erratic lifetime. Some last several months while others break the first time gas is put in the counter. Also, the film leakage rate increases each time the counter is refilled. Formvar films have a considerably longer life and a much smaller leakage rate and have lasted 1–1½ years with no detectable leakage through the window. A different technique is required for making the Formvar films which has been thoroughly described by Henke.[14]

V. EXCITATION OF SOFT X-RAYS

A. Electron Excitation

Until recently electron excitation was the only method of producing soft X-rays without vaporizing the solid. Recently with improved instrumentation X-ray excitation has become practical. However, electron excitation is by far the more efficient method. When photographic films were used as detectors, electron beam currents from 50–100 mA were used with a target potential of about 2000 V. For a grating spectrometer with the target behind the entrance slit, using a flow proportional counter, a 3600 grooves per millimeter grating and 25 μ slits, currents of 1–2 mA at 4000 V give adequate intensities. If high resolution is not required, gratings with fewer grooves per millimeter and larger slit widths can be used, which require only several hundred microampere beam current. In Section III-C, it was shown that when the target is behind the entrance slit a crystal spectrometer gives a greater intensity than a grating spectrometer. Consequently, a smaller power input will be required when using a crystal analyzer. But when a fine focus beam is in the place of the entrance slit, a 0.1 μA beam current will be adequate for the grating spectrometer.

In order that only the ruled portion of the grating "sees" the X-ray source, it is important that the electron beam be focused and positioned so that it falls within the area indicated by the solid lines (area a) in Figure 12. If X-rays are produced outside this boundary, they will be scattered randomly by the unruled portion of the grating analyzer, increasing the intensity in the zero order and the background. After the electron beam has been focused so that it has a width equal to or smaller than the area indicated in Figure 12, it must be positioned within this area. This can be done by deflecting the electron beam magnetically, electrostatically, or by mechanically moving the target and gun until the X-ray intensity is at a maximum.

For those interested in quantitative analysis, the exciting voltage should be adjusted to give the highest product of P/B times peak intensity. This is because the least detectable weight percent for a given element is a spectral peak, which has an intensity above background of about three times the standard deviation of the noise background. The P/B as a function of voltage for the C K band at 90° takeoff angle is shown in Figure 13. The background is taken on the low-energy side at a distance of three times the width

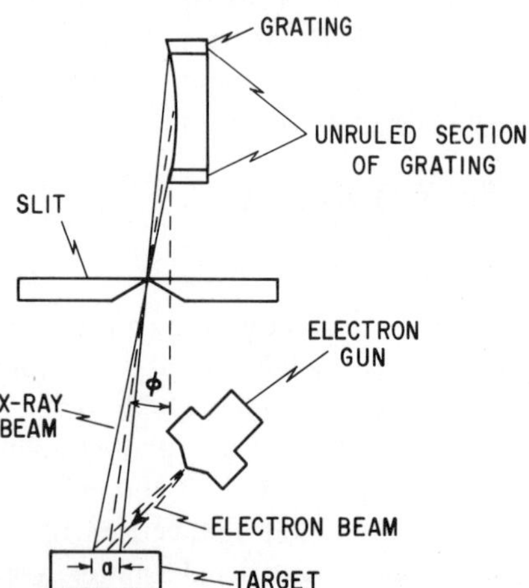

Fig. 12. Area of target seen by ruled surface of grating. For maximum efficiency the electron beam spot must stay with the boundaries indicated by *a*.

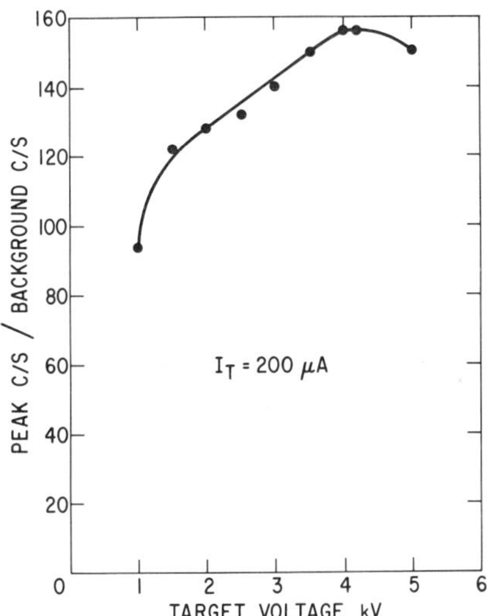

Fig. 13. Peak to background ratio for C K band as a function of target voltage, takeoff angle 90°. Reprinted by permission, J. E. Holliday, *J. Appl. Phys.*, **33**, 3259 (1962).

at half maximum. It can be seen that the maximum P/B occurs at approximately 4 kV for the C *K* band. For smaller takeoff angles the maximum P/B occurs at higher voltages. In order to avoid surface contamination and focusing problems in some microprobes, some investigators have found it necessary to operate with electron beam potentials of 10–20 kV. At the higher voltages there will be a greater penetration with increased absorption of the soft X-rays. The work of Ong[16] who measured counts/sec per 0.1 μA/target current as a function of target voltage is shown in Figure 14. For a takeoff angle of 15° the maximum occurs at 8 kV, while for a 48° takeoff angle it occurs at about 10 kV. The effect on P/B at the higher voltage can be seen by comparing the work of Nicholson and Hasler,[10] and Ong.[16] In Table 5 Nicholson and Hasler reported a P/B of 100 for the C *K* band at 5 kV, while Ong reported a P/B of 70 for the C *K* band at 17 kV, both using a lead stearate crystal. Thus, the increased self-absorption and white radiation at higher excitation voltages does not appreciably reduce the intensity and P/B. Self-absorption does become a more significant problem when measuring light element spectra in a heavy element matrix. This is shown by the work of Anderson[34] in Figure 15. For a 52° takeoff angle the maximum intensity for the C *K* band occurs at 17.5 kV for graphite and 7.5 kV for Fe_3C.

When comparing the fine structure of emission bands it is important that they all be measured at the same voltage. Changes in shape and wavelength

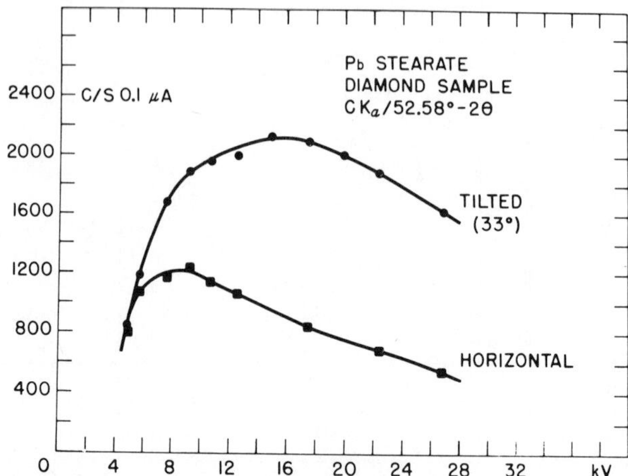

Fig. 14. Counting rate per 0.1 μ A target current as a function of the anode voltage for C K band. Takeoff angle 15° (horizontal) and 48° (tilted). Reprinted by permission, Poen Sing Ong, in *The Electron Microprobe,* T. O. McKinley, K. F. J. Heinrich and D. B. Wittry, eds., Wiley, New York, 1966, p. 47.

of emission bands when the excitation voltage is varied have been observed by Chopra,[35] Liefeld,[36] and Holliday.[22] Fischer and Baun[37] have also found changes in intensity distribution of emission bands with takeoff angle. Liefeld and Chopra have observed changes in the peak wavelength and in the emission edge of the Ni L_{III} band from Ni with excitation voltage. They attributed these changes to self-absorption resulting from the overlap of the absorption and emission edges. The amount of the self-absorption would supposedly change with the depth of electron penetration which is a function of the excitation voltage. However, Holliday[1] has shown, by placing a 1000 Å Ti film between grating and detector, that even when the L_{III} absorption edge overlaps the Ti L_{III} emission edge radiation there is no change in the slope of the Ti L_{III} edge or in peak wavelength. These experiments did show that the Ti $L_{\text{II}}/L_{\text{III}}$ intensity ratio is affected by self-absorption. However, the Fe $L_{\text{II}}/L_{\text{III}}$ intensity ratio was not affected by self-absorption. Although self-absorption can in some cases explain the changes in $L_{\text{II}}/L_{\text{III}}$ intensity ratio with excitation voltage, it does not appear to change the L_{III} emission edge slope and peak wavelength. Since the contribution of a surface oxide or carbide (due to carbon contamination) will change with excitation voltage and takeoff angle it is quite possible that most of the observed changes with

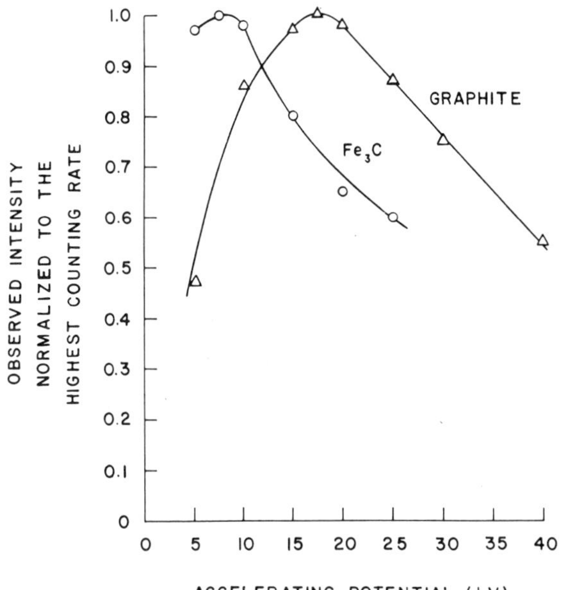

Fig. 15. The C *K* band peak intensity as a function of target voltage for graphite and Fe_3C. Takeoff angle was 52°. Reprinted by permission, Christian A. Anderson, in *The Electron Microprobe,* T. D. McKinley, K. F. J. Heinrich and D. B. Wittry, eds., Wiley, New York, 1966, p. 66.

excitation voltage are due to the increasing contribution of surface contamination with decreasing excitation voltage. The author[1] has shown that surface contamination can contribute significantly to the pure metal emission band even at 4 kV excitation voltage.

Differences in the shape and wavelength of emission bands have also been observed between X-ray and electron excitation. In Figure 16 is shown the B *K* emission band of $Na_2B_4O_7$ excited by X-rays and electrons. The two peaks, one on the high and the other on lower energy side of the main peak, are satellites. Although there are differences on the high energy side, the change of intensity distribution of the main peak is more pronounced on the low energy side, which would not be expected if self-absorption were the cause.

In the majority of materials excited by both excitation methods Mattson observed no change in the band intensity distribution. Faessler[38] has stated that the only time he observed a difference between electron and X-ray

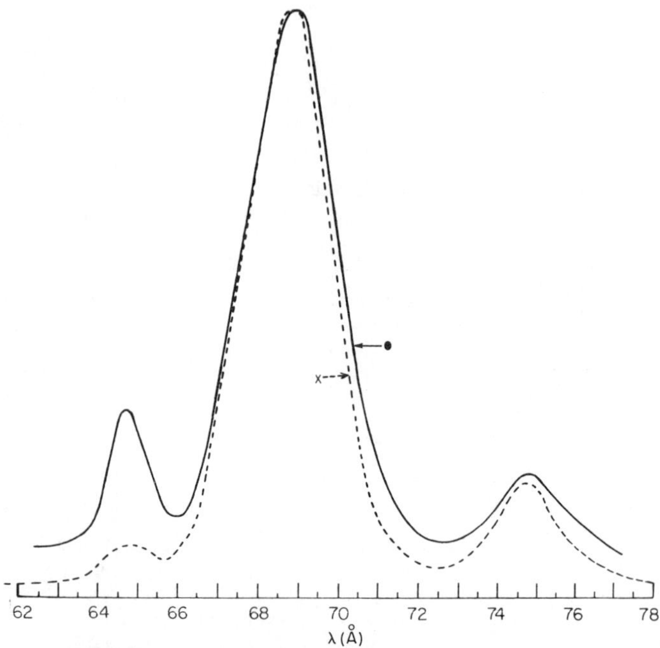

Fig. 16. B *K* emission spectrum from $Na_2B_4O_7$ excited by electrons (e) and C *K* band X-rays (x), using lead stearate crystal. Reprinted by permission, R. A. Mattson and R. C. Ehlert in *Advances in X-ray Analysis*, G. R. Gavin, M. Fay, W. M. Mueller, eds. Vol. 9, Plenum Press, New York, 1966, p. 463.

excitation was when the action of the electron beam changed the chemical state of the target. However, Cauchois[39] has reported that the intensity distribution of Mg *K* band From Mg changed when going from electron (3 kV) to Al *K* X-ray excitation. The reason for this difference could be the same as that given for band changes when the excitation voltage is varied. Since an Al *K* X-ray will penetrate to a greater depth than *a* 3 kV electron, there will be a greater contribution from surface oxide with the electron excitation than for the Al *K* X-ray excitation.

These results show that when measuring emission bands, chemical changes in the target due to the electron beam must be avoided, and the target surface should be free of contamination.

Other factors to consider in electron excitation are the contamination produced by the electron beam and the chemical changes in the surface as a result of electron bombardment. To keep contamination at a minimum the target must be in a clean vacuum and there must be provision in the X-ray chamber for cleaning the target under vacuum. Ennos,[40] and Heide[41] have shown that the principle source of contamination in the vacuum system is from an absorbed hydrocarbon layer on the surface of the target. Under the action of the electron beam the hydrocarbons are cracked, leaving a layer of carbon, and the hydrogen is pumped away. In order to keep hydrocarbon vapors at a minimum, there should be no oil diffusion pumps and only metal O-rings should be used in the X-ray chamber. Since mercury pumps require forepumps, which contribute contaminating oils, it is best to use ion-getter pumps. Because spectrometers require a certain amount of lubrication, it is almost impossible to eliminate all sources of oil in the spectrometer chamber completely. It is best to isolate the X-ray chamber from the spectrometer chamber by a very thin (~500 Å) plastic window. Even with the isolating window it is best to remove nonmetal seals and oil diffusion pumps in the spectrometer chamber to prevent contamination of the analyzer. When the fine focus beam is in the place of the entrance slit, it is next to impossible to isolate the X-ray chamber from the spectrometer chamber. In addition, most fine-focus electron beams are formed in a microprobe column with a light optical system which requires nonmetal seals. In this case it is best to isolate the X-ray target from the contaminating vapors of the vacuum by surrounding the target with a cold chamber. Campbell and Gibbons[42] have described such a device and reported that at $-135°$ the contamination rate had been reduced by 3 orders of magnitude compared to what it was at room temperature.

In the case of the microprobe, Duncomb, and Melford[43] report that a focused electron beam produces a ring of carbon contamination about 8 μ o.d. which deposits peripheral to the point of electron impact. The center of the ring where the electron beam strikes remains relatively free of carbon. If the probe is defocused, the area where the beam strikes becomes contaminated. This is shown in Figure 17 which plots the C K band intensity as a function of time for a defocused beam (curve a) and a focused beam (curve b). There is no carbon build-up for the focused beam, but there is for the defocused beam.

Besides the contamination due to the residual vapors in the vacuum, Holliday[2,44] has shown that there is contamination that comes from the target itself. This can best be seen by referring to Figures 18 and 19. Figure 19 is the mass spectrum in the X-ray chamber before the Mo target was heated to 600°C by electron bombardment. It is seen from Figure 18 that there were no hydrocarbon vapors in the system. The target had been cleaned by ion bombardment to assure that there were no hydrocarbons on the surface of the Mo. During heating by electron bombardment, the only evidence of hydrocarbons was a very small increase in mass 13. Mass 28 had increased 30

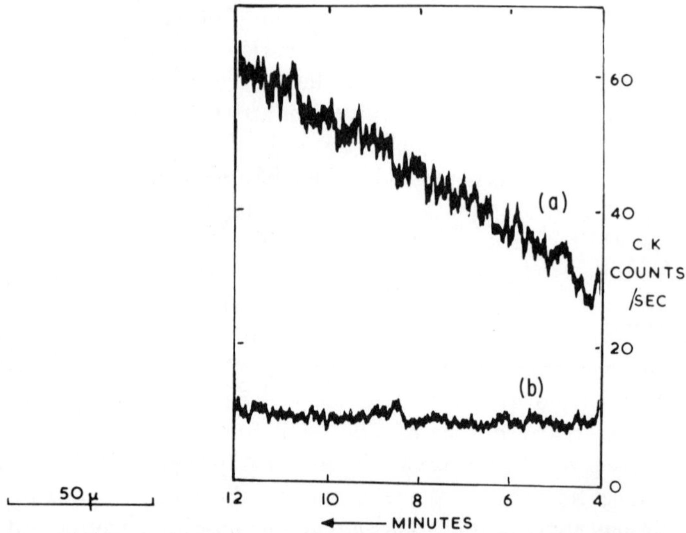

Fig. 17. Carbon contamination build up by action of an electron microprobe beam on a pure iron target. Curve *a* is the C K band intensity as a function of time for a defocused beam and curve *b* is for a focused beam. The photograph to the left shows the contamination visually. Reprinted by permission, P. Duncumb and D. A. Melford, International Symposium on X-ray Optics and X-ray Microanalysis. 4th, Orsay, 1965. Optique des rayons X et microanalyse. X-ray Optics and microananalysis. Ed. by R. Castaing, P. Deschamps, and J. Philibert. Paris, Herman, 1966, p. 241.

times, however, which could be either nitrogen or CO. Since there was a larger increase in mass 12 (carbon) than mass 14 (nitrogen), it can be assumed that mass 28 is mostly CO. Also, there was a large increase in CO_2. After heating the target it was found that the C K band from the Mo target had increased in intensity. The increase in carbon on the Mo surface can be explained by the following reaction:

$$2CO \rightleftharpoons C + CO_2$$

At elevated temperatures and in the presence of a suitable catalyst (the electron beam and metal surface) CO disproportionates to carbon and carbon dioxide. It has also been found that when an oxide is being formed by the action of the electron beam there is no contamination.

The above results show that it is necessary to thoroughly degas the surface before measuring X-ray spectra. The target must also be cleaned in the vacuum after degassing. It has been found that as soon as the target is exposed to air and put back in the X-ray chamber the rate of carbon contamination will increase over what it was just before it was taken out. The most effec-

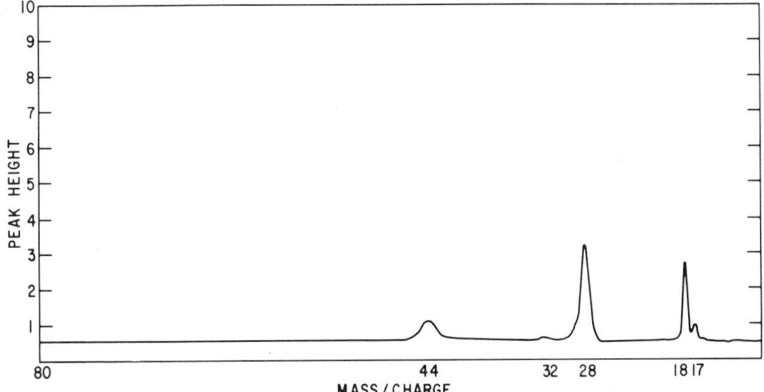

Fig. 18. Typical mass spectrum of residual gases in X-ray chamber after filling liquid nitrogen trap, vacuum 1 × 10–⁷ torr. Reprinted by permission, J. E. Holliday, in *Handbook of X-rays,* Emmette F. Kaelble, ed., McGraw-Hill, New York, 1967, p. 38-13.

tive cleaning method in vacuum is ion bombardment[2] using argon at a pressure of 10^{-2} to 10^{-3} torr. By this method it is possible to remove completely a visible layer of carbon deposited by the action of the electron beam.

Changes in chemical state as a result of electron bombardment have not been thoroughly studied. For the group IV, V, and VI transition metals, heating with the electron beam causes oxides to form on the surface in a vacuum of 10^{-7} torr. Certain unstable carbides such as Nb_2C, MoC, and W_2C will be converted to the stable form by the heating action of the electron beam. Cooling the target will help prevent some chemical change by

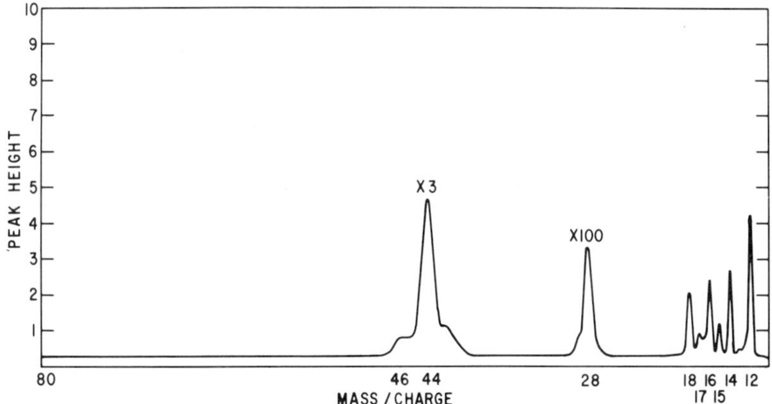

Fig. 19. Mass spectrum of gases evolved from Mo target when heated to 600°C. Pressure rose to 1 × 10^{-5} torr. Reprinted by permission, J. E. Holliday, in *Handbook of X-rays,* Emmett F. Kaelble, ed. McGraw-Hill, New York, 1967, p. 38-13.

the electron beam. For example, it has been shown by the author[22] that if martensite is cooled during electron bombardment it will not convert to Fe_3C. However, if it is not cooled the entire target will be transformed to Fe_3C + ferrite in 2 or 3 min by an electron beam with 4 W of power.

B. X-ray Excitation

Before the recent improvements in instrumentation, it was impossible to generate soft X-rays by fluorescence with sufficient intensity to be detected. With the use of crystals and gas counters in the long-wavelength region fluorescent excitation is now practical. Although the efficiency is considerably less than that of electron excitation, X-ray excitation offers certain advantages over electron excitation. Henke was the pioneer in soft X-ray fluorescent excitation. In order to obtain adequate X-ray intensities and to obtain low background counting rates, the target must be excited by an intense soft X-ray radiation not much shorter in wavelength than the emission spectra being measured.

The Henke[15] tube for fulfilling these conditions is shown in Figure 20. The high intensity X-ray beam for exciting the target T is obtained by a close coupling between the anode and the target. The power limit of the Henke tube for a copper anode is 4 kW. In order to prevent the copper anode from melting, it is water cooled. The copper anode is generally operated at 6 kV which is the optimum voltage for generating Cu L_{III} (13.3 Å) radiation at the takeoff angle indicated in Figure 20. At 4 kW this voltage would give an anode current of approximately 666 mA. To insure that no contamination reaches the target, such as evaporated material from the anode, a thin plastic window (W in Fig. 20) is placed between the anode and the target. The entire soft X-ray generating tube is in a seaparate housing and has its own pumping system. These tubes are available commercially from Philips Electronics.

Another type of tube for soft X-ray fluorescence is shown in Figure 21. This tube was developed by Mattson and Ehlert.[21] It has the advantage that soft X-rays can be generated either by X-rays (Fig. 21*a*) or electrons (Fig. 21*b*). Also, gases can be excited by electron bombardment, Figure 21*d*. If desired, a wire cathode can be used (Fig. 21*c*). The anode is biased positively for X-rays and negatively for electron excitation. It is important in X-ray excitation, because of the large current, that the anode be at positive with respect to ground and the filament at ground. When the filament is at ground the backscattered electrons will not strike the crystal and other parts of the spectrometer giving rise to high background and possibly damaging the crystal. When the anode is positive it is necessary to use a high dielectric fluid rather than water for cooling the anode. Fluorochemical or F-77 made by 3M Company is a good fluid for this purpose. Because of the large currents

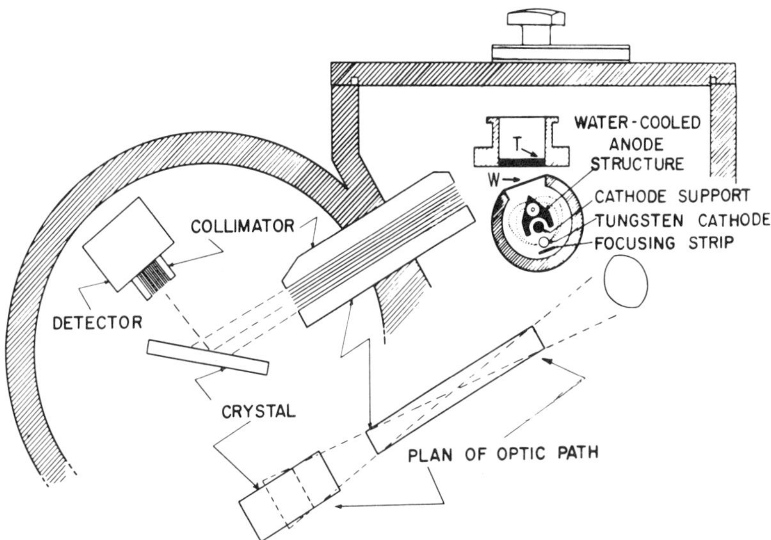

Fig. 20. X-ray optics of the vacuum spectrograph with a soft X-ray excitation source. *T* is the target and *W* a thin plastic window. Reprinted by permission, Burton L. Henke, in *Advances in X-ray Analysis,* G. R. Mallett, M. Fay and W. M. Mueller, eds. Vol. 7, Plenum Press, New York, 1964, p. 461.

involved in X-ray excitation it is important that there is adequate baffling to prevent the secondary X-rays from reaching the analyzer and detector. Also there are a large number of ions generated in the residual gas in the vacuum that must be kept away from the detector and analyzer.

It was indicated earlier that the exciting X-rays must have an energy only slightly greater than the soft X-rays being measured. In Figure 22 is shown the work of Mattson[45] on various anode materials which were used for exciting C K radiation from graphite. It can be seen that the Fe and Cu L_{III} bands give a C K efficiency of 18 counts sec^{-1} W^{-1} but at different voltages. For Fe L_{III} the maximum is at 5 kV and for the Cu L_{III} band it is at about 9 kV. The fact that the maximum for copper occurs at a higher voltage is understandable since the Fe L_{III} peak is at 17.54 Å (706.8 eV) and Cu L_{III} at 13.3 Å (932 eV). Copper is a better anode material than iron for several reasons: one is that copper is a better heat conductor and the other is that the higher operating voltage of copper requires less beam current for a given power input than iron.

It might be expected that since the Ti L_{III} band (27.45 Å) has a wavelength close to the C K band (44.85 Å) it should be a more effective generator of C K radiation than either the Fe L_{III} or Cu L_{III} bands. From Figure 22

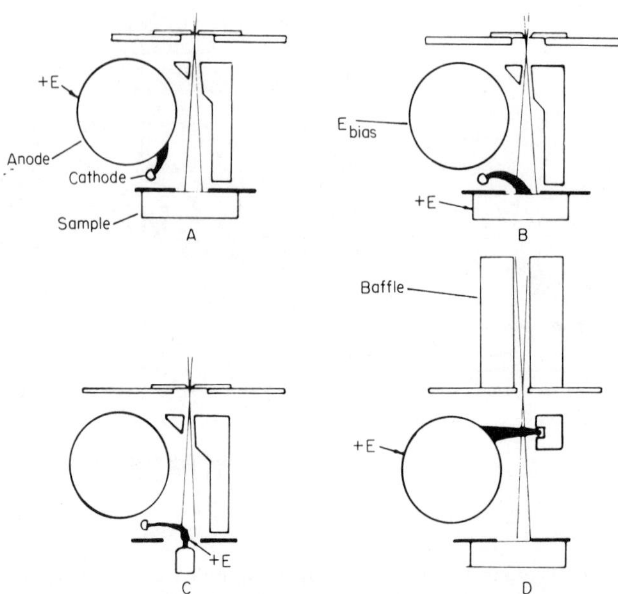

Fig. 21. Schematic representation of X-ray source. (*a*) X-ray excitation, (*b*) electron excitation, (*c*) electron excitation of heated wire, (*d*) electron excitation of gaseous sample. Reprinted by permission, R. A. Mattson and R. C. Ehlert, in *Advances in X-ray Analysis,* G. R. Mallett, M. Fay and W. M. Mueller, eds. Vol. 9, Plenum Press, New York, 1966, p.472.

it can be seen that it is less efficient than Fe L_{III} or Cu L_{III}. The reason is that for a given power input the Ti L_{III} band has about one fourth the intensity of either the Fe or Cu L_{III} bands as seen from measurements by the author in Table 7.

Table 8 shows the peak intensity over background for the light elements as obtained by Henke[11] using X-ray fluorescence analysis. The anode voltage was 6 kV and the beam current was 300 mA. For generating X-rays that have wavelengths longer than the C K band, carbon is used as the X-ray anode rather than copper. Although it is possible to obtain intensities similar to those obtained by electron excitation, the P/B is poorer and considerably more power is required. Since there is no white radiation with X-ray excitation the poorer P/B is an instrumental problem. Ehlert and Mattson[12] have generated the C K bands by X-ray and electron excitation in the same instrument. For electron excitation the efficiency is 5100 counts sec^{-1} W^{-1} while for X-rays it is 18 counts sec^{-1} W^{-1}, showing that a spectrometer using electron excitation is 280 times more efficient than one using X-ray excitation.

Table 7. L_{III} emission band parameters for first series transition metals, using a 3600 grooves/mm 1° blazed Pt surface[a] grating

Metal	Counts sec^{-1}	P/B	$W_{1/2}$	λ Å[c]
Ti	205	8	3.2	27.45
V	400	13	5.5	24.47
Cr	560	20	4.7[b]	21.64
Mn	690	13	4.5[b]	19.6
Fe	1000	30	4.0[b]	17.54
Ni	1340	22	3.4[b]	14.6
Cu	2000			13.3

[a] V_T = 4 kV and 1 mA beam current.
[b] 2nd order.
[c] Improved accuracy from that reported by author in handbook of X-rays.

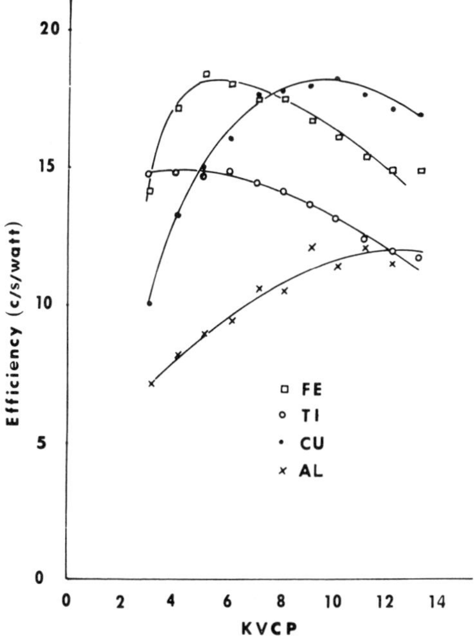

Fig. 22. C *K* band excitation efficiency excited by X-rays from Fe, Ti, Cu and Al as a function of excitation energy of the electrons which cause fluorescence of Fe, Ti, Cu and Al targets. Reprinted by permission, R. A. Mattson, in *Advances in X-ray Analysis,* G. R. Mallett, M. Fay and W. M. Mueller, eds. Vol. 8, Plenum Press, New York, 1965, p. 338.

Table 8. Light element X-ray fluorescence analysis. Reprinted by permission of Burton. L. Henke, *Advances in X-ray Analysis*, G. R. Mallet, M. Fay and W. M. Mueller, eds. Vol. 9, Plenum Press, New York, 1966, p. 432

					Analysis sensitivity		
					Fluorescent sample		Peak Int./Background
Element	Band	X-ray tube anode	Crystal analyzer	Proportional counter voltage	Element	Sample	counts sec^{-1}
O	O K (23.6 Å)	Cu	PbS	2325	53.3% O	SiO_2	$\frac{13297}{759} = 17$
N	N K (31.6 Å)	Cu	PbS	2370	56.4% N	BN	$\frac{3840}{462} = 8$
C	C K (44.85 Å)	Cu	PbS	2410	100% C	Graphite	$\frac{21681}{696} = 31$
B	B K (67.8 Å)	C	PbS	2490	43.6% B	BN	$\frac{12444}{108} = 11$
Be	Be K (114 Å)	C	PbL	2600	100% Be	Beryllium foil	$\frac{1855}{132} = 14$

X-ray excitation has the advantage that the carbon contamination problem is not present. Also insulator and powdered materials can be investigated without heating the sample. It is almost impossible to bombard insulators without intense local heating. With certain materials, as indicated earlier, this heating will change the characteristics of the material under investigation. Also, evaporation of material from insulator targets due to electron bombardment will deposit on the analyzer, resulting in poorer response from the analyzer. However, spectrometers using electron excitation have the advantage that a lower minimum detectable limit is possible than with spectrometers using X-rays excitation because of the better P/B, as shown by comparing results in Tables 8 and 5. Thus lines and bands that have a P/B only slightly greater than one would be impossible to detect by X-ray excitation. In addition, it is not possible to obtain a fine focus X-ray beam for microanalysis.

C. Penetration Depth

The electron penetration and the depth from which X-rays come from within the target is different for electrons than for X-ray excitation. The range of electrons in aluminum as measured by Holliday andSternglass[46] together with that of several other investigators[47–49] is shown in Figure 23. The experimental depth from which X-rays come by electron excitation, corresponding to 50% of X-ray production, as obtained by Anderson[34] on SiO_2 films is also shown in the same figure. It will be seen from Figure 23 that the depth from which 50% of the X-rays come is about one-half the range of 2.5 keV electrons. For 20 keV electrons it is about one-fourth the range of electrons. No detailed work has been published comparing the range of X-rays in solids and the depth from which the fluorescent X-rays originate. As long as the X-rays being generated have a wavelength close to the wavelength of the exciting X-rays, the depth from which the X-rays come will be about the same as the depth to which the exciting X-rays penetrate, and this depth can be determined from the mass absorption coefficient. The difference between the penetration depth of X-rays and electrons can be seen from the following examples. According to Figure 23 a 2500 V electron will be completely absorbed by 0.025 mg/cm^2 Formvar film. The range energy curve in Figure 23 can be used to determine the stopping power of Formvar, because in mg/cm^2 it should be about the same as Al. However, a 284 eV C *K* band X-ray will only be 10% absorbed in a 0.025 mg/cm^2 Formvar film.

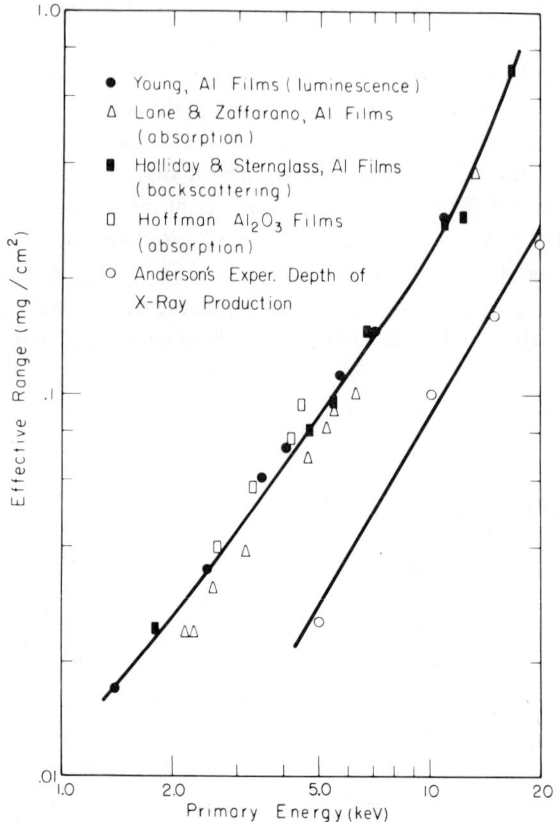

Fig. 23. Effective range of electrons in Al and Al_2O_3 as a function of primary electron energy compared to Anderson's[34] experimental depth of X-ray production.

VI. EMISSION LINES AND BANDS

In the previous section reference has been made to emission lines and emission bands. In this section the basic physics of emission lines and bands in the 10–150 Å wavelength region will be discussed.

The energy level diagram of Mo for the transitions that are in the 10–150 Å region is shown in Figure 24. The energy position of the levels are in close agreement with those given by Sandstrom.[3] The energy widths of the levels are only approximate. The X-ray level notation and their corresponding quantum numbers are given to the right, where n is the total quantum number, l is the angular momentum, and j is due to electron spin. The quantum number j is the vector sum of l and s (electron spin). When two or more

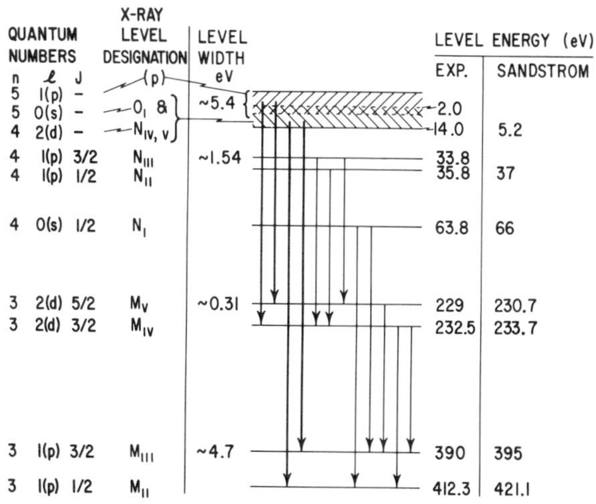

Fig. 24. Conventional energy level diagram for Mo showing *M* and N series electron transitions. Both X-ray and quantum mechanical designation for the levels are shown. (*a*) J. E. Holliday; (*b*) A. E. Sandstrom.[3]

electrons contribute to the total angular momentum it is represented by J. The X-ray emission spectra is usually given in X-ray notation rather than quantum numbers because it was historically developed before quantum mechanics. When quantum numbers are used l is designated by s, p, d . . . rather than by 0, 1, 2, 3. . . The electron transitions that give rise to emission bands are shown by the heavy arrows and transitions that give rise to lines as shown by the light arrows. The allowed transitions determined by dipole selection rules,[50] are those in which l changes by $\pm$ 1 and j changes by 0 or $\pm$ 1. Forbidden transitions sometimes occur that do not obey dipole selection rules but the resulting emission line is very weak.

A. Emission Lines

Emission bands are transitions between a broad outer level with a complex electron distribution (called a band) and an inner narrow level. An emission line is produced by a transition between two inner levels, which usually have Lorentzian type electron distributions. A typical emission line is the transition between the M_V and N_{III} levels or $4d_{3/2} \rightarrow 3p_{3/2}$. The Mo M_V N_{III} line at 64.2 Å is shown in Figure 25 with a definitely Lorentzian shape. There are two smaller peaks: the M_{IV} N_{III} line at approximately 63 Å and the M_{IV} N_{II} line at 63.5 Å. The differences in intensity result from differences in transition probabilities. The intensity ratio of the two lines is given by the ratios of the statistical weight

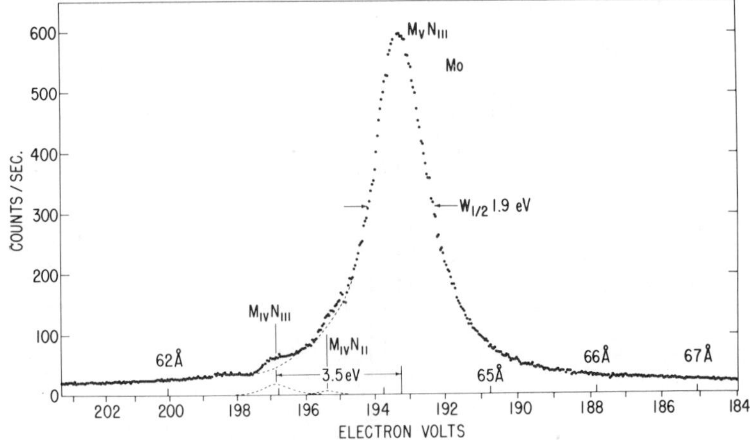

Fig. 25. Mo$M_{IV}N_{III}$,$M_{IV}N_{II}$ and M_VN_{III} emission lines of bulk Mo using Al grating with 2160 grooves/mm and 25 μ slits. Target potential 4kV, beam current 1 mA. Reprinted by permission, J. E. Holliday in *Handbook of X-rays,* Emmett F. Kaelble, ed. McGraw-Hill, New York, 1967, p. 38-22.

of the level represented by $(2J + 1)$. According to this relation the $M_V\,N_{III}$ liue should be 1.5 times more intense than the $M_{IV}\,N_{III}$. However, in Figure 25 it is seen that the Mo $M_V\,N_{III}$ line is about 50 times more intense than the Mo $M_{IV}\,N_{III}$ or $M_{IV}\,N_{II}$ lines. In Table 9 is shown the hydrogenetic transition probabilities calculated by Bethe and Salpeter.[51] The $M_V\,N_{III}$ line should be the weakest of all M series lines but Holliday[2] has shown it to be the strongest. This shows that hydrogenetic transition probabilities are of little value in solids. The $N_V\,N_{III}$ line occurs in the soft X-ray region for the elements Y–Sn. The intensity and wavelength of these lines have been published by the author.[2] Davidson and Wyckoff[52] have also measured some of these M series lines using a lead stearate crystal.

From Figure 24 it will be seen that not all the inner levels are narrow. The N_{III} level for example has a width of approximately 4.7 eV and some inner levels are broader than bands. When measuring the fine structure of emission bands, it is important that the inner level is narrow so that the fine structure is not smeared out. The M_V level of Mo has a width of approximately 0.31 eV, which is sufficiently narrow that fine structure can be observed in the Mo M_V band.

When a neutral atom is ionized in an inner shell, a transition may occur in which the inner vacancy is filled by an electron from a higher level. The transition results in the emission of an X-ray as indicated above. Alternatively, the transition may occur without emission of radiation. In this case the X-ray energy is transferred to another electron of the same atom and this electron is ejected. This process is called the Auger process, and the ejected electron is called an Auger electron. Radiationless reorganization of an atom ionized in an inner shell has three important consequences in X-ray spectra. It influences the breadth of X-ray emission lines and absorption edges, the intensities of the X-ray emission lines, and it is one cause of satellite lines. The satellite line arises from transitions involving atoms doubly ionized in an inner shell. Referring to the energy level diagram in Figure 24, this can occur in the following manner. If there is a hole in the Mo L_{II} level created by an incident electron into which an electron from the Mo M_V level falls, and the X-ray created by this transition ejects an electron from the Mo M_V band by the Auger process, then the Mo M_V level will be doubly ionized. Double ionization of the Mo M_V level causes a greater energy difference between the Mo M_V and Mo N_{III} levels. Electron transitions between these new energy levels will result in higher energy quanta being emitted and a satellite line occurring at a wavelength shorter than the Mo M_V N_{III} line at 64.2 Å. In most cases the intensity of the satellite will be less than that of the parent line, since the probability of a doubly ionized level is less than that for a singly ionized level. An M_V N_{III} satellite is shown in Figure 26. The satellite line has considerably less intensity than the M_V N_{III} line. A doubly ionized level can also be produced when a single bombarding electron ejects two electrons simultaneously. A transition from the conduction band to the doubly ionized Mo M_V level would be called a Mo M_V band satellite. Band satellites have been found to be strongly influenced by chemical combination.

As indicated above, transitions from outer levels of atoms in a solid give rise to bands. Different X-ray notation is used for emission lines and emission bands. For example, a transition from the L_{II} to the K level which is an inner

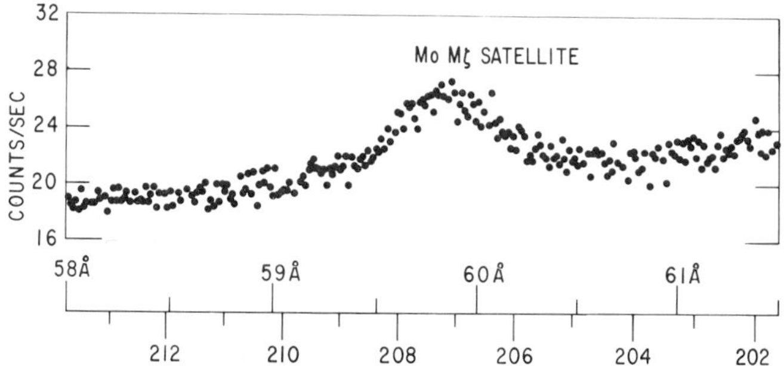

Fig. 26. The MoMζ (M_V N_{III}) line satellite. The target potential is 4 kV, using a 2160 grooves/mm 1° blazed replica grating.

Table 9. Transition probabilities for hydrogen in 10^8 sec^{-1}. Reprinted by permission of Hans A. Bethe and Edwin E. Salpeter, *Quantum Mechanics of One and Two Electron Atoms*, Academic Press, New York, 1957, p. 266

Initial	Final	$n = 1$	2	3	4	5	Total	Lifetime in 10^{-8} sec
$2s$	np	—	—	—	—	—	0	∞
$2p$	ns	6.25	—	—	—	—	6.25	0.16
2	mean	4.69	—	—	—	—	4.69	0.21
$3s$	np	—	0.063	—	—	—	0.063	16
$3p$	ns	1.64	0.22	—	—	—	1.86	0.54
$3d$	np	—	0.64	—	—	—	0.64	1.56
3	mean	0.55	0.43	—	—	—	0.98	1.02
$4s$	np	—	0.025	0.018	—	—	0.043	23
$4p$	ns	0.68	0.095	0.030	—	—	0.81	1.24
	nd	—	—	0.003	—	—		
$4d$	np	—	0.204	0.070	—	—	0.274	3.65
$4f$	nd	—	—	0.137	—	—	0.137	7.3
4	mean	0.12_8	0.083	0.089	—	—	0.299	3.35

5s	np	—	0.012	0.008_5	0.006_5	—	0.027_7	36
5p	ns	0.34	0.049	0.016	0.007_5	—	0.415	2.40
	nd	—	—	0.001_5	0.002	—		
5d	np	—	0.094	0.034	0.014	—	0.142	7.0
	nf	—	—	—	0.000_5	—		
5f	nd	—	—	0.045	0.026	—	0.071	14.0
5g	nf	—	—	—	0.042_5	—	0.042_5	23.5
5	mean	0.040	0.025	0.022	0.027	—	0.114	8.8
6s	np	—	0.007_3	0.0051	0.0035	0.0017	0.0176	57
6p	ns	0.195	0.029	0.0096	0.0045	0.0021	0.243	4.1
	nd	—	—	0.0007	0.0009	0.0010		
6d	np	—	0.048	0.0187	0.0086	0.0040	0.080	12.6
	nf	—	—	—	0.0002	0.0004		
6f	nd	—	—	0.0210	0.0129	0.0072	0.0412	24.3
	ng	—	—	—	—	0.0001		
6g	nf	—	—	—	0.0137	0.0110	0.0247	40.5
6h	ng	—	—	—	—	0.0164	0.0164	61
6	mean	0.0162	0.0092	0.0077	0.0077	0.0101	0.0510	19.6

level is referred to as the $K_{\alpha 2}$ line. If the L_{II} level happens to be the outer level as is the case in carbon, then it is a band and the transition from L_{II} band to the K level is no longer referred to as $K_{\alpha 2}$ but as the C K band. The $\alpha 2$ notation implies that the electrons in the L_{II} level have distinct quantum numbers of $n = 2$, $l = 1$ and $j = \frac{1}{2}$. In a band, quantum numbers do not have the same significance as they do for electrons in inner levels. There is a great deal of admixture of the s, p, d, . . . electrons caused by band overlap as indicated in Figure 24, where the outer s and p bands overlap the $4d$ band. The admixture is more pronounced in metals than in insulators. For Be and Li there is a large overlap between the $2s$ and $2p$ bands so that the outer levels of these elements will be an admixture of s and p electrons. If all of the outer shell electrons had $2s$ type symmetry, as they do in the free atom, there would be only a weak intensity from Be and Li in the soft X-ray region because a $2s \rightarrow 1s$ transition is forbidden. Since the $2s$ and $2p$ levels of the nonmetals for the second period elements are separated, they have more significance for X-ray measurements.

The depth within the atom to which the levels are affected by neighboring atoms depends on how well the inner levels are screened by the outer electrons. For Mo, the M_{III} levels are not changed by the presence of neighboring atoms in a solid, and the M_V N_{III} line does not change in shape and wavelength with chemical combination. The M_V N_{III} line of Y was effected by changes in chemical combination. The M_V N_{III} line from Y has an asymmetry index* of 1.0 and a $W_{1/2}$ of 0.45 eV (uncorrected) compared with the M_V N_{III} line from Y_2O_3 which has an asymmetry index of 1.5 and a $W_{1/2}$ of 0.8 eV. The reason the M_V N_{III} line is changed for Y and not for Mo in chemical combinations is that Mo has 6 screening electrons above the M_{III} level, while Y has only 3. Thus, the N_{III} level of Mo has more screening from neighboring atoms than does the N_{III} level of Y.

B. Emission Bands

The interest in soft X-ray emission bands by metallurgists and metal physicists began with the idea that emission bands would give the shape and width of the occupied portion of the density of states curve. The level density is a function of the energy E and is defined as the number of states per unit volume of the solid in the energy range E to $E + dE$. The number of states per unit volume is denoted by $N(E)dE$. It is not the purpose of this chapter to derive the density of states function or a detailed mathematical relation between the density of states function and emission spectra. The

* The index of asymmetry is defined as the ratio of the part of the full width at half maximum lying to the long-wavelength side of the maximum ordinate to that on the short-wavelength side.

solution is only summarized and the reader is referred to Tomboulian[53] for more detailed discussions of this subject in his *Handbuch der Physik* article.

For free electrons the density of states curve is given by

$$N(E) = (2\pi/h)(2m)^{3/2}E^{1/2} \tag{13}$$

where h is Plank's constant, m is mass of the electron, and E is the energy. The above expression does not consider lattice structures and is not valid for the real crystal, where the zone structure has a strong influence on $N(E)$. To determine $N(E)$ for the real crystal, the shape of the energy surface in k space and its behavior near zone boundaries must be known. For the real crystal the $N(E)$ curve is given by the following relation:

$$N(E) = (1/8\pi^3)\int_A (dA/|\text{ grad } E\,|) \tag{14}$$

where the integration is over the surface A on which the energy has a given value, and the gradient is with respect to the wave vector k.* The $E^{1/2}$ dependence for the $N(E)$ curve still holds in a crystal if the energy surface is not touching the Brillouin zone boundaries. As the energy surface approaches the zone boundary, the $N(E)$ curve departs from the $E^{1/2}$ dependence, reaches a peak corresponding to a discontinuity when the zone boundary is touched, and then if the electron concentration is small compared with that required to fill the first zone the $N(E)$ curve drops to zero. This is shown in Figure 27. The fact that the integration for the $N(E)$ curve is over the entire surface shows that the $N(E)$ curve is not dependent on direction. Thus, the $N(E)$ curve will be the same from any given plane of a single crystal, as it is from a polycrystal.

It has been shown by Tomboulian[53] that X-ray emission intensity aside from frequency variation depends on the following quantity:

$$\int_A |M_{0K}|^2/|\text{grad } E|dA \tag{15}$$

where $|M_{0K}|^2$ is $\rho[\int_i \psi^*\delta/\partial x\psi_K d\tau]^2$, which is the transition probability, and ψ_0 and ψ_K denote the wave functions of the atom in the initial and final states. Aside from a constant the expression under the integral (quantity 15) after $|M_{0K}|$ is removed is the density of states. The quantity 15 can be written

$$|M_{0K}|^2N(E) \tag{16}$$

* Bloch has shown that wavefunction of an electron can be represented as a modulated plane of the form

$$\psi\, R^{(r)} = U_R{}^{(r)}\, C^{-k,r}$$

where k is the propagation vector.

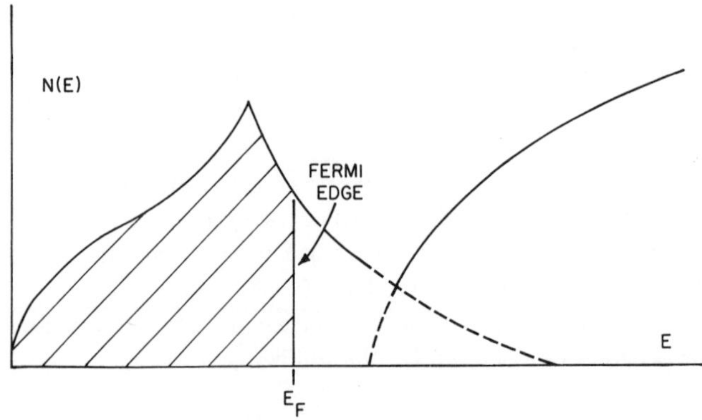

Fig. 27. The form of the $N(E)$ in the Brillouin zone theory where the first and second zones overlap which is shown by the dashed line. The shaded area is the filled portion of the band.

When the emission band intensity is measured in photons/sec. the relation between X-ray bands and $N(E)$ is

$$I(E) \; \alpha \; \nu |M_{0K}|^2 N(E) \tag{17}$$

where ν is the frequency of a given state. Thus, it is seen that emission bands do not give direct information concerning the density of states, but the density of states times transition probability. The transition probability varies with the energy and depends on the electronic level structure of the solid. Since $|M_{0K}|^2$ is a function of ν^2, relation 17 can be written in the following form after Skinner

$$I(E) \; \alpha \; \nu^3 \; F(E) \; N(E) \tag{18}$$

Where $F(E)$, according to Skinner,[54] is an atomic transition probability in contrast to $|M_{0K}|$ which is thought to be influenced by the neighboring atoms. The only difference between equations 17 and 18 is the factor ν^2. Some investigators [53,55] have partially corrected their emission bands for transition probability by dividing by ν^2.

If it were possible to completely correct $I(E)$ for transition probability it would still not be possible to obtain the $N(E)$ curve. There are several additional considerations that are not indicated in equation 16. The first consideration is that because of the above-mentioned dipole section rules the $I(E)$ curve shows only a partial density of states curve. For example a transition from the conduction band to an inner level with p-type symmetry would reflect electrons with only s- and d-type symmetry in the band. For transitions to an inner level with d-type symmetry only electrons with p-type symmetry would be revealed by the resulting emission band. In order to obtain a complete picture of the band several emission bands which reflect different symmetry in the band would have to be measured. The relation between the partial density of states and the total occupied states is given by

$$N(E) = N(E)_{s+d} + N(E)_p \tag{19}$$

In addition to the density of occupied states obtained from the emission bands there is a density of states for the unoccupied states which can be obtained from absorption spectra. In Figure 27 is shown the total density of states with the occupied portion shaded. The boundary between the occupied and unoccupied states is called the Fermi edge, and the energy at which the edge occurs is labeled the Fermi energy. The $I(E)$ curve obtained from the instrument must also be corrected for the broadening and energy dependence of the inner level. In addition, the emission band obtained from the spectrometer must be corrected for instrumental broadening, overlapping bands and widening of the emission edge due to temperature. Since a method for correcting the bands for these factors has been given by the author[2] in the *Handbook of X-rays*, they will not be discussed fruther. Also, Parratt[56] discusses the above-mentioned corrections and methods of making them.

C. Emission Bands from Transition Metals and Alloys

Figure 28 shows the Mo M_{IV} and M_{V} emission bands. By dipole selection rules this $5p$–$3d$ transition would give a partial density of states curve with p type symmetry. Since there are no $5p$ electrons in the free atom there must be an overlap between the $5p$ and $4d$ bands. Because the M_{IV} and M_{V} bands are transitions from the same band only the M_{V} band will be considered in detail. The M_{V} band of Nb showing p type symmetry in the band is indicated in Figure 29. These bands show a typical intensity distribution which is observed in soft X-ray spectroscopy, i.e., a pronounced leading edge, strong asymmetry, and tailing to the bottom of the band. Some of the bands show more peaks than others. The M_{V} bands in Figures 28 and 29 show two distinct peaks. Wiech[57] has also observed a double peak in the Nb M_{V} band but it is less resolved than for the Nb M_{V} band in Figure 29. The M_{V} bands were corrected for background, M_{IV} overlap, instrumental error and width of the M_{V} level and are shown in Figure 30. The instrumental error is a Gaussian curve with a $W_{1/2}$ of 0.6 cV and the M_{V} level width is 0.31 eV and 0.28 eV for Mo and Nb, respectively. The method of correcting the emission bands using the analog computer has been described by the author[2] in the *Handbook of X-rays*.

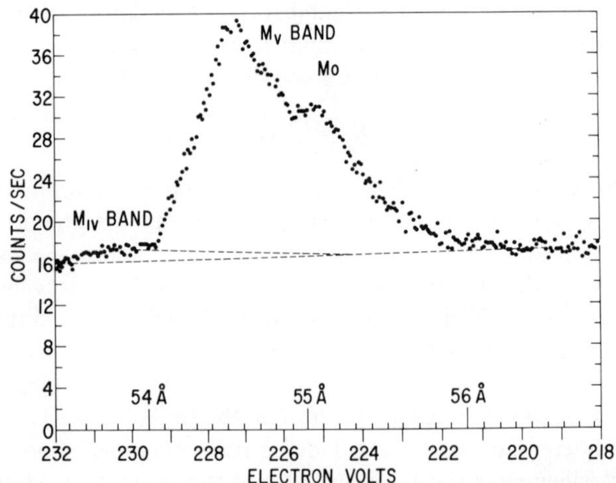

Fig. 28. The M_{IV} ($5p \rightarrow 3d\ 3/2$) and M_V ($5p \rightarrow 3d\ 5/2$) emission bands of Mo obtained using a 2160 grooves/mm 1° blazed replica grating with a Pt surface. The target voltage is 4 kV and a 1.4 ma beam current. Reprinted by permission, J. E. Holliday, *The Electron Microprobe,* T. D. McKinley, K. F. J. Heinrich and D. B. Wittry, eds. Wiley, New York, 1966, p. 3.

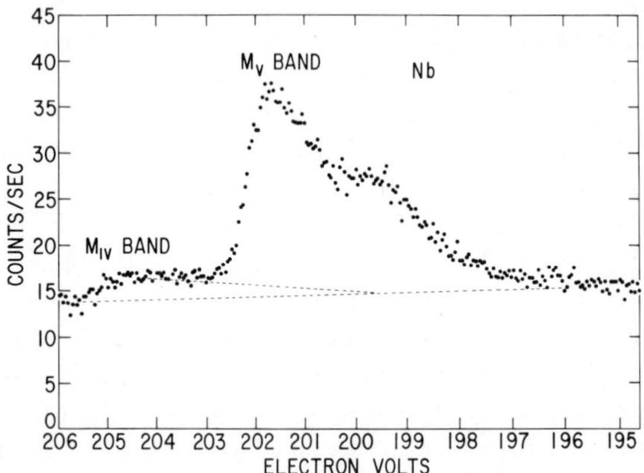

Fig. 29. M_{IV} ($5p \rightarrow 3d\ 3/2$) and M_V ($5p \rightarrow 3d\ 5/2$) emission bands of Nb using Al grating with 2160 grooves/mm. Target potential 4 kV, beam current 1.4 mA. Reprinted by permission, J. E. Holliday, *The Electron Microprobe,* T. D. McKinley, K. F. J. Heinrich and D. B. Wittry, eds. Wiley, New York, 1966, p. 3.

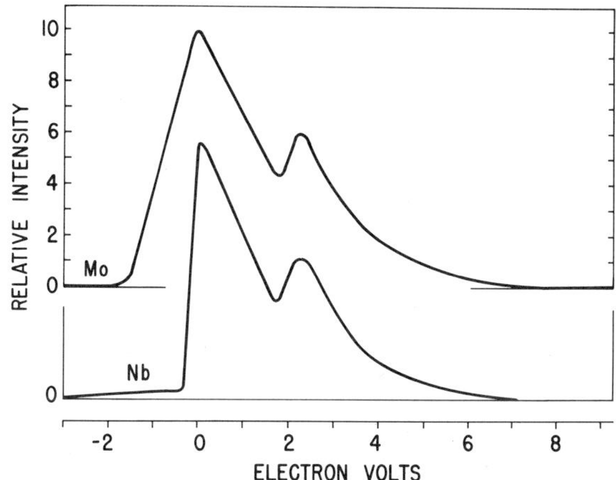

Fig. 30. The M_V emission bands of Mo and Nb metal corrected for instrumental error, inner level width, M_{IV} overlap and background. The intensity distribution of both M_V bands are identical beyond the peak. (Peaks normalized). Reprinted by permission, J. E. Holliday, in *Proc. Conf. Soft X-ray Spectrometry and the Band Structure of Metals and Materials,* Academic Press, New York and London, 1968 (in press.)

The width of the Mo emission edge at the base after correction is 1.5 eV. The temperature contribution to this edge width according to Skinner[54] is 6 kT where k is Boltzman's constant and T is the absolute temperature. The target temperature was approximately 350°C which gives an edge width of 0.32 eV. After all corrections are made, the emission edge width is approximately 1.2 eV. If the emission edge corresponds to the Fermi edge, it should have zero width after making the above corrections. One explanation for the emission edge of the Mo M_V band not having zero width can be obtained from the total $N(E)$ curve in Figure 31, for the second series transition metals as obtained from specific heat measurements by Ostenburg et al.[58] It can be seen that the total $N(E)$ curve is a maximum at Nb. Thus, the apparent width of the edge for the Mo M_V band is the rise in $N(E)$ between 1 and 0 eV. The Fermi edge at Mo (Fig. 31) is quite small in comparison to the height of the emission edge and apparently is not observable in the measurement. Since the peak of the total $N(E)$ curve occurs at Nb, it would be expected that the Nb M_V band emission edge would have zero width after corrections and the shape of the Nb and Mo M_V bands would be identical beyond the band peak. Both the Mo and Nb M_V bands have the same intensity distribution beyond the peak (Fig. 30). The second peak on the low-energy side of the total $N(E)$ curve in Figure 31 is also seen for the Nb and Mo M_V bands. The Nb M_V band after being corrected for background, M_{IV} band overlap, inner level width, and instrumental error has an edge width

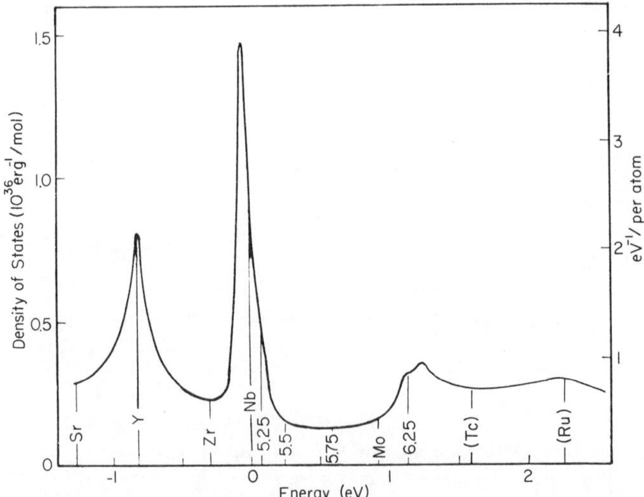

Fig. 31. Density of states obtained from the low temperature specific heat data for the *4d* band of transition metals and their alloys. The Fermi levels of these metals and alloys are shown. (Alloys are specified by the number of electrons per atom, e/a = 5 for Nb metal and 6 for Mo metal.) Reprinted by permission, D. O. Van Ostenburg, M. J. Lan, M. Shmizu, and A. Katsuki, *J. Phys. Soc. Japan,* **18,** 1748 (1963).

of 0.32eV. The temperature of the target was also 350°C giving an edge width of 0 eV after correction. Thus, the general shape of the Nb and Mo M_V emission band is predicted by the total $N(E)$ curve as determined by specific heat measurements. However, quantitatively the comparison between the corrected M_V bands in Figure 30 and the total $N(E)$ curve is not as good. The separation between the two peaks of the M_V band according to $N(E)$ in Figure 31 should be 0.8 eV but it is actually 2.3 eV.

Even after the corrections are made there is still a considerable amount of tailing to the bottom of the band and some tailing at the emission edge. As indicated above, this tailing at the end points of the band has made determination of the energy width of the band difficult. A number of correction methods have been applied. Some investigators[53–56, 59] have tried to fit an $E^{3/2}$ or $E^{1/2}$ dependence to the tail, depending on whether the inner level was K or L respectively. An $E^{1/2}$ or $E^{3/2}$ dependence, however, does not appear to fit the experimental band. Skinner[54] proposed that the tailing effect is due to Auger transition in the band which broadens the levels, and is more probable at the bottom of the band than near the top. Thus, the broadening effect of these transitions would be more pronounced at the bottom of the band. With a high resolving power spectrometer the author[1] has shown that many of these tails are made up of component peaks which are probably satellites. However, in order to establish a lower limit to the bottom of the band, a linear extrapolation of the band was employed. Table 10 shows the extrapolated band width, the edgewidth, $W_{1/2}$ and the energy at which the emission edge occurs, for the corrected M_V bands of Y, Zr, Nb and Mo. Temperature corrections were only applied to emission edges which corresponded to the Fermi edge. The calculated bandwidth as determined by Altman[60] for Y and Zr are also shown in Table 10. It will be seen that the experimental band width of Y is smaller than predicted by Altmann's calculation while it is just the opposite for Zr.

The first series transition metals are of interest in metal physics. The L_{II}, L_{III} bands ($4s + 3d \rightarrow 2p$ transition) occur in the 10–30 Å region for the first series transition metals. The L_{II} band is a ($4s + 3d \rightarrow 2p\ J_{1/2}$) transition and the L_{III} band is a ($4s + 3d \rightarrow 2p\ J_{3/2}$) transition. The $L_{II,III}$ emission bands for Ti, Fe and Ni using a 3600 grooves per millimeter, 1° blaze, Pt grating and 25 μ slits as measured by the author, are shown in Figures 32–34. The peak intensity, wavelength and $W_{1/2}$ for the first series transition metals L_{III} bands are shown in Table 7.

Skinner, et al.,[61] Fischer and Baun, [62–64] Bonnelle,[19a] Liefeld,[65] and Lukirskii and Brytov[66] have also measured some of the $L_{II,III}$ emission bands of the first series transition metals. Since the original measurements by Skinner et al., a number of arguments have been presented as to why the L_{III} bands did not show the double peak indicated by the theoretical band calculations. The L_{III} bands measured by the author are shown in Figures 32–34 and indicate that there is more than one peak in the L_{III} band, although there is some question as to which peaks are part of the band and which are satellites. The peak A on the low-energy side of the Ti L_{III} band (Fig. 32 and 40) is considered to be a cross transition which increases in intensity relative to the Ti L_{III} peak for certain type of compounds. This peak is more prominent for the Ti L_{III} band from TiO in Figure 40 than TiC in Figure 32 because for TiO peak A is an $02p \rightarrow$ Ti $2p$ cross transition, while for TiC it is a C $2s \rightarrow$ Ti $2p$ cross transition. In the case of the Fe L_{III} band (Fig. 33) there are two peaks which have been labeled *a*, *b* for identification purposes. Fischer and Baun, as well as Cauchois and Bonnelle have also observed peaks *a* and *b*. The three peaks for the Ni L_{III} band in Figure 34 have also been labeled *a*, *b*. Liefeld[65] considers hump *a* to be a satellite because it is on the high-energy side of the Ni L_{III} absorption edge. Other investigators[19,36,65] who measured the Ni L_{III} band observed hump *a* but they did not report any of the fine structure in hump *a* or on the low energy side of peak *b*. The dashed curve drawn by Cuthill et al.[67] in Figure 35 is typical of previous measurements of the Ni L_{III} band. They did not include hump *a*

Table 10. Corrected M_V emission band parameters for some 2nd series transition metals

Metal	Emission edge energy position	Edgewidth[a] eV	Basewidth[a] eV	$W_{1/2}$ eV	Calculated[b] Bandwidth eV	M_V level width eV
Y	157.1 eV	.18[c]	~2.18	1.6	3.3	0.23
Zr	179.85 eV	0.75	4.95	2.95	3.9	0.25
Nb	202.3 eV	0. [c]	4.2	2.75		0.28
Mo	229.0 eV	1.5	5.7	3.5		0.31

[a] Extrapolated and includes edgewidth.
[b] Altmann[60].
[c] Corrected for temperature.

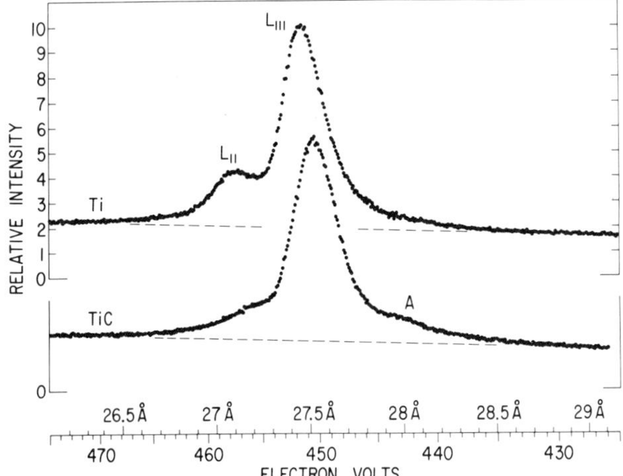

Fig. 32. The Ti $L_{II,III}$ ($3d + 4s \rightarrow 2p$) band from $TiC_{0.95}$ and from Ti metal. The Ti L_{III} peak from $TiC_{0.95}$ has shifted towards lower energy (peaks normalized). Reprinted by permission, J. E. Holliday, *J. Appl. Phys.*, **38**, 4727 (1967).

because this hump is not part of the band. However, Cuthill et al.[67] observed fine structure on the low-energy side of the Ni M_{III} band ($3d + 4s \rightarrow 3p$ transition) shown in Figure 35 which occurs at about the same eV from the main peak as does the fine structure in Figure 34. They have made some progress in correlating this fine structure with theoretical band calculations. A comparison of the first series transition metal L_{III} bands with the total $N(E)$ curve from specific heat measurements will be instructive. In Figure 36 is shown Cheng et al.'s[68] total $N(E)$ curve for the d band of the first series transition metals. This curve is similar in shape to that for the second series in Figure 31. The total $N(E)$ curve in Figure 36 predicts an $N(E)$curve for Fe that has a low intensity hump on the high-energy side, a high intensity peak in the middle of the band followed by a lower intensity peak. It will be seen that the Ni Fe L_{III} bands in Figure 33 have a similar intensity distribution. Since it is not certain whether peak a is part of the Fe L_{III} band it may be premature to make this comparison. However, the $N(E)$ curve in Figure 36 does show as in the case of the Mo M_V band that sharp emission edges are not to be expected for the L_{III} bands of the first series except possibly V.

Although some correlation has been made between the transition metal emission bands and the band model there is less agreement with the emission band from alloys. Curry[69] and his group at the University of Leeds have measured the emission band from a large number of binary alloys which are:

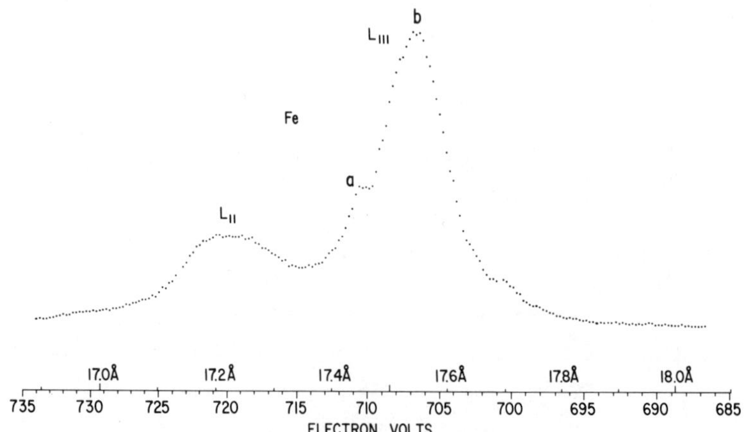

Fig. 33. Fe $L_{II,III}$ ($3d + 4s \rightarrow 2p$ transition) emission band from Fe using high resolving power spectrometer with a 3600 grooves/mm 1° blazed grating.

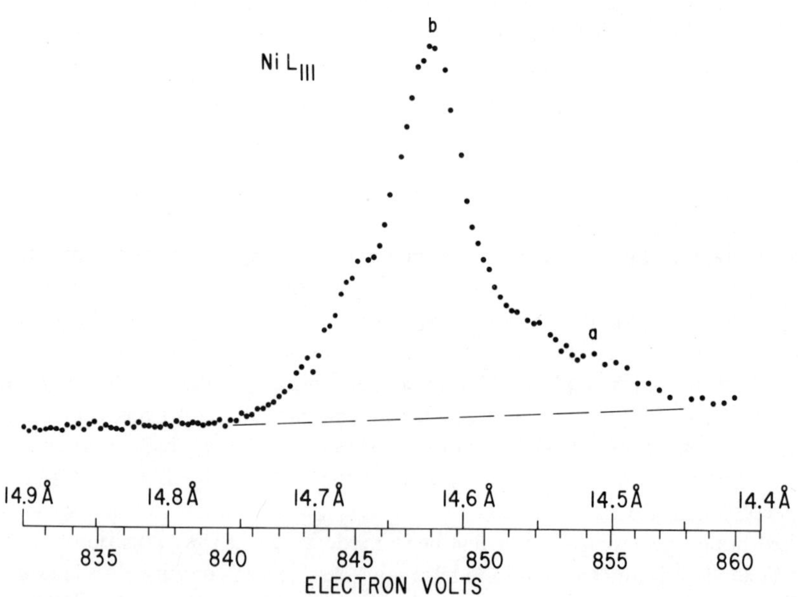

Fig. 34. Ni L_{III} ($3d + 4s \rightarrow 2p$ transition) emission band from Ni using a high resolving power spectrometer with a 3600 grooves/mm grating.

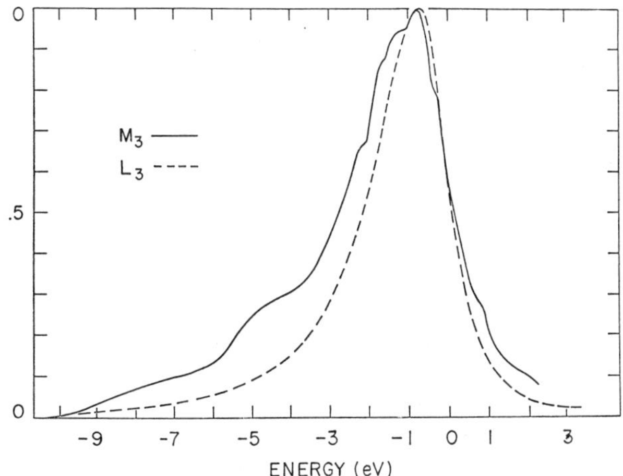

Fig. 35. Ni M_{III} ($3d + 4s \rightarrow 3p$ transition) band from Ni; the dashed curve is typical of previous measurements of the Ni L_{III} band. Reprinted by permission, J. R. Cuthill, R. A. McAlister, M. L. Williams and R. E. Watson, *Phys. Rev.* 164, 1012 (1967).

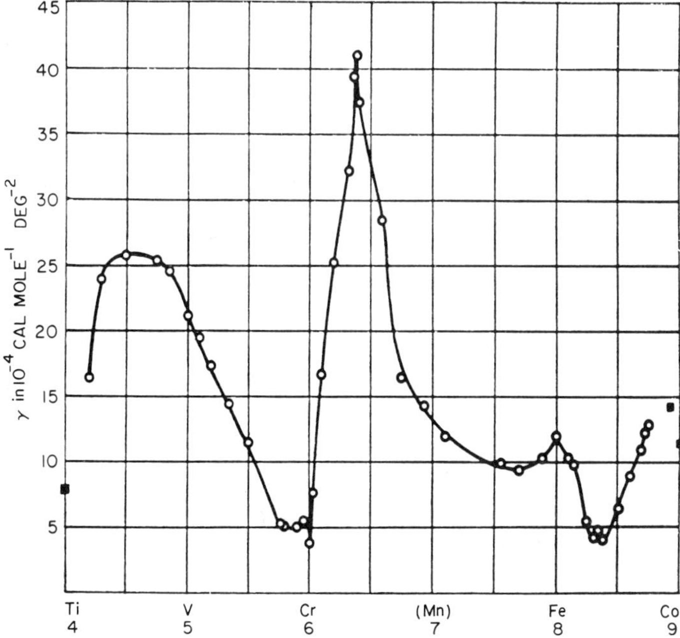

Fig. 36. Electronic specific heat coefficient of b.c.c. alloys of the first series transition elements. Reprinted by permission, C. H. Cheng, K. P. Gupta, E. C. Van Reuth, and Paul Beck, *Phys. Rev.*, **126**, 2030, (1962).

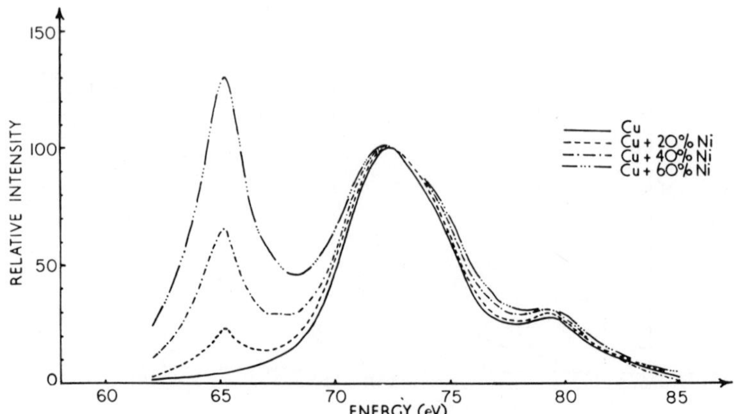

Fig. 37. Cu and Ni $M_{II,III}$ emission bands ($3d + 4s \rightarrow 3p$ transition) bands from 0, 20, 40 and 60% Ni in Cu. Reprinted by permission, J. Clift, C. Curry, and B. J. Thompson, *Phil. Mag.*, **8**, 593 (1963).

Cu/Ni, Ni/Zr, Al/Fe, Cu/Si, Al/Cu, Al/Mn, and Mg/Si ($L_{II,III}$ emission bands of Al, Mg, and Si, $M_{II,III}$ bands of Zr, Cu, Fe, Ni) and Mn. From these measurements they have made some rather interesting conclusions which are given below. An examination of the Al K emission spectra for a number of Al alloys which were measured by Fischer and Baun[63] indicates that their results are in agreement with Curry's conclusions. The statements made below also incorporate some of Fischer and Baun's conclusions and the authors preliminary measurements of the $L_{II,III}$ bands of the Ti/Fe and Ti/Ni intermetallic compounds.

(*1*) Many of the emission bands of the above alloys did not change perceptibly on alloying. The Ti $L_{II,III}$ band did change when combined with Ni. The amount of change appeared to be related to the electronegativity difference of the two interacting metals.

(*2*) No change in emission bands was observed in going from the order to disorder state. Catterall and Trotter[70] also reported that there was no effect of order–disorder on emission bands. There was no observed dependence on structure. A definite dependence on composition was observed.

(*3*) Band widths do not change substantially on alloying (rarely by more than 10% and usually much less). In addition the tailing at the emission edge and bottom of the edges as observed for the pure metal makes it difficult to measure changes with any degree of accuracy. However, shifts in the peak were observed on alloying and these can be measured to a high accuracy.

(*4*) There was little tendency toward equality of emission bands as would be expected on a simple common valence band model. This is illustrated by Clift et al.'s[71] work on Cu/Ni in Figure 37. The difference that existed between the metal bands before alloying was maintained after alloying and in some cases the difference increased on alloying.

(*5*) Curry[69] stated that a strong tendency for preferential grouping of valence electrons around the two atoms was indicated by the results in (*4*).

It is clear that the above conclusions are not in agreement with the rigid band model that pictures a common band in which the Fermi edge moves toward higher or lower energy depending on whether electrons are added or subtracted from the band.

Some X-ray spectroscopists argue that satellites are the reason band widths cannot be measured accurately and that they also prevent the changes in band width on alloying from being observed. Liefeld[65] claims that in order to avoid satellites the bands must be measured with an excitation voltage that is only 1.1 times the threshold energy of the inner level. It is also claimed that transition probabilities change the shape of the band more than has been realized and this is the reason that common bands are not observed in alloys. However, the author[1] in his introductory remarks to Part II of the proceedings of the Conference on Soft X-ray Spectroscopy and the Band Structure of Metals and Alloys has stated that important experimental information is obscured by the attempt to make emission bands fit the band model through extensive corrections and by measuring emission bands under adverse experimental conditions (most bands have negligible intensity for excitation voltages 1.1 times threshold). Although there is justification for correlating pure metal emission bands with the band model, this is not true of alloys and compounds. There are a number of observed changes in emission bands with alloying and chemical changes that are not easily related or explained by the band model. These changes and their significance will be discussed in the next section.

VII. CHANGES IN SHAPE AND WAVELENGTH OF THE EMISSION BANDS OF ELEMENTS WHEN CHEMICALLY COMBINED

The most significant changes in emission bands with alloying and chemical combination are peak energy shift and intensity distribution changes. Unlike band width the peak shift can be measured to a high accuracy. Although these parameters are not easily related to the band model, it has been shown[22, 72–74] that they are related to changes in electron distribution and chemical bonding concepts. In general, the greater the electron transfer or electronegativity difference between the two elements in the alloy or compound the greater the peak shift and change in intensity distribution on alloying or when the compound is formed. When the amount of electron tranfers is a factor in bond strength as measured by heat of formation it has been shown[74] that intensity distribution and wavelength shift can be correlated with bond strength.

Thus soft X-ray spectroscopy can provide information regarding the amount of electron transfer or ionic character to the bond. This would be valuable information in a number of metallurgical processes. In addition the measurement of emission band fine structure using a microelectron probe as the X-ray source could provide information as to the nature of the bond in small phase second particles and inclusions in the metal matrix. This technique would be a valuable supplement to X-ray diffraction for studying glassy materials or where the particles are too fine to have long range order.

Since the interaction of the 2nd period elements with metals is of importance in a number of metal processes, and because changes in bands are most pronounced for these reactions, the changes in metal and nonmetal emission bands when oxides, nitrides, carbides and borides are formed will be considered. If more background is desired as to peak intensity, wavelength and intensity distribution of the 2nd period elements, the reader is referred to the author's[2] *Handbook of X-rays* article.

A. Oxides

O'Bryan and Skinner[75] measured the emission band shape of a large number of oxides using a grating analyzer. Recently Fisher and Baun[76] and Mattson and Ehlert[21] have used lead stearate and KAP crystals to measure the O K bands from oxides. Mattson and Ehlert[21] have made some interesting measurements of the O K band from gases which are shown in Figure 38. As would be expected the O K band from O gas has a narrower $W_{1/2}$ and has less structure than when oxygen is combined with another element. Although it was shown in Section III-C that there will be additional peaks in the O K band when the above crystals are used as analyzers, the possibility of being able to determine from fine structure measurements whether the gas is combined or free in the metal is of considerable value in metallurgy. This is especially true for very low percentages of gases in metals.

The O K bands measured by the author, for Ti + 25% O, $TiO_{1.02}$ and $SrTiO_3$ are shown in Figure 39. There is no pronounced change in shape between the O K band for 25% O in solution and $TiO_{1.02}$. The hump on the low energy side appears to be more pronounced for $TiO_{1.02}$ than for 25% O in solution. There is a shift towards higher energy in the O K peak in going from Ti + 25% O to $SrTiO_3$. The index of asymmetry has changed from 0.45 to 2 in going from Ti + 25% O to $SrTiO_3$.

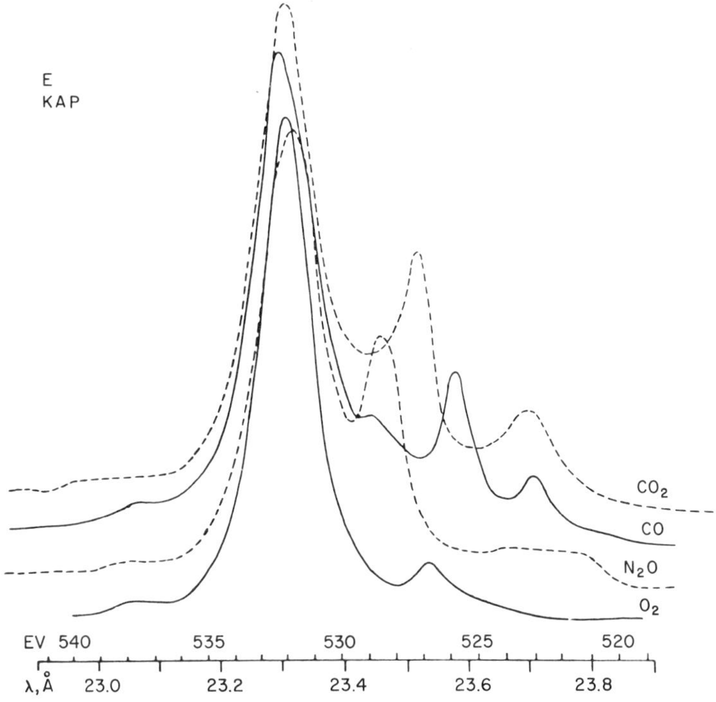

Fig. 38. O *K* emission bands from N_2O, CO_2, CO and O_2 using a KAP crystal. Reprinted by permission, R. A. Mattson and R. C. Ehlert, in *Advances in X-ray Analysis*, G. R. Mallet, M. Fay, and W. M. Mueller, eds. Vol. 9, Plenum Press, New York, 1966, p. 482.

Changes in shape and peak wavelength are also observed for the Ti $L_{II,III}$ bands from Ti + 25% O, $TiO_{1.02}$ and Ti metal in Figure 40. Before discussing these bands it is important to acknowledge the work of Blochin and Shuvaev,[77] Vainshtein and Chrikov,[78] Memnonov and Kolobova,[79] and Meisel and Nefedow[80] who have done a considerable amount of work on titanium compounds in the *K* emission bands in the hard X-ray region. Lukirskii and Brytov[66] have reported some good work on the $L_{II,III}$ bands of titanium and chromium oxides using a grating analyzer. Because of poor intensity in the fourth order, only the band parameters of the oxide were published and not the spectra. There are several obvious changes in the Ti $L_{II,III}$ emission band in going from the metal to the oxide. The L_{II}/L_{III} intensity ratio* has

*It was shown earlier that in addition to chemical changes the Ti L_{II}/L_{III} intensity ratio is affected by self absorption.

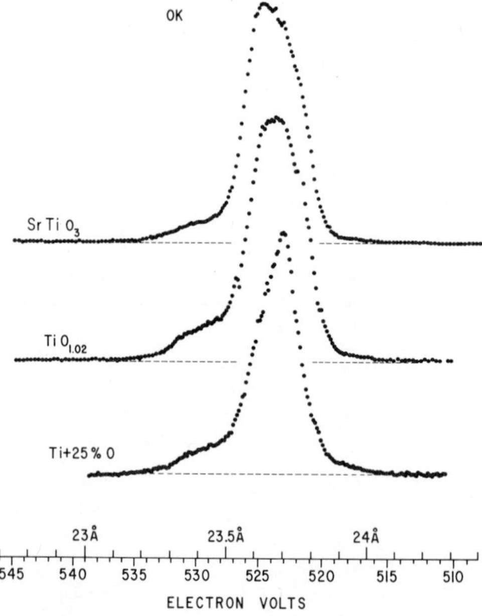

Fig. 39. O K bands ($2p \rightarrow 1s$ transition) normalized from Ti + 25% O, $TiO_{1.02}$ and $SrTiO_3$ using a high resolving power spectrometer with a 3600 grooves/mm 1° blazed grating. The target potential was 4 kV.

increased, the Ti L_{III} peak energy has shifted toward higher energy and there is an increase in the intensity of the satellite labeled A. As indicated in Table 11 (Holiday's meas.) the L_{II}/L_{III} ratio increases and L_{III} peak shifts towards higher energy for the other first series transition metals relative to the pure metal. The satellite A was found to increase relative to the Ti L_{III} peak with the Ti oxidation number, as shown in Figure 41. From Table 12 it will be seen that the A/L_{III} intensity ratio also increases with the Ti L_{III} energy shift. From a comparison of the Ti L_{III} shifts and the A/L_{III} intensity ratios for TiO_2 and $SrTiO_3$ in Table 12, it can be seen that these parameters are not a function of the amount of oxygen in the target.

B. Nitrides

The N K band from ZrN and BN is shown in Figure 42. The N K band from TiN is not shown due to the overlap of the Ti L_I line. The peak wavelengths of the N K bands are the same. It is of interest to note that the relation between the shapes of the N K bands from BN and ZrN are similar to that between the C K band of graphite and ZrC (Fig. 43). It is planned to remeasure the above N K bands and other N K bands with a higher

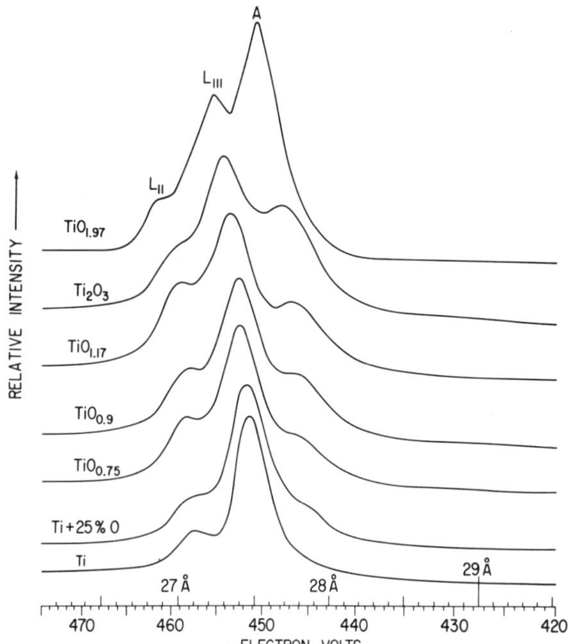

Fig. 41. The Ti $L_{II,III}$ bands (normalized to Ti L_{III} band from $TiO_{0.75}$) intensity distribution for titanium, Ti + 25% O and titanium oxides, TiO_x. The potential was constant at 4 kV and percentage deviation was = 1.5. Reprinted by permission, J. E. Holliday in *Soft X-ray Band Spectra and Electronic Structure of Metal and Alloys,* Derek J. Fabian, ed., Academic Press, London and New York, (1968) p.128.

IV transition metal carbides has a $W_{1/2}$ of about 1/3 that of the C K band from carbon contamination. Thus, the surface must be thoroughly cleaned and there must be no carbonaceous contamination deposited by the electron beam during the course of the measurements.

The C K bands of diamond graphite, TiC, VC and Fe_3C* which were measured by Holliday[74] are shown in Figure 44. The peak heights of the C K bands of diamond and the carbides have been normalized to the peak height of graphite. When going across the first series transition metal carbides (groups IV–VIII) the C K band changes from a single nearly symmetrical peak to a more complex band structure similar to the C K band of graphite. This has also been shown[70] to be the case for the second and third series transition metal carbides. Because of the correspondence of the peaks of the

* The C K bands of graphite and diamond have been compared to other investigators' measurements in a recent publication.[74]

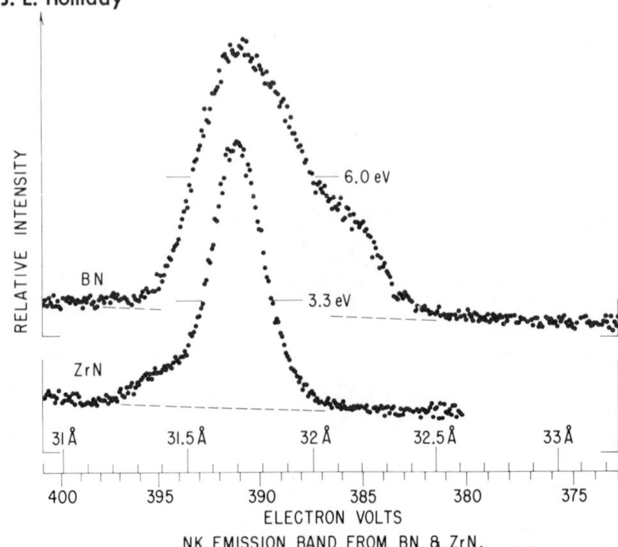

Fig. 42. Comparison of N K emission bands (2*p* →1*s* transition) from BN and ZrN. The BN target had a thin coating of carbon to prevent charging. The peak heights have been normalized. Reprinted by permission, J. E. Holliday in *The Advances in X-ray Analysis*, G. R. Mallett, M. Fay and W. M. Mueller, eds. Vol. 4, Plenum Press, New York, 1966, p. 372.

carbide C *K* bands with the humps labeled *a*, *b*, *c*, *d*, *e*, and *f* on the C *K* bands of diamond, graphite, it has been shown[1] that these bands and carbide C *K* bands can be resolved into components peaks. In Figure 46 the intensity distributions of the C *K* bands of graphite and diamond have been reproduced using a Dupont curve analyzer by assigning the proper weighting to 7 Gaussian peaks that correspond in wavelength to humps *a*, b' (Fig. 47), *b*, *c*, *d*, *e*, and *f*. The intensity distribution of the carbide C *K* bands of TiC, VC and Fe_3C have been reproduced by keeping the wavelength position of the subpeaks *a*, b', *b*, *c*, *d*, *e*, and *f* constant and varying the relative peak heights as shown in Figure 46. In the future when the emission band is resolved into Gaussian peaks which are indicated by the humps on the envelope of the band they will be referred to as subpeaks. This term is used since these peaks suggest subbands.

It was found necessary to have an additional subpeak between subpeaks *b* and *a*, labeled peak b', in order to reproduce all of the C *K* band shapes in Figures 45 and 46. Confirmation of the reality of this peak was obtained when the C *K* band of TiC was remeasured on the higher resolving power spectrometer. Another peak is then seen on the high energy side of *b* between peaks *a* and *b* (Fig. 47). Also, peak *c* is much more prominent in the C *K*

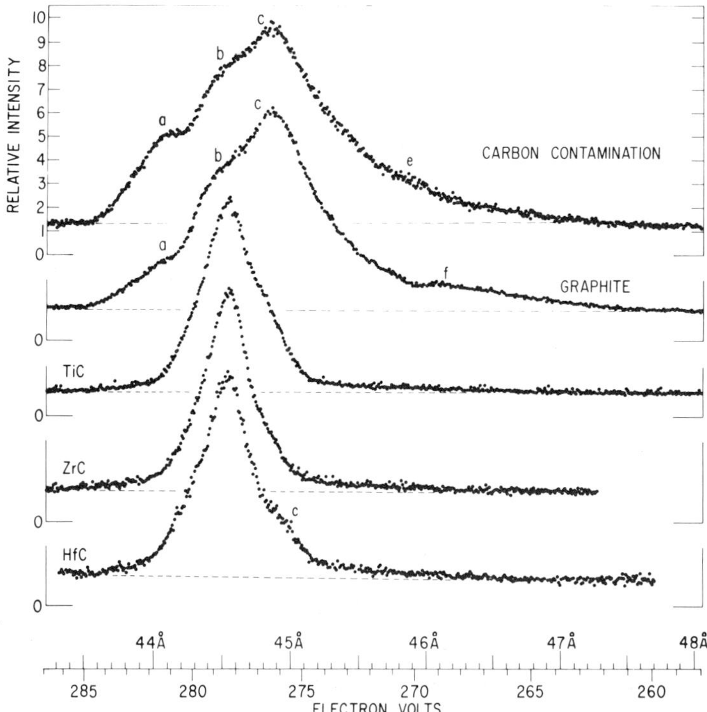

Fig. 43. The carbon *K* emission bands (2*p* ⟶ 1*s* transition) from carbon contamination deposited by the electron beam, graphite, and group IV carbides, TiC, ZrC, and HfC (peaks normalized to peak c of electron deposited carbon).

band of TiC in Figure 47 than it was in Figure 44. Further justification for the physical reality of subpeaks *a*, *b*′, *b*, *c*, *d*, *e*, and *f* is that it was possible to also reproduce the peak wavelengths of the C *K* bands in the synthesized bands without changing the position of the subpeaks.

For group IV (Figs. 43 and 46) carbides, Gaussian peak *b* is predominant while less weighting is given to subpeaks *a*, *b*′, and *c*. The low-energy tail of group IV carbides is due to the very small intensity peaks *d* and *e*. In group V carbides (Figs. 44 and 46), more weight is given to peaks *a* and *c*. For group VI and higher carbides (Figs. 44 and 46), the weighting is more equally divided among the various subpeaks. Thus, the differences in peak wavelengths shown for the C *K* bands in Table 13 and the changes in intensity distribution shown in Figures 43 and 44, appear to be due to changes in the weighting of the Gaussian subpeaks and not in a shift in the inner level as is the case for the Ti L_{III} bands from TiO_x, Figure 41.

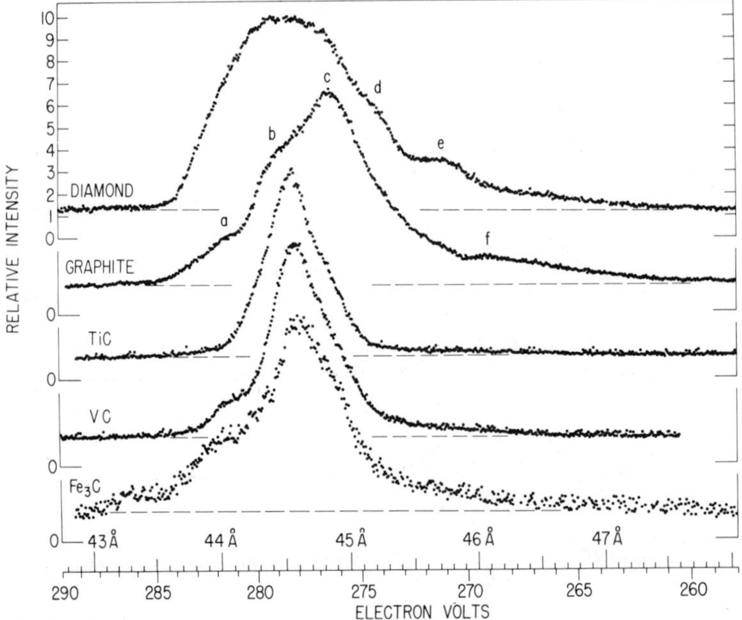

Fig. 44. Carbon *K* emission band ($2p \rightarrow 1s$ transition) (peaks normalized) from diamond, graphite, TiC, VC and Fe_3C. Reprinted by permission, J. E. Holliday, *J. Appl. Phys.*, **38**, 4722, 1967.

It would be of interest to apply the above analysis to the C *K* bands from three different states of carbon in an iron matrix shown in Figure 48. Curve (*a*) is the C *K* band from graphitized iron. The micrograph of the graphitized Fe-1.83 wt % C alloy is shown in Figure 49. The micrograph shows that all of the carbon has been precipitated as free carbon. The C *K* band from graphitized iron has the same peak wavelength as the graphite C *K* band (Table 13), but the shape is different. The a/c intensity ratio is 0.65 for "graphitized" iron compared to 0.3 for electrode graphite. As a result of the increase in the a/c intensity ratio, the $W_{1/2}$ for the graphitized iron C *K* band is 8.3 eV compared to 6.0 eV for the C *K* band from electrode graphite. These changes indicate that the electronic structure of the carbon precipitated in iron is different from that of electrode graphite.

Curve *b* is the C *K* band from Fe_3C which was obtained by electron bombarding a Fe-1.83 wt % C alloy. Before bombardment the carbon was in solution in iron as austenite and martensite. The conversion from austenite and martensite to Fe_3C + ferrite shows the effectiveness of heat from the electron beam in producing chemical changes. A thermocouple measurement

Table 13. C *K* band parameters from transition metal carbides. Compared to heats of formation and melting temperatures

Material	Peak shift[a] v (eV)	Peak shift[a] $\Delta\lambda$ Å	Peak wavelength	Halfwidth[b] $W_{1/2}$ (eV)	Asymmetry[b]	ΔH°_{298}/C atom heat of formation	Melting[c] point °C
Graphite	—		44.85 Å	6.0	0.83		
Diamond	+2.1	−0.33	44.52	8.1	1.25		
Group A							
TiC	+1.95	−0.31		3.0	1.1	−43	3200
VC	+1.8	−0.29		3.3	1.45	−28	2850
Cr_3C_2	+1.9	−0.30		3.3	1.6	−10.5	1870
ZrC	+2.05	−0.325		2.4	0.85	−44	3530
NbC	+1.9	−0.30		2.4	1.05	−33	3500
HfC	+2.0	−0.32		3.0	0.80	−50	3800
TaC	+1.2	−0.19		3.0	0.80	−32	3880
$TaC_{0.5}$	+1.7	−0.27		2.7	1.7		3400
Group B							
$(MnCo)_4C$	+1.2	−0.19		5.2	0.9		
Fe_3C [d]	+1.8	−0.29		4.4	1.25	+5.98	1650
80% [e] Martensite, 20% Austenite	+1.0	−0.105		5.0	0.7	+4.2	(1200)
Mo_2C	+1.2	−0.19		4.4	1.2	+4.2	2600
WC	+1.8	−0.29		6.7		+8.4	2850

[a] Peak shift relative to graphite. [b] Not corrected for instrumental error. [c] Taken mostly from Peter T. B. Shaffer, *High Temperature Materials*, Plenum Press, New York 1964 and Lawrence S. Darken and Robert W. Curry, *Physical Chemistry of Metals*, McGraw-Hill, New York, 1953, p. 364 and Edmund K. Storms, *The Refractory Carbides*, Academic Press, New York and London, (1967). [d] As second phase in Fe-1.83 wt. % C alloy. [e] Fe-1.83 wt. % C alloy.

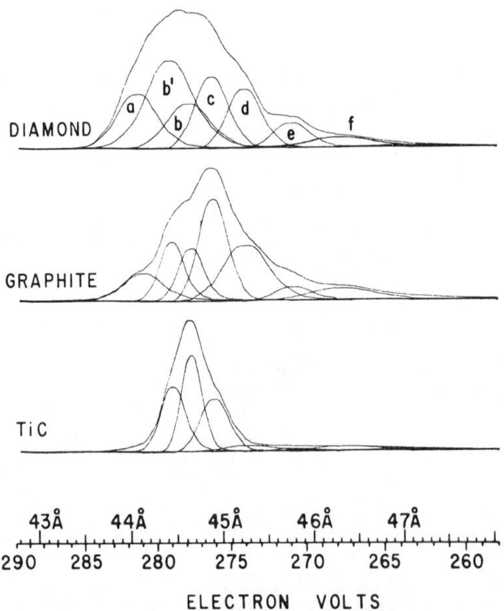

Fig. 45. The C K emission bands from diamond, graphite and TiC resolved into Gaussian curves *a*, *b′*, *b*, *c*, *d*, *e*, and *f*. These curves correspond to humps *a*, *b′*, *b*, *c*, *d*, e and *f* on the C K bands of diamond, graphite, and TiC (Fig. 47). Reprinted by permission, J. E. Holliday in *Soft X-ray Band Spectra and the Electronic Structure of Metals and Materials,* Derek J. Fabian, ed., Academic Press, New York and London, 1968 p. 117.

indicated that the target temperature during bombardment was about 400°C. A comparison of the micrographs of the Fe-1.83 wt % C alloy before and after electron bombardment is shown in Figure 50. Both microscopic examination and X-ray diffraction showed that the austenite had been converted to Fe_3C. Comparing the shape of curve *c* with that of the Fe_3C C *K* band in Figure 44 shows that they both have the same shape and wavelength.

To prevent the austenite + martensite from transforming to cementite by electron bombardment the target was cooled with a chilled high dielectric fluid. Curve *c* in Figure 48 is the C *K* band from carbon in solution in austenite + martensite in the Fe-1.83 wt % C alloy. Both optical microscopic inspection and X-ray diffraction measurements showed that the carbon was still in solution as austenite + martensite after electrical bombardment. It can be seen (curve *c*, Fig. 48) that the shape of the C *K* band from austenite + martensite is close to the shape of the C *K* band from Fe_3C but there are important differences between the bands. Relative to graphite, the peak of the C *K* band from austenite + martensite is shifted 1.0 eV as compared to 1.8

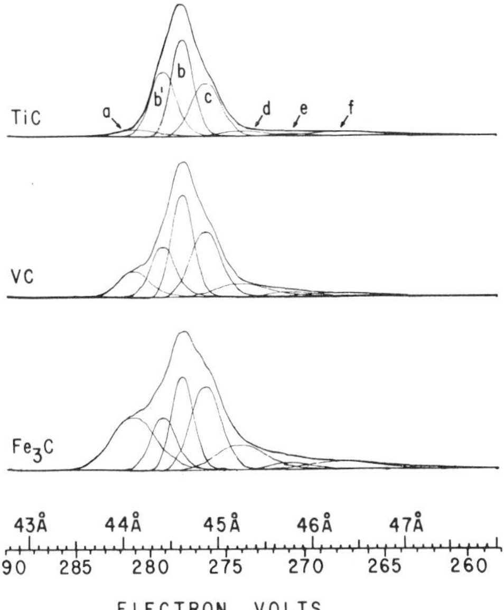

Fig. 46. The C K emission bands from TiC, VC, and Fe_3C resolved into Gaussian curves *a*, *b′*, *b*, *c*, *d*, *e*, and *f*. Reprinted by permission, J. E. Holliday in *Soft X-ray Band Spectra and the Electronic Structure of Metals and Materials,* Derek J. Fabian, ed., Academic Press, New York and London, 1968, p. 121.

eV for Fe_3C and it is steeper on the low-energy side whereas the C *K* band from Fe_3C is steeper on the high-energy side (Table 13). The change in intensity distribution and energy shift of the C *K* band from austenite + martensite relative to that of Fe_3C appears to be due to an increase in the intensity of peak *c* relative to *b*. This indicates that the C *K* band from carbon dissolved in Fe is closer to the distribution of graphite than is the C *K* band from Fe_3C.

The above results show that the three different states of carbon in iron can be identified by the shape of the C *K* band. Although this method of identification is not as efficient as X-ray diffraction or optical methods on a macro basis, it is of considerable value for the microprobe.

The author's measurements of the Ti $L_{II,III}$ emission band from $TiC_{0.95}$ compared to the Ti $L_{II,III}$ band from the pure Ti is shown in Figure 32. The Ti $L_{II/III}$ intensity ratio has decreased and the Ti L_{III} peak has shifted towards lower energy which is opposite to that observed for TiO. The L_{II}/L_{III} intensity ratio and the L_{III} peak energy shifts for the first series transition metal carbides and their respective metals and oxides is shown in Table 11. It can be seen that the peak of the L_{III} band has shifted toward lower energy in TiC and VC, while for Cr_3C_2, and 50% Fe_3C + Ferrite the peak has shifted

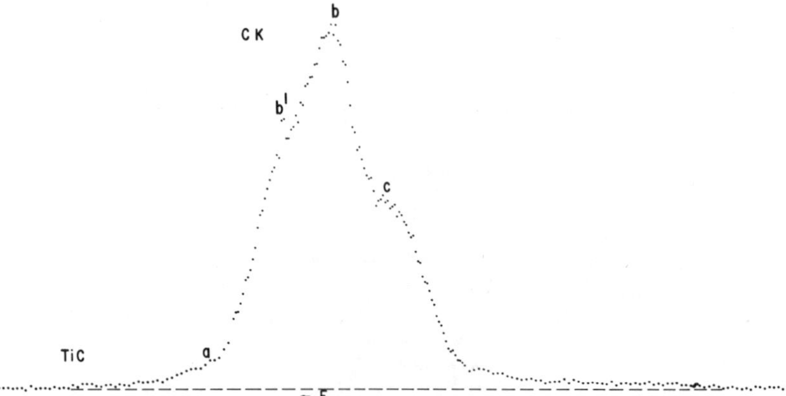

Fig. 47. The C *K* emission band TiC obtained on a higher resolving power spectrometer with a 3600 grooves/mm 1° blazed grating. Reprinted by permission, J. E. Holliday in *Soft X-ray Band Spectra and the Electronic Structure of Metals and Materials,* Derek J. Fabian, ed., Academic Press, New York and London, 1968, P. 121.

slightly towards higher energy relative to the L_{III} band of the respective pure metal. The L_{II}/L_{III} ratio is less for TiC and VC, but there is no change in the ratio with respect to that of the pure metal for Cr_3C_2 or Fe_3C + Ferrite. The data in Table 11 indicates that there is a change in the L_{II}, L_{III} band parameters relative to the pure metal at group VI carbides.

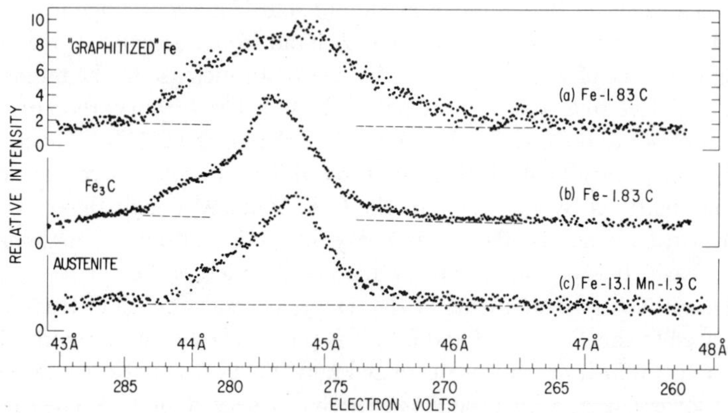

Fig. 48. Normalized C *K* emission bands (*a*) C *K* band from "graphitized" Fe-1.83C alloy; (*b*) C *K* band from Fe_3C in Fe-1.83 wt % C alloy; (*c*) C *K* band for carbon in solution as 80% martensite and 20% austenite in Fe-1.83 wt % C alloy. The percentage deviation is ± 2.5. Target voltage is 4 kV.

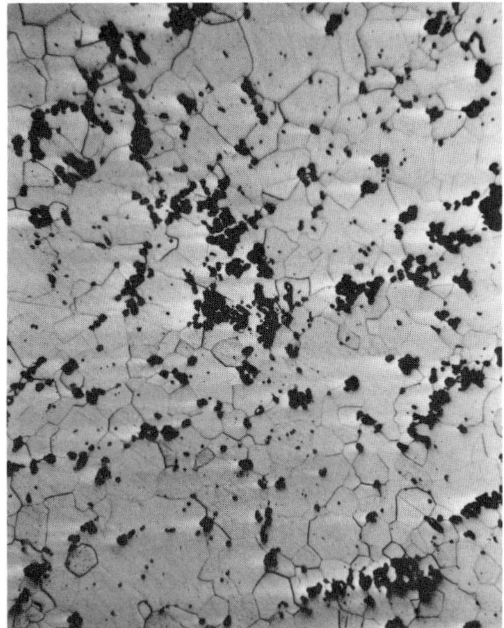

Fig. 49. "Graphitized" Fe formed from Fe-1.83 wt % C alloy. X500. Reprinted by permission, J. E. Holliday, *J. Appl. Phys.*, **38**, 4720 (1967).

D. Diborides

The B K bands from boron, ZrB_2, TiB_2 and VB_2 with peak intensities normalized to the boron B K band measured by Holliday[22] using a grating analyzer are shown in Figure 51. The boron B K band has a relatively sharp emission edge and a long tail to the low energy side of the band. There is a peak, at 69.5 Å that appears to be present for all the B K bands shown in Figure 52. Using a grating, Crisp and Williams',[81] and Sagawa and Aita's[82] measurements of the boron B K band shows this peak to be more pronounced. Fischer and Baun,[83] using a lead stearate analyzer, also show this peak at 69.5 Å for the boron B K band (Fig. 52). However, their spectrum of the B K band from TiB_2 in Figure 51 does not show a peak at 69.5 Å, apparently because of the poorer resolving power of the lead stearate analyzer. The grating analyzer shows evidence of another peak at about 67 Å for ZrB_2, TiB_2 and it is especially prominent for VB_2. The $W_{1/2}$ of the B K bands of the diboride compounds shown in Table 14 are nearly equal, while the B K band from B N is about 1.5 times broader. The differences in $W_{1/2}$ between the B K band from boron and the diborides is smaller than between the C K

Table 14. Ti $L_{II,III}$ and B K Band Parameters

Ti $L_{II,III}$ Bands

Material	Peak λ Å	L_{III} shift ΔeV	$W_{1/2}$ of L_{III}[a] ΔeV	L_{II}/L_{III}
Ti	27.45	—	3.2	0.25
TiB_2		−0.5 eV	6.0	0.22

B K Bands

	Peak λ Å	K Shift ΔeV	$W_{1/2}$ of K[a] ΔeV	Asymmetry[a] K
Boron	67.56	—	4.3	1.8
TiB_2	56.9	−0.4	3.8	2.2
ZrB_2	57.2	−0.4	3.35	1.4
VB_2	67.89	−1.15	4.25	1.7
BN	68.65	−2.92	5.5	0.9

[a] Not corrected for instrumental error.

band of graphite and group IV and V carbides. The smaller changes in $W_{1/2}$ with chemical combination would make calibration easier for quantitative analysis of the borides than of the carbides. The peak energy shift for the compounds in Table 12 is in the same direction as that of the B K band for B_2O_3. The shift for VB_2 relative to the shift of the B K band from boron is about three times that of the ZrB_2 and TiB_2.

Figure 53 shows the author's measurements of the metal Ti $L_{II,III}$ emission bands from TiB_2 compared to that from the pure Ti. Like the case of the Ti L_{III} emission band of TiC, there is a shift in the peak of the L_{III} band towards lower energy and a decrease in the L_{II}/L_{III} intensity ratio relative to the pure metal. In general, the changes in band characteristics for the Ti $L_{II,III}$ emission band in going from pure metal to the diboride is similar to the case of the Ti $L_{II,III}$ band for TiC. However, the Ti L_{II}/L_{III} intensity ratio for $Ti_{52}Fe_{48}$ has increased relative to that of Ti metal. This similar to the Ti L_{II}/L_{III} intensity ratio change observed for the oxides of Ti.

VIII. ELECTRON DISTRIBUTION AND BONDING

It is clear from the examples given in the previous section that peak energy shift and intensity distribution changes are the significant types of

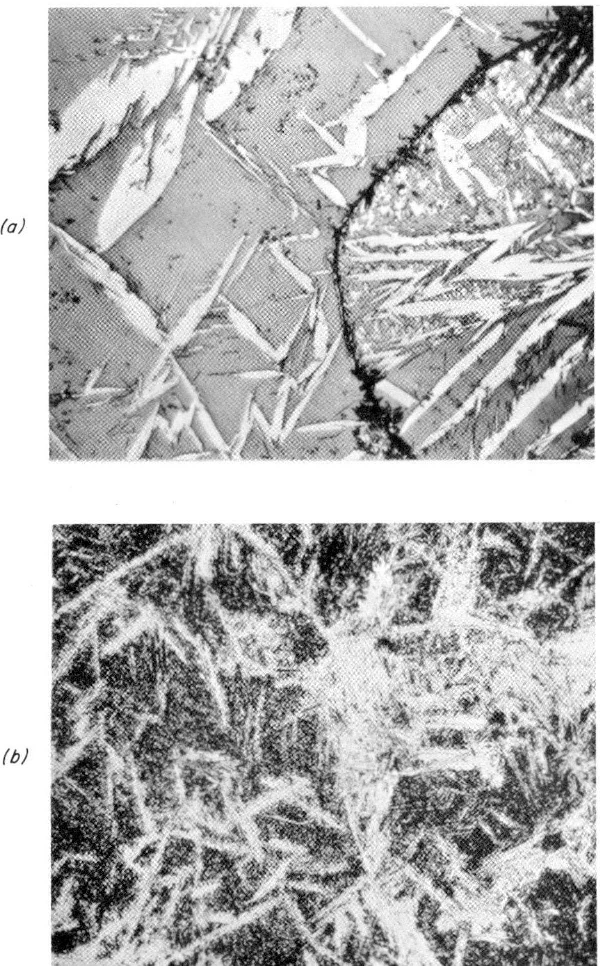

Fig. 50. Micrograph of Fe-1.83 wt% C alloy quenched from the completely austenitic conditions: X500. (*a*) Target before electron bombardment: 75% austenite and 25% martensite. (*b*) Target after approximately 5 min of electron bombardment. Austenite and martensite converted to ferrite and Fe_3C. Reprinted by permission, J. E. Holliday, *J. Appl. Phys.*, **38**, 4720 (1967).

alterations in the emission bands with chemical combination. Faessler[72] and Goehring[84] were the pioneers in relating peak energy shift (this does not include peak shifts resulting from changes in the intensity distribution) to changes in electron distribution with chemical combination. They showed

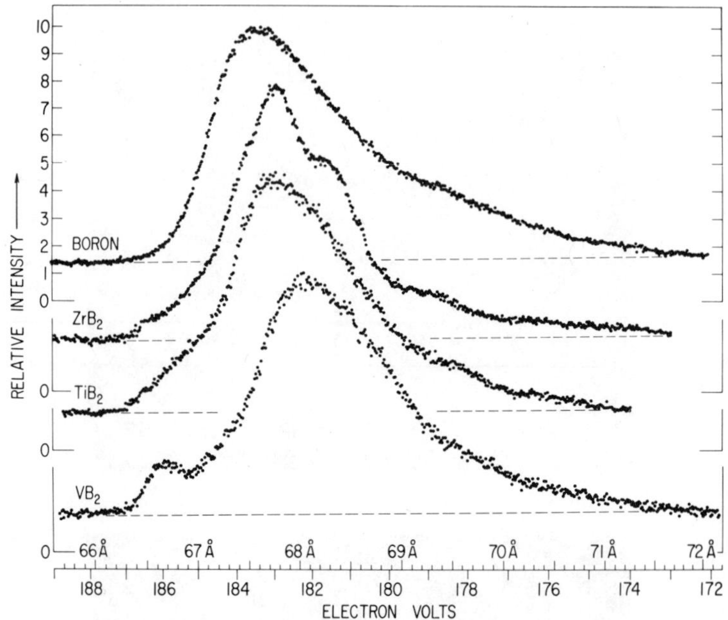

Fig. 51. Comparison of the B *K* bands (peaks normalized to boron B *K*) from boron ZrB_2, TiB_2 and VB_2. Excitation potential is 4 kV.

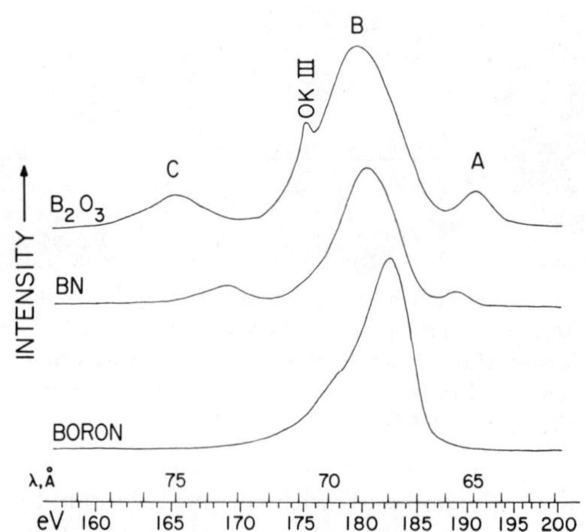

Fig. 52. Boron *K* emission bands from boron, BN and B_2O_3. Reprinted by permission. David W. Fischer and William L. Baun, in *Advances in X-ray Analysis*, W. M. Mueller, G. R. Mallett and M. Fay, Eds., Vol. 9, Plenum Press, New York, 1966. p. 335.

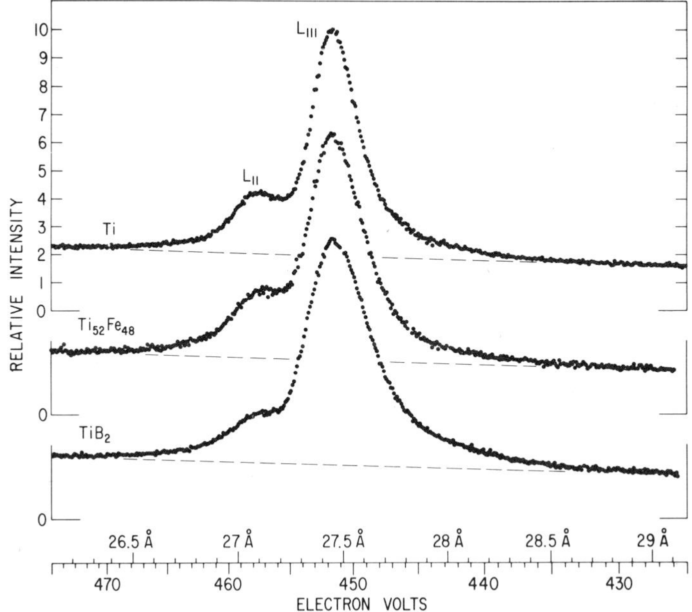

Fig. 53. Comparison of the Ti $L_{II,III}$ emission bands from Ti, TiB_2 and $Ti_{52}Fe_{48}$. Excitation voltage is 4 kV.

that for a given oxide or sulfide series the peak shift increased with the degree of ionization of the oxygen of sulfide atoms. Fischer and Baun[83] have reported a relation between peak energy shift of the Cl $L_{II,III}$ band and the electronegativity difference of the Cl and metal ion for a number of metal chlorides. Pauling[86] states that the amount of ionic character of the bond is a function of the electronegativity difference. Clift et al.[71] indicate that the amount of change in the emission band of alloys relative to the respective pure metal depends on the amount of electron transfer between solute and solvent atoms. The author[1] has recently published results showing that the peak energy shift of the Ti L_{III} band from TiO_x compounds is a function of electrons transferred from Ti to the oxygen atom and the degree of ionic character to the bond.

A. Ionic Bonding

Figure 54 shows the energy shift of the Ti L_{III} band as a function of x in TiO_x compounds. Since the Ti L_{II} band and the satellite A move in the same way as the L_{III} peak the Ti L_{III} peak shift is due to a shift in the level and not a change in weighting of the subpeaks. Although three different oxide

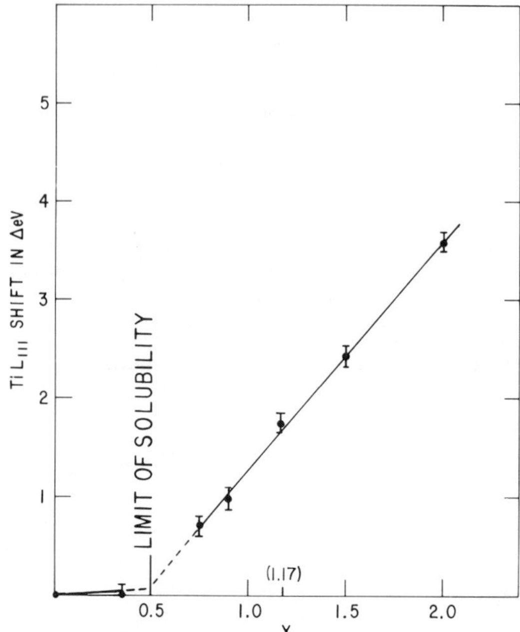

Fig. 54. The peak energy shift of the Ti L_{III} band from titanium oxides, TiO_x and Ti + 25% O relative to the Ti L_{III} band from titanium metal. The excitation potential was constant at 4 kV. The experimental error is indicated by the straight line through the points. Reprinted by permission, J. E. Holliday in *Soft X-ray Band Spectra and the Electronic Structure of Metal and Materials,* Derek J. Fabian, Ed., Academic Press, London and New York (1968) p. 101.

phases are represented ($x = 0.85$ to $x = 1.17$ NaCl, $x = 1.5$ trigonal, and $x = 1.97$ rutile) in Figure 54, within experimental error a straight line can be drawn. However, there is a break in the straight line when going from the compound to the solution. It would then appear that wavelength shift is not a function of structure, but is a function of chemical state. It has already been shown in Section VI-C that emission bands are not a function of alloy structure. From simple valence considerations the amount of electron transfer per titanium atom and the amount of ionic character to the Ti—O bond increase with the titanium oxidation number. As can be seen from Figure 54, the energy shift of the Ti L_{III} band from TiO_x compounds relative to that of titanium, also increases with the titanium oxidation number and is a maximum for TiO_2 which is about 100% ionically bonded. It is generally considered that about 4 electrons are transferred from the titanium atom to oxygen atom in TiO_2 and about one electron per titanium atom in TiO. This would indicate that the 0.9 eV energy shift of the Ti L_{III} band is equivalent to a transfer of approximately one electron to the oxygen atom. The change in the Ti $L_{II,III}$ (Fig. 41) intensity distribution relative to that of titanium metal becomes greater with the amount of ionic character of the Ti—O bond. Also, the height of peak A in Figure 41 increases directly with energy shift of the Ti L_{III} peak. The relation of the shape and energy shift is well illustrated by comparing the intensity distribution of the Ti $L_{II,III}$ emission bands from $SrTiO_3$ and TiO_2 in Figure 55. The Ti $L_{II,III}$ emission bands from these oxides have nearly identical shapes and the same L_{III} peak wavelengths. Although they have different structures and different percentage of oxygen the common parameter between the two compounds is that the Ti atom has a valence of approximately 4 in each case.

If shape is related to the degree of electron transfer or valence, then it would be expected that the emission band of an element A when combined with an element B of a given group x in the periodic table would not change when combined with other elements in group x. It has been shown that the C K band has approximately the same shape from carbides of a given transition metal group. This is seen in Figure 43 for group IV transition metal carbides. Fischer and Baun[63] have reported that elements of the same subgroup each have virtually the same effect on the Al K band.

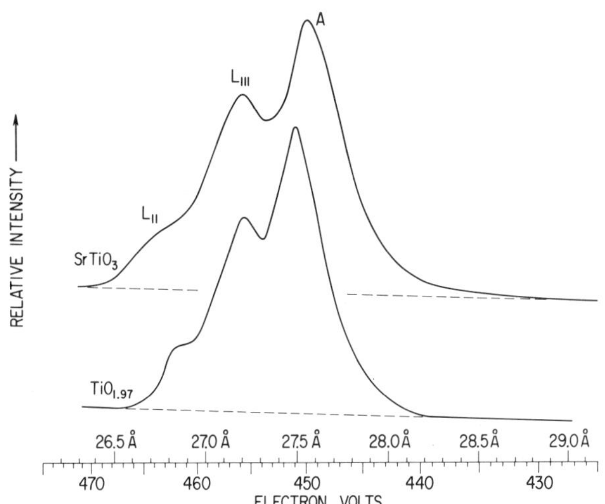

Fig. 55. Comparison of the Ti $L_{III,II}$ bands from TiO_2 and $SrTiO_3$. Reprinted by permission, J. E. Holliday in *Soft X-ray Band Spectra and the Electronic Structure of Metal and Materials,* Derek J. Fabian, Ed., Academic Press, London and New York (1968) p. 101.

B. Metallic Bond

From the above mentioned relation between electron transfer and band changes there should be little alterations in band shape and wavelength when two elements are combined that have a small electronegativity difference. This is probably the reason why there are no observable changes in the Ni and Cu M_{III} bands on alloying. However, the metal and nonmetal emission bands of transition metal borides, carbides and nitrides and intermetallic compounds such as $Ti_{52}Fe_{48}$ (Fig. 53) which have larger electronegativity differences, do show alterations in the emission bands when the compound is formed. Preliminary measurements of the Ti $L_{II,III}$ band from TiNi, which has a larger electronegativity difference than $Ti_{52}Fe_{48}$, shows a greater change relative to the Ti $L_{II,III}$ band from pure Ti than does the Ti $L_{II,III}$ band from $Ti_{52}Fe_{48}$. Even though all of these compounds have metallic properties, the intensity distribution and peak energy changes indicate electron transfer between the two elements of the metal compound. For example, the shift in peak energy of the L_{III} bands for TiB_2, $TiC_{0.95}$ and $VC_{0.95}$ (Tables 11 and 14) towards lower energy relative to the respective pure metal shows that there is electron transfer between the two elements of the metal compound. Since this shift is in a direction that is opposite to the oxides of Ti and V, it is possible that the electron transfer is from the nonmetal to the metal atom.* The idea that charge is transferred from carbon to the metal *d* band in transition metal carbides has been advanced by Kiessling,[87] Robins,[88] Dempsey,[89] Costa and Conte,[90] Lye[91] and Logothetis.[92] Their conclusions are based on both theoretical and experimental consideration. However, Ern and Switendick's[93a] band calculations show the C $2p$ band below the Ti $3d$ band indicating electron transfers from Ti to carbon. A transfer of electrons between the metal and carbon atoms would indicate a "semi-ionic" character for the carbon–metal bond, which is discussed by Williams[93] for TiC. The idea of ionic character in a metallic compound with an electron transfer in a direction opposite to that indicated by the electronegativity difference is considered by some to be completely unreasonable. However, Pauling[94] has explained this type of electron transfer in transition metal compounds that have metallic properties by the "electroneutrality principle." For example, if atom A is more electronegative than atom B the normal transfer would be from B to A which would result in ionic character to the bond with a positive

*Recent electron spectroscopy measurements (ESCA) by Lars Ramquist et al. (101) have indicated that the charge transfer in the transition metal groups iv and v carbides is from the metal to the carbon atom. Regardless of the final out come of the direction of charge transfer, the ESCA meansurement confirms the ionic character of the bond in the transition metal carbides.

charge on B and a negative charge on A. In a metallic bond charge neutrality has to be maintained and this can be accomplished if the electron transfer is in the reverse direction from A to B which reduces the charges on the atoms and stabilizes the compound. It will be noted that this reverse charge transfer is in the same direction as that described above for group IV and V transition metal carbides and borides. It will be noted from Table **11** that the shift of the metal L_{III} peak for Cr_3C_2 and higher carbides is in the same direction as the oxides of these metals. This shows an electron transfer from metal to nonmetal for Group VI and higher carbides which is the normal direction for the electronegativity difference of the metal and nometal atoms. Pauling predicts a change in the significance of the electron neutrality principle at group VI transition metal carbides.

C. Bond Strength

If the electron transfer is a factor in determining bond strength as measured by melting temperature and heats of formation it has been shown by Holliday[74] that it is possible to relate bond strength to changes in emission band parameters. Das Gupta[95] has attempted to show a mathematical relation between heat of formation and peak energy shift. In Table 13 the transition metal carbides have been divided into two groups according to their melting temperatures and heats of formation. With the exception of Cr_3C_2, group A consists of group IV and V carbides which have high negative heats of formation and melting temperatures higher than the pure metals. Group B consists of group VI and higher carbides which have positive heats of formation and lower melting temperatures that are about the same as the pure metal. It was shown in the section on ionic bonding that the intensity distribution is related to electron transfer. This shape relation for the C K bands from the transition metal carbides is that the greater the electronegativity difference between the carbon and metal atom the more a single subpeak predominates in the C K bands (Figs. 43 and 46). Faessler[96] has found that for the Si K band from Si compounds a single subpeak also predominates for the greatest electronegativity difference. Thus, the $W_{1/2}$ of the C K bands in Table 13 indicate the same thing as the metal L_{III} peak shifts of the carbides discussed above. The highest bonded carbides (group A) are those with the largest electron transfer which results in a very stable compound. This high stability is reflected in the high melting temperature and negative heats of formation of group A carbides. At group VI carbides the electron configuration and smaller electronegativity difference results carbide. Although Cr_3C_2 is listed with group A carbides it should be placed between group A and B carbides. For group VI and higher carbides the C *K* band intensity distribution is closer to that of graphite than from group IV and V carbides. This indicates strong carbon–carbon bonds which are probably stronger than the carbon–metal bonds.

IX. QUANTITATIVE ANALYSIS

Although the peak X-ray intensity is the most common parameter for quantitative analysis, it can be seen from the results in the previous sections that the wavelength shift, the L_{II}/L_{III} and A/L_{III} intensity ratios, and the emission band intensity distribution can also be used as parameters in quantitative analysis. The fact that the depth from which soft X-rays come is very near the surface as shown in Figure 23, can give an advantage to soft X-ray spectroscopy over other quantitative methods such as hard X-rays, gamma rays or neutron activation analysis. Bunshah[97] has suggested that this fact could be used to study diffusion of one metal in another by successive sectioning of the target and making quantitative mesurements of the target after each sectioning. Because of the small depth from which soft X-rays come, a curve of the amount of material as function of depth could be obtained.

The two basic considerations in light-element quantitative analysis are the minimum detectable limit and the accuracy with which these elements can be measured. Very little work has been published on actual measure-

Table 15. Carbon concentration in an iron matrix. Reprinted by permission from John P. Moskal, Poen Sing Ong, Patrica Cream, *Norelco* Application Data Report, **144,** (1965)

Sample	Weight percent carbon
1	pure iron
2	0.054
3	0.146
4	0.218
5	0.305
6	0.406
7	0.567
8	0.810
9	1.07
10	1.40
11	1.64

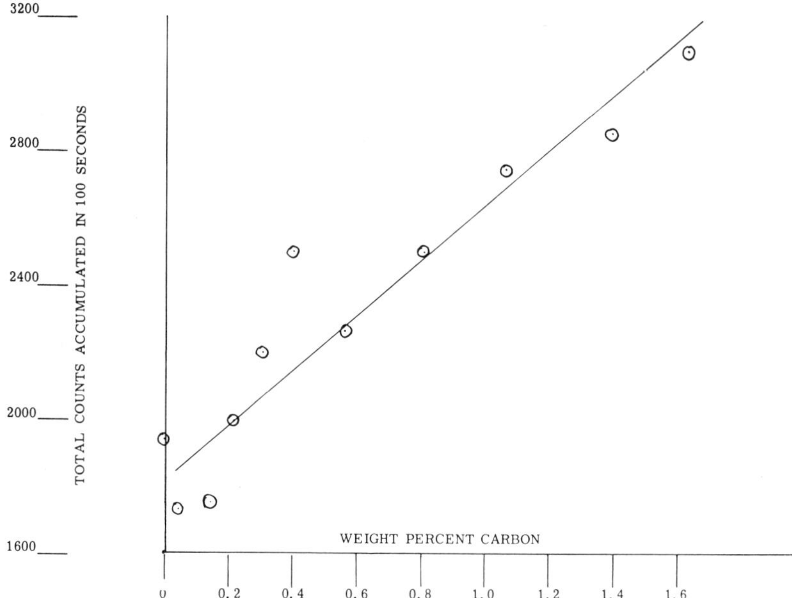

Fig. 56. Intensity of C K band as a function of wt % carbon in a iron matrix for an excitation voltage of 10 kV. Reprinted by permission, John P. Moskal, Poen Sing Ong and Patrica Cream, Norelco Application Data Report, **144**, (1965).

ments of low percentages of the second period elements in a matrix. Moskal, Ong, and Cream[98] have published a curve of the C *K* band intensity in counts sec^{-1} as a function of weight percent carbon in an iron matrix for a 10 kV electron beam. This curve is shown in Figure 56. The weight percent carbon used for these measurements is shown in Table 15. The ordinate was adjusted so the C *K* band intensity from the pure iron target was zero. Actually there is a C *K* band intensity of approximately 6 counts sec^{-1} from the pure iron. The actual amount of carbon in the pure iron was not given. However, the C *K* band intensity was probably from either the residual carbon on the iron surface or carbon contamination due to electron bombardment. Since a straight line can be drawn through the points, the surface carbon contamination must have been approximately the same for each sample. The linear plot also indicates that the shape and wavelength of the C *K* band was the same for all of the samples. These results show that with the existing instrumentation, carbon can be detected to about 0.05 wt % carbon in an iron matrix. Before serious work is done on measuring low percentage of carbon with a probe, some means must be provided for cleaning in the vacuum.

Moskal, Ong and Cream[98] were not able to obtain a linear relation between counts sec^{-1} and weight percent below 0.6 wt% carbon using a 30 kV

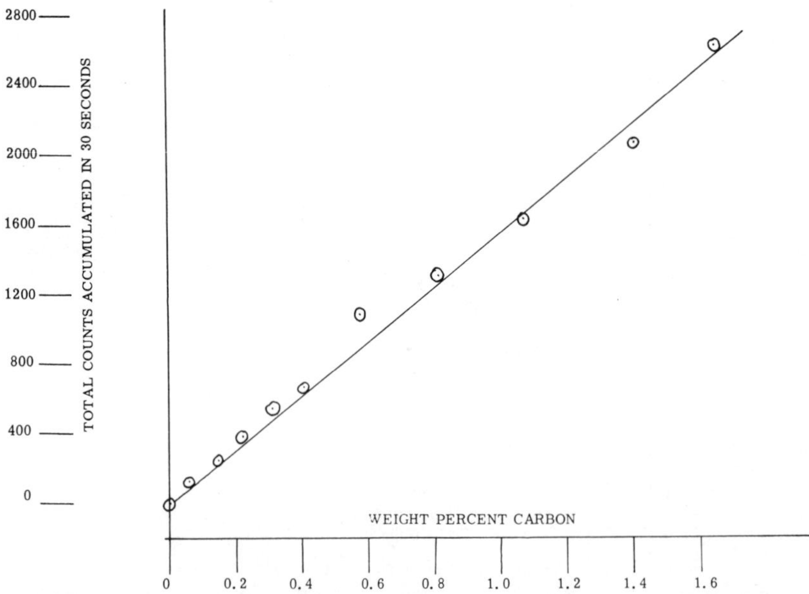

Fig. 57. Intensity of C K band as a function of wt % carbon in an iron matrix for an excitation voltage of 30 kV. Reprinted by permission, John P. Moskal, Poen Sing Ong and Patrica Cream, Application Data, *Norelco Report,* **144**, (1965).

electron beam (Fig. 57). They attributed this difficulty to the fact that a larger sample volume is represented in the X-ray intensity for the 30 kV electron beam than for the 10 kV electron beam. Since the quantitative carbon values in Table 15 are from bulk analysis, it is to be expected that the 30 kV electron beam would give a more linear plot than the 10 kV electron beam. The factor that must be considered in measuring carbon at higher voltages is that the maximum counting rate for carbon in iron is 7.5 kV shown in Figure 15 for a 52° takeoff angle. Ranzetta and Scott[99] showed that the wt% carbon, as determined by X-ray intensity using C *K* radiation from diamond as a standard, declined from 18 wt% at 2 kV to 1 wt% at 20 kV. Their takeoff angle was 20°. Although Moskal et al.[98] should still be able to obtain a straight line relation below 0.6 wt% carbon using a 30 kV electron beam, the accuracy will be greatly reduced because of the lowered intensity above background being measured. This inaccuracy would be especially high for a sample with a low percentage of carbon and would explain the large scatter of the points below 0.6 wt% carbon. Ransetta and Scott also showed that for an Fe_3C target, 6.7 wt% (equilibrium value of carbon in cementite) was obtained for a 2 kV electron beam.

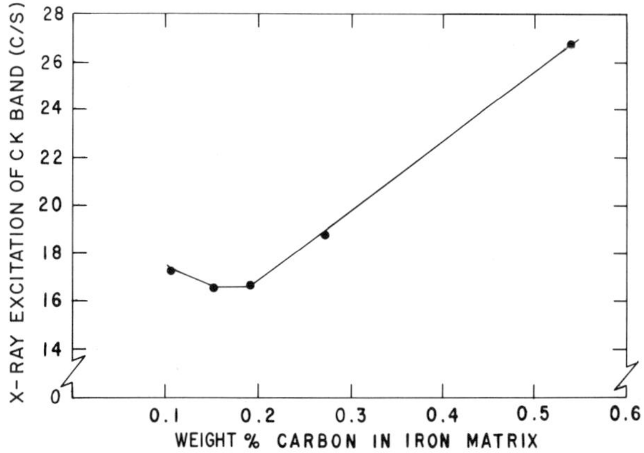

Fig. 58. Intensity of C K band, excited by Cu L_{III} band (13.3 A), as a function wt% carbon in an iron matrix which was plotted from Henke's[15] data.

Low percentages of carbon determined by X-ray excitation should be more accurate than those determined by electron excitation, since there is no carbon contamination due to X-ray bombardment. Figure 58 shows the counts sec^{-1} as a function of weight percent carbon in iron determined by Henke[15] using fluorescent analysis on National Bureau of Standards standards. The C K band was excited with Cu L_{III} radiation at 13.3 Å. The target surfaces were thoroughly cleaned before they were put into the vacuum system and additionally cleaned by ion bombardment while they were in the vacuum system. The plot shows that the counts sec^{-1} is not a linear function of wt% carbon below 0.19 wt% carbon. This is similar to the difficulty that Moskal et al.[98] had using the 30 kV electron beam which was indicated above. However, the penetration depth of the Cu L_{III} band radiation is approximately the same as a 10 kV electron beam. Consequently it does not appear that the reason for nonlinearity below 0.19 wt% C in Figure 58 is the same as that in Figure 57. It is possible that the bulk quantitative values given by the National Bureau of Standards are not representative of the carbon near the surface below 0.19 wt% carbon. In some iron-carbon alloys the amount of carbon near the surface is different from the amount deeper within the iron. This fact shows that considerable amount of cleaning of the surface is required before measurements from near the surface can be representative of the bulk. For example, it was found that approximately 2 hrs of ion cleaning were required before the C K band from a Fe-13.1 Mn-1.3 C alloy was representative of the bulk.

With the exception of carbon, very few measurements have been made on low percentages of the second period elements. Holliday[7] has reported that as little as 0.1 wt% nitrogen in niobium could be detected with soft X-rays. The quantity of nitrogen in niobium was determined by chemical analysis. The lowest detectable (L.D.) wt% is usually defined where the peak intensity is about three times greater than the standard deviation of the noise background. On this assumption the L.D. wt% can be approximated from the following equation,

$$\text{L.D. wt\%} = 3w/[N_x(N_x/N_b)]^{1/2} \tag{20}$$

where N_x is the total counts of a small weight percent, w, of the element being measured in a given sample matrix and N_b is the X-ray background from this sample. Since the L.D. wt% is based on the total number of counts, N_x, it will be a function of the counting time. The maximum allowable time depends on such factors as stability of X-ray source, detector and associated electronics, and rate of contamination of the target. The above equation leads to overly optimistic values for the L.D. wt% because the equation does not consider the experimental difficulties encountered in measuring very low percentages of an element in a matrix. Some of these experimental difficulties have been indicated above. Using the values of N_x and N_b reported by Henke[15] for 0.19 wt% C in steel and a 100 sec counting time, the L.D. wt% would be 0.01 wt% C in iron. Since spectra measured with spectrometers using electron excitation have shown about (Table 16) 10 times better P/B than for spectrometers using X-ray excitation (Table 8), the L.D. wt% would be approximately 0.001 wt% carbon in an iron matrix. However, as shown in Figure 56 the L.D. wt% for carbon in iron that can actually be measured with the present instrumentation is about 0.05 wt% C in iron.

Another factor not considered in equation 20 is accuracy. Some problems of accuracy have already been discussed. Investigators who work with the microprobe are well aware of the corrections required to convert conventional X-ray line intensities into actual percentages of an element in a given matrix. However, when dealing with emission bands of the light elements, the intensity measurements are further complicated by changes in shape and wavelength with chemical combination. Also the surface properties, both physical and chemical, are a more significant factor in the soft X-ray region than they are in the hard X-ray region.

Some of the problems involved in the accuracy of intensity measurements of the light elements will be illustrated by comparing two different sets of intensities measurements of the C K band from a number of carbides in Table 17. The first column of CK band intensity measurements were measured

Table 16. Intensity and P/B of K Emission band parameters from second period elements.[a] Reprinted by permission, J. E. Holliday, in *Electron Microprobe*, T. D. McKinley, K. F. J. Heinrich and D. B. Wittry, eds., Wiley, New York, 1966, p. 7

Radiation	Source	Peak int. counts sec^{-1}	P/B	I_TmA	Voltage	Blaze
B K	LiF	17,800[b] 16,100	130	1	3500	1°
O K	FeO	17,400[b] 16,000	40	1	3500	1°
N K	BN	17,000[b] 15,300	125	1	3500	1°
C K	Graphite	27,200[b] 24,000	180	1	3500	1°
B K	BN	2,730[b] 2,050 2,050	55	1	3500	1°
Be K	Be	4,150[b] 1,500	42	1	4000	7°.55′

[a] 600 grooves/mm 1° blazed, Al grating and 40 μ slits.
[b] Corrected for counter window.

by Manzione and Fornwalt[100] using a microprobe and lead stearate crystal. The second column of intensity measurements were obtained by Holliday using a grating spectrometer. The grating was a 3600 grooves/mm 1° blaze with an aluminum surface.

Because of the different instruments used, it is not expected that the two sets of data will have the same absolute intensity measurements, but the relative intensity should be the same. However, from the table it will be seen that Holliday's value of the C K band intensity from ZrC is greater than NbC, HfC, and TaC, while Manzione and Fornwalt[100] show just the opposite results. Since ZrC has a greater weight percent carbon than NbC, HfC, and TaC, it appears that Holliday's value for ZrC is the correct one. There could be several reasons why Manzione and Fornwalt's C K intensity from ZrC is too low. Zirconium carbide can have a large defect structure and it is possible that the ZrC target was nonstoiciometric. In addition Manzione and Fornwalt[100] corrected for background by using the background of the pure metal rather than the carbide. This is poor practice, because in the soft X-ray region the carbon atom contributes about twice as much to the background as does the metal atom, consequently the background from the

Table 17. Comparison of carbide C K band intensities obtained from a grating spectrometer and a lead stearate crystal spectrometer

Specimen	Carbon content, w/o	Manzione and Fornwalt lead stearate, microprobe 10 kV	Holliday 3600 groove/mm grating, 4 keV, 1.4 mA
		Counts sec $^{-1}$ [a]	Counts sec $^{-1}$ [b]
Diamond	100	488.4	—
B_1C	21.7	5.3	—
SiC	30.0	11.1	—
TiC	20.0	43.9	100
VC	19.1	30.1	75
Cr_3C_2	13.3	12.6	45
ZrC	11.6	7.6	50
NbC	11.4	10.5	27
Mo_2C	5.9	4.1	25
HfC	6.3	9.7	29
TaC	6.2	7.8	25
WC	6.1	5.7	—
UC	4.8	50.7	—
$UC_{1.5}$	7.0	64.1	—
UC_2	9.2	74.9	—

[a] Corrected for background taken on the pure metal.
[b] Corrected for background taken on carbide.

carbide will be almost double that for the respective pure metal. For example, in TiC the background is 33 counts sec^{-1} at 26.5 Å while for Ti metal it is only 18 counts sec^{-1}. It can also be noted that there is a greater difference between the C *K* band intensity of diamond and those of the carbides for Manzione and Fornwalt's data than that reported by Holliday. The intensity from UC is rather interesting since the C *K* band from it is greater than from any other carbide. They state that there is a U $N_{VI,VII}O_{IV,V}$ line at 43.2 Å whose interference was reduced by pulse height discrimination. The peak of the C *K* band occurs at 44.85 Å, and the tails of emission lines can have measurable intensity approximately 3 Å on either side of the peak, which can be seen for the Mo M_VN_{III} line in Figure 25. As indicated in Section IV on detectors, the variation of pulse heights in the soft X-ray region makes it extremely unlikely that it is possible to resolve 10.5 eV (equivalent to 1.65 Å at 44 Å) by pulse height analysis. Before their intensity values of the

C K band from UC can be taken seriously, the intensity of the U $N_{IV,VII}O_{IV,V}$ from pure U will have to be measured to determine what effect it has on the intensity of the C K band from UC.

With the exception of Manzione and Fornwalt's[100] value for the C K band from ZrC, both sets of data show that for a given series there is a decrease in the intensity of the C K band with increasing group number. For a given group the C K band decreases in intensity with increasing transition metal series. However, the percentage change in intensity is considerably different from the percentage change in wt%. For example, there is approximately a 5% reduction in wt% in going from group IV to group V transition metal carbides, while for the C K band peak intensity there is approximately 30% reduction for the first series, 40% for the second and 15% for the third (Holliday's data). The fact that the respective metal of the carbides are adjacent in the periodic table rules out self-absorption as the reason for the discrepancy. Since there is a change in the intensity distribution of the carbide C K bands in going across a given series, (Fig. 44) the area under the curve may be more representative of the wt% of carbon than the peak intensity. In addition to alterations in peak wavelength and intensity distribution, changes in electron distribution with chemical combination may also effect the peak intensity.

The author[9] has published results which show that there are changes in peak intensity of some metal emission bands with chemical combination that cannot be related to changes in the amount of material. This was determined by showing that an emission line, which had a wavelength close to the emission band did not have the same percentage change as the emission band with chemical combination. In Table 18 are shown peak intensity changes for the Ti L_{III} and Zr M_V emission bands. For the Ti L_{III} band the Ti L_l line at 31.45 Å was used as the reference standard, while the Zr $M_V N_{III}$ line

Table 18. Relative peak intensity changes of emission bands due to bonding changes

Material		% int. Peak change in Band
Ti	Ti L_{III}	0
TiC		+15
TiO		— 5
	Zr V	M
Zr		0
ZrC		+40

at 81.5 Å was used as the reference standard for the Zr M_V emission band. In the case of TiC there was a 10% reduction in the peak intensity of the Ti L_l line in going from Ti to TiC but there was a 5% increase in the peak intensity of the Ti L_{III} band for the same chemical change. For the Zr M_V band there was a 40% relative increase in the peak intensity. Since the above changes in peak intensity are not due to physical changes, it appears that they are due to changes in the electron distribution in the valence band. This shows the importance of using standards whose elements have the same bonding and electron distribution as the elements being measured.

The above results show that a considerable amount of work is required in order to determine the L.D. wt% and the accuracy with which low percentages of the second period elements can be measured. In summary the following points must be observed in light element quantitative analysis:

(*1*) The target must be cleaned in the vacuum. When measuring C *K* bands the target must be cleaned until the C *K* radiation remains constant with cleaning time.

(*2*) Standards must be obtained in which the quantity of the light element near the surface is known.

(*3*) The electron beam voltage should be 10 kV or less.

(*4*) The bonding and electron distribution of the element being measured must be the same as the standard.

General References

H. W. B. Skinner, "The Soft X-ray Spectroscopy of Solids," *Phil. Trans. Roy. Soc. London*, **A239**, 95 (1940).

D. H. Tomboulian, "The Experimental Methods of Soft X-ray Spectroscopy and the Valence Band Spectra of the Light Elements," *Handbuch der Physik*, Vol. 30, S. Flugge, Ed., Springer-Verlag, Berlin, 1957, pp. 246–304.

L. G. Parratt, "Electronic Band Structure of the Solids, by X-ray Spectroscopy,"*Rev. Mod. Phys*., **3**, 616 (1959).

M. A. Blochin, *The Physics of X-rays*, (translated by Office of Technical Information, from a publication of the State Publishing House of Technical-Theoretical Literature, Moscow, 1957), 2nd ed.

J. E. Holliday, "Soft X-ray Emission Spectroscopy in the 10 to 150 Å Region," E. F. Kaelble, Ed., McGraw-Hill, New York, 1967, pp. 38–1 to 38–41.

Soft X-ray Ban Spectra and Electronic Structure of Metals and Materials, Derek J. Fabian, Ed., Academic Press, London and New York (1968).

Special Issue "Ultrasoft X-rays," *Norelco Rep*., **14**, 75 (1967).

References

1. J. E. Holliday, in *Soft X-ray Band Spectra and the Electronic Structure of Metals and Materials*, Derek J. Fabian, ed. (University of Strathclyde, Glasgow, Sept. 18–21, 1967), Academic Press, London and New York (1968) pp. 101–32.

2. J. E. Holliday, in *Handbook of X-rays*, E. F. Kaelbe, Ed., McGraw-Hill, New York, 1967, Chapt. 38, pp. 38-1 to 38-41.
3. A. E. Sandstrom, in *Handbuch der Physik*, Vol. 30, S. Flugge, Ed., Springer-Verlag, Berlin, 1957, pp. 226–227.
4. J. B. Nicholson, C. F. Mooney, and G. L. Griffin, in *Advances in X-ray Analysis*, Vol. 8, W. M. Mueller, G. R. Mallett, and M. Fay, Eds., Plenum Press, New York, 1966, p. 306.
5. D. Richardson, Bausch & Lomb Co., private communication (1958).
6. J. E. Holliday, *Rev. Sci. Instr.*, **31**, 891 (1960).
7. J. E. Holliday, *J. Appl. Phys.*, **33**, 3259 (1962).
8. J. E. Holliday, *J. Opt. Soc. Am.*, **52**, 1312, WB17 (1962).
9. J. E. Holliday, in *The Electron Microprobe*, T. D. McKinley, K. F. J. Heinrich, and D. B. Wittry, Eds., Wiley, New York, 1966, pp. 3–22.
10. J. B. Nicholson and M. G. Hasler, in *Advances in X-ray Analysis*, Vol. 9, W. M. Mueller, G. R. Mallett, and M. Fay, Eds., Plenum Press, New York, 1966, pp. 420–29.
11. B. L. Henke, in *Advances in X-ray Analysis*, Vol. 9, W. M. Mueller, G. R. Mallett, and M. Fay, Eds., Plenum Press, New York, 1966, pp. 430–40.
12. R. C. Ehlert and R. A. Mattson, in *Advances in X-ray Analysis*, Vol. 9, W. M. Mueller, G. R. Mallett, and M. Fay, Eds., Plenum Press, New York, 1966, pp. 456–470.
13. P. S. Ong, private communication (1965).
14. B. L. Henke, in *Advances in X-ray Analysis*, Vol. 8, W. M. Mueller, G. R. Mallett, and M. Fay, Eds., Plenum Press, New York, 1965, pp. 269–84.
15. B. L. Henke, in *Advances in X-ray Analysis*, Vol. 7, W. M. Mueller, G. R. Mallett, and M. Fay, Eds., Plenum Press, New York, 1964, pp. 460–88.
16. P. S. Ong, in *The Electron Microprobe*, T. D. McKinley, K. F. J. Heinrich, Eds.,Wiley, New York, 1966, pp. 43–57.
17. P. Fisher, *J. Opt. Soc. Am.*, **44**, 665 (1954).
18. R. C. Ehlert, General Electric Co., private communication (1966).
19. C. Bonnelle, "Contribution a L'Etude des Metaux de Transition du Premier Groupe, du Cuivre, et de Leurs Oxydes par Spectroscopie X dans de le Domaine de 13-22 Å," Theses de Docteur Es-Sciences Physiques, L'Universite de Paris, 1964.

19a. C. Bonnelle, in *Soft X-ray Band Spectra and the Electronic Structure of Metals and Materials*, Derek J. Fabian ed., Academic Press, London and New York, 1968, p. 163.

20. C. Bonnelle, University of Paris, private communication (1967).
21. R. A. Mattson and R. C. Ehlert, in *Advances in X-ray Analysis*, Vol. 9, W. M. Mueller G. R. Mallett, and M. Fay, Eds., Plenum Press, New York, 1966, pp. 471–85.
22. J. E. Holliday, *Norelco Rep.*, **14**, 84 (1967).
23. P. Fisher, R. S. Crisp, and S. E. Williams, *Opt. Acta*, **5**, 31 (1958).
24. S. E. Williams, *J. Quant. Spectr. Radiative Transfer*, **2**, 621 (1962).
25. J. A. Catterall and J. Trotter, *Phil. Mag.*, **3**, 1424 (1958).
26. J. R. Cuthill, R. A. McAllster, and M. L. Williams, *Phys. Rev. Letters*, (USA) **16**, 993 (1966).
27. D. W. Fischer and W. L. Baun, in *Advances in X-ray Analysis*, Vol. 7, W. M. Mueller, G. R. Mallett and M. Fay, Eds., Plenum Press, New York, 1964, pp. 489–96.
28. A. P. Lukirskii and I. A. Brytov, *Instr. Exptl. Tech. (USSR)*, No. **5**, 1083 (1965).
29. P. S. Ong, in *Advances in X-ray Analysis*, Vol. 8, W. M. Mueller, G. R. Mallett, and M. Fay, Eds., Plenum Press, New York, 1964, pp. 341–351.

30. A. P. Lukirskii, M. A. Rumsk, and C. A. Karpovich, *Opt. i Spektroskopiya*, **9**, 343 (1960).
31. R. Wilson, *Phil. Mag.*, **41**, 66 (1950).
32. A. P. Lukirskii, V. A. Formichev, and I. A. Brytov, *Opt. Spectr.* (*USSR*), **20**, 202 (1966).
33. B. L. Henke, R. L. Egin, R. E. Lent, and R. B. Ledingham, *Norelco Rep.*, **14**, 112 (1967).
34. C. A. Anderson, in *The Electron Microprobe*, T. D. McKinley, K. F. J. Heinrich and D. B. Wittry, Eds., Wiley, New York, 1966, pp. 58–74.
35. D. Chopra, Doctoral Thesis (New Mexico State University, University Park, New Mexico, 1964).
36. D. Chopra and R. Liefeld, *Bull. Amer. Phys. Soc.*, **B62**, 404 (1964).
37. D. W. Fischer and W. L. Baun, *J. Appl. Phys.*, **38**, 4830 (1967).
38. A. Faessler, in *Soft X-ray Band Spectra and the Electronic Structure of Metals and Materials*, Derek J. Fabian, ed. Academic Press, London and New York, 1968, pp. 93–98.
39. Y. Cauchois in *Soft X-ray Band Spectra and the Electronic Structure of Metals and Materials*, Derek J. Fabian, ed. *Proc. Conf. Soft X-ray Spectroscopy and the Band Structure of Metals Alloys*, Academic Press, New York and London, 1968, p. 71–79.
40. A. E. Ennos, *Brit. J. App. Phys.*, **4**, 101 (1953).
41. H. G. Heide, *Z. Angew Phys.*, **15**, 116 (1963).
42. A. J. Campbell and R. Gibbons, in *The Electron Microprobe*, T. D. McKinley, K. F. J. Heinrich, and D. B. Wittry, Eds., Wiley, New York, 1966, pp. 75–82.
43. P. Duncumb and D. A. Melford, International Symposium on X-Ray Optics and X-ray Microanalysis. 4th, Orsay, 1965. Optique des rayons X et microanalyse. X-ray Optics and Microanalysis. R. Castaing, P. Deschamps, and J. Philibert. Eds., Paris, Hermann 1966. pp. 240–253.
44. J. E. Holliday, in *Developments in Applied Spectroscopy*, Vol. 5, L. R. Pearson and E. G. Grove, Eds., Plenum Press, N. Y., 1966, pp. 77–105.
45. R. A. Mattson, in *Advances in X-ray Analysis*, Vol. 8, W. M. Mueller, G. R. Mallett, and M. Fay, Eds., Plenum Press, New York, 1965, pp. 333–40.
46. J. E. Holliday and E. J. Sternglass, *J. Appl. Phys.*, **30**, 1428 (1959).
47. J. R. Young, *J. Appl. Phys.*, **27**, 1 (1956).
48. R. O. Lane and D. J. Zaffarano, *Phys. Rev.*, **94**, 960 (1954).
49. O. Hoffman, *Z. Physik*, **143**, 147 (1952).
50. G. Herzberg, in *Atomic Spectra and Atomic Structure*, Dover Publications, New York, 1944, p. 50.
51. H. A. Bethe and E. E. Salpeter, "Quantum Mechanics of one-and-two Electron Atoms," Academic Press, New York, 1957, p. 266.
52. F. D. Davidson and R. W. G. Wyckoff, in *Advances in X-ray Analysis*, Vol. 9, W. M. Mueller, G. R. Mallett and M. Fay, Eds., Plenum Press, New York, 1966, pp. 344–53.
53. D. H. Tomboulian, in *Handbuch der Physik*, Vol. 30, S. Flugge, Ed., Springer-Verlag, Berlin, 1957, pp. 246–304.
54. H. W. B. Skinner, *Phil. Trans. Roy. Soc.* (*London*), **A239**, 95 (1940).
55. T. Sagawa, The Science Reports of the Tohoku University, **44**, 115 (1960).
56. L. G. Parratt, *Rev. Mod. Phys.*, **3**, 616 (1959).
57. G. Wiech, Dissertation zur Erlangung der Doktorwürde, (der Ludwig-Maximilians-Universität, München), December, 1964.

58. D. O. Van Ostenburg, D. J. Lan, Masao Shimizu, and Atsushui Katsuki, *J. Phys. Soc. Japan*, **18**, 1744 (1963).
59. R. S. Crisp and S. E. Williams, *Phil. Mag.*, **6**, 365 (1961).
60. S. L. Altmann, in *Soft X-ray Band Spectra and the Electronic Structure of Metals and Materials*, Derek J. Fabian, ed. Academic Press, New York and London, 1968, pp. 265–278.
61. H. W. B. Skinner, T. Bollen, and J. E. Johnston, *Phil. Mag.*, **45**, 1070 (1954).
62. D. W. Fischer, *J. Appl. Phys.*, **36**, 2048 (1965).
63. D. W. Fischer and W. L. Baun, AFML-TR-191 (1966).
64. D. W. Fischer and W. L. Baun, *J. Appl. Phys.*, **38**, 229 (1967).
65. R. Liefeld, in *Soft X-ray Band Spectra and the Electronic Structure of Metals and Materials*, Derek J. Fabian, ed., Academic Press, New York and London, 1968 pp. 133–149.
66. A. P. Lukirskii and I. A. Brytov, *Akad. Sci. USSR Bull. Phys. Ser.*, **28**, 841–52 (1964).
67. J. R. Cuthill, A. J. McAlister, M. L. Williams, and R. E. Watson, Phys. Rev. 164, 1006 (1967).
68. C. H. Cheng, K. P. Gupta, E. C. Van Reuth, and P. Beck, Phys. Rev. *126*, 2030 (1962)
69. C. Curry, in *Soft X-ray Band Spectra and the Electronic Structure of Metals and Materials*, Derek J. Fabian, Ed., Academic Press, New York and London, 1968 pp. 173–184.
70. J. A. Catterall and J. Trotter, *Proc. Phys. Soc.*, **79**, 691 (1962).
71. J. Clift, C. Curry and B. J. Thompson, *Phil. Mag.*, **8**, 593 (1963).
72. A. Faessler, in *Colloq. Spectr. Int'l. 19th Univ. Maryland*, 1962, E. R. Lippincot and M. Margoshes, Eds., Spartan Books, Washington, D. C. p. 307.
73. J. E. Holliday, in *Advances in X-ray Analysis*, Vol. 9, W. M. Mueller, G. R. Mallett, and M. Fay, Eds., Plenum Press, New York, 1966, p. 365.
74. J. E. Holliday, *J. Appl. Phys.*, **37**, 4720 (1967).
75. H. M. O'Bryan and H. W. B. Skinner, *Proc. Royal Soc. (London)*, **176**, 229 (1940).
76. D. W. Fischer, AFML-TR-65-58 (1965).
77. M. A. Blochin and A. T. Shuvaev, *Bull. Acad. Sci. U.S.S.R. Phys. Ser.*, **2q**, 429 (1962).
78. E. E. Vainshtein and V. I. Chrikov, *Soviet Phys. "Doklady"*, **7**, 724 (1963).
79. S. A. Nemnonov and K. M. Kolobova, Fitz. Metal. i, *Metalloved*, **22** 680 (1966).
80. A. Meisel and W. Nefedow, *Z. Phys. Chem.*, **219**, 194 (1962).
81. R. S. Crisp and S. E. Williams, *Phil. Mag.*, **6**, 365 (1961).
82. T. Sagawa and D. Aita, *Jap. J. Appl. Phys.* (in press).
83. D. W. Fischer and W. L. Baun, in *Advances in X-ray Analysis*, Vol. 9, W. M. Mueller, G. R. Mallett and M. Fay, Eds., Plenum Press, New York, 1966, pp. 329–343.
84. A. Faessler and M. Goehring, *Naturwiss.*, **39**, 169 (1952).
85. D. W. Fischer and W. L. Baun, in *Advances in X-ray Analysis*, Vol. 9, W. M. Mueller, G. R. Mallett and M. Fay, Eds., Plenum Press, New York, 1966, pp. 329–343.
86. L. Pauling, *The Nature of the Chemical Bond*, 3rd ed., Cornell University Press, Ithaca, N. Y., 1960, p. 97.
87. R. Kiessling, *Met. Rev.*, **2**, 77 (1957).
88. D. A. Robins, *Powder Met.*, **1/2**, 172 (1958).
89. E. Dempsey, *Phil. Mag.*, **8**, 285 (1963).
90. P. Costa and R. R. Conte, *Met Soc. Am. Inst. Mining Met. Petrol. Eng. Inst. Metals Div. Spec. Rept. Ser.* (AIME New York 1964), No. 13, pp. 3–27.
91. R. G. Lye, *J. Phys. Chem. Solids*, **26**, 407 (1965).
92. R. G. Lye, and E. M. Logothetis, *Phys. Rev.*, **147**, 622–635 (1966).

93. W. S. Williams, *Science*, **152**, 41 (1966).
93a. V. Ern and A. C. Switendick Phys. Rev., *137*, A1927 (1965).
94. L. Pauling, *The Nature of the Chemical Bond*, 3rd. ed. Cornell University Press Ithaca, N.Y., 1960, p. 432.
95. K. Das Gupta, *Phys. Rev.*, **80**, 281 (1950).
96. A. Faessler, University of Munich, private communication (1967).
97. R. F. Bunshah, University of California, private communication (1968).
98. J. F. Moskal, P. S. Ong, and P. Cream, 14th Annual Conference on Applications of X-ray Analysis, University of Denver, 1965, *Norelco Application Data Report*. No. 144
99. G. V. T. Ranzetta and D. V. Scott, *Brit. J. Appl. Phys.*, **15**, 263 (1964).
100. A. V. Manzione and D. E. Fornwalt, *Norelco Rep.*, **12**, 3 (1965).
101. Lars Ramquist, Kjell Hamrin, Gunilla Johansson, Anders Fahlman and, Car Nordling, Uppsala University Institute of Physics Rep., No., UVI p. 609, (1968)

Author Index

Numbers in parentheses are reference numbers and indicate that the author's work is referred to although his name is not mentioned in the text. Numbers in italics show the pages on which the complete references are listed.

Abakumov, G. I., 63(167), *67*
Abresch, K., 98, *113, 182*
Acs, L., 138, *141*
Adams, R. N., 153(13), *183*
Adler, I., 298(43), *321, 323*
Agazzi, E. J., 126(15), *139,* 173
Aimoto, Y., 255(34), *272*
Aita, D., 397, *417*
Albrecht, W. M., 84(35), 85(37), *113*
Alder, A. B., 14(11), 20(11), *23*
Alimarin, I. P., 41(68), 51(68), *65*
Alkemade, C. T. J., 187, 188(11), *220*
Allison, S. K., 282(3), 285(4), 310(3), *320, 323*
Altmann, S. L., 377, 378, *417*
American Society for Testing and Materials, 6(5), *23,* 71, 80(27), 82(31), 84(27,36), 85(27), 92(59), 93, 99, 103, 106, *112–114,* 116(1), 134(1), 137, *139,* 179(58), *184,* 295(22), *321*
Ammann, R., 177(53), *184*
Amos, M. D., 204, 207(97), 209, 211, 216, 217(97), *222*
Amphlett, C. B., 56(133), *66*
Analytical Chemistry, 226, *272*
Andermann, G., 298(42), 301, 303, *321, 322*
Anderson, C. A., 351, 353, 363, 364, *416*
Anderson, J. R. A., 271(109), *274*
Anderson, J. W., 217(108), 218, *223*
Anson, F. C., 152(8), *183*
Aoyama, T., 35(37), *64*
Apple, R. F., 45
Argauer, R. J., 258(79), *274*
Arnesen, R. T., 256(57), *273*
Aronin, L. R., 136, *140*
Arthur, P., 176
Ashby, W. D., 295(26), *321*
Ashley, S. E. Q., 50(97), *65*
Askworth, W. J., *323*
Aspinal, M. L., 74(6,6a), 76(6), *112*
Aztec Instruments, 188(58), *221*

Baamen, A., 255(36), *273*
Babko, A, K., 36(45), *64,* 271(106), *274*
Baghurst, H. C., 256(50), *273*
Bahnstedt, U., 256(52), *273*
Baird, A. K., 298(37), *321*
Ball, R. G., 167
Banks, C. V., 107(81), *114*
Barabas, S., 138, *141*
Bard, A. J., 176, 177(53), *184*
Bardin, M. B., 164
Barieau, R. E., 312, *323*
Barker, G. C., 168(38,39), *183*
Barnes, R. B., 187(5), *220*
Barnes, W. J., 47(79), *65*
Bartholomé, E., 197
Bassett, J., 255(27), *272*
Bastian, R., 251(13), *272*
Bate, L. C., 62(162), *67*
Bates, R. G., 145(1), *182*
Bauer, G. A., 256(51), *273*
Bauer, H. H., 168(36), *182, 183*
Bauman, F. A., 136(51), *140*
Bauman, R. P., *223,* 230, *272*
Baun, W. L., 286(7), *320,* 346, 352, 377, 382, 384, 397, 400, 401, 403, *415–417*
Bausch and Lomb, 257(77), *274*
Bawden, A. T., 172(49), *184*
Beach, A. L., 72, 80(19), 81(29), *112*
Bearden, J. A., 295(20,21), *321*
Beck, E., 107, *114*
Beck, P., 379(68), 381, *417*
Beckman, A. O., 246(7), *272*
Beckman Instruments, 188(57), 194(65), 216(57), 217(57), *221, 222*
Bedi, R. D., 178
Beeghly, H. F., 316(92), 317, *323*
Beernsten, D. J., 81(28), *112*

Beilby, A. L., 153(14), *183*
Belew, W. L., 154(17), 167(17), 169(17), *183*
Bell, G. F., 218, *223*
Bellanca, S., 31(5), *63*
Bender, G. T., 136(52), 137(53), *140*
Benedetti-Pichler, A. A., 13, *23*
Benson, V. M., 255(35), *272*
Berg, E. W., 244(84), 264(84), *274*
Berlandi, F. J., 55(121,122), *66*
Bermijo Martinez, B., 188(43), *221*
Bernstein, F., 297, *321*
Berry, J. W., 187(5), *220*
Bertilsson, G. O. B., 194(66), *222*
Bertin, E. P., 313, *323*
Bethe, H. A., 366, 368, *416*
Biecher, D. G., 256(63), *273*
Bingham, C. D., 133(40), *140*
Birks, F. T., 61(154), *67*
Birks, L. S., 289, 290, 299, *320, 321, 323*
Blake, C. A., Jr., 37(65), *65*
Blocher, J. M., Jr., 79(13), *112*
Blochin, M. A., *323,* 326, 385, *414, 417*
Blom, L., 131, *140*
Blosser, E. R., 1, 35(26), *64*
Blurton, K. F., 171(47), *184*
Boiteux, H., 187, 188(13), 193(13), 197(13), 199(81), 201(13), 214(104), *220, 222, 223*
Boling, E. A., 204, 215(106), *222, 223*
Bollen, T., 377(61), *417*
Boltz, D. F., 133(43), *140,* 225, 252(14,15), 253(19,22), 254(68), *272, 273*
Boltz, G., 255(33), *272*
Bonnelle, C., 343, 377, *415*
Bonsels, W., 35(24,25), 45, *64*
Booman, G. L., 177(54), 178, *184*
Booth, E., 78, 88, *112, 113*
Bortsova, V. A., 256(49), *273*
Bowman, H. R., 293(19), 294(19), *321*
Box, G. F., 188(23), *221*
Boyd, B. R., 298(38), *321*
Boyle, W. G., Jr., 115, 122(9), 126(17), 127(17), 132, 134(9), 138, *139, 140*
Bozhevol'nov, E. A., 270(96), 271(96,108), *274*
Bradenstein, M., 306(62), *322*
Bradshaw, G., 35(36), *64*
Brady, J., 79(14), *112*
Branco, J. J. R., 300(48), *322*
Bresle, A., 169(40), *183*
Breyer, A. C., 55(126), 59(126), *66*
Breyer, B., 168(36), *182, 183*
Bricker, C. E., 51(105), 53(105,107), *66*
Bridge, E. P., 36(38), 45, *64*
Bright, H. A., 80(21), *112*
British Iron and Steel Research Associates, 87, *113*
Brody, J. K., 62(158), *67*
Brooks, E., 290, *321*
Brooks, F. R., 126(15), *139*
Brooks, M. S., 6(4), *23*
Brooks, R. R., 58(139), *67*
Brooks, W., 153(14), *183*
Brophy, V. A., 35(28), *64*
Brown, D. M., 298, *321*
Brown, K. B., 37(65), *65*
Brown, R. A., 176
Bruins, E. H., 131, *140*
Bryant, F. S., 88, *113*
Bryson, T. C., 129(20), *140*
Brytov, I. A., 346, 349(32), 377, 385, *415–417*
Buchanan, J. D., 31(5), *63*
Buchanan, R. F., 60(144), 62(158), *67*
Budenz, R., 256(52), *273*
Buhrke, V. E., 295(26), *321*
Bunsen, R., 187, *220*
Bunshah, R. F., 406, *418*
Burger, J. C., 188(25), 205(91), *221, 222*
Burke, K. E., 134(47), 138, *140*
Burkhalter, T. S., 181(61), *182, 184*
Burnett, B. B., 48(83), *65*
Burr, A. F., 295(21), *321*
Burson, K. R., 167
Bush, E. L., 35(33), *64*
Bush, G. H., 36(41), *64*
Busker, D., 271(97), *274*
Butler, E. B., 31(9), *63*
Bystroff, R. I., 185

Cain, D. A., 218(116), *223*
Cameron, J. F., 319, *323*
Campanile, V. A., 126(15), *139*
Campbell, A. J., 355, *416*
Campbell, I. E., 79(13), *112*
Campbell, M. C., 61(157), *67*
Campbell, M. E., 35(27,29), 45, *64,* 255(26), *272*
Campbell, W. J., 286(8), 298(45), 305, *320–322*

Carlson, T. A., 56(132), *66*
Carney, D. J., 86(40), 87, *113*
Carson, R., 45
Cartwright, J., 34(17), *63,* 188(27), 217, *221*
Carver, R. J., 313(87), *323*
Cary Instruments, *223*
Case, O. P., 255(28), *272*
Cass, D. E., 317, *323*
Casto, C. C., 53(108), *66*
Catterall, J. A., 346, 382, 389(70), *415, 417*
Cauchois, Y., 290, 321, 354, *416*
Caugherty, B., 319, *323*
Cederlund, J., 287(13), *320*
Čejhan, O., 138, *141*
Center, E. J., 53(116), *66*
Chadwick, J., 287(10), *320*
Chambers, W. E., 206(93), *222*
Chao, M. S., 153(10), *183*
Chase, D. L., 45, 53(116), *66*
Chemical and Engineering News, 146(2), *182*
Cheng, C. H., 379, 381, *417*
Cherkesov, A. I., 270(93), 271(103), *274*
Chipman, J., 86(40), 87, 90, *113*
Chirnside, R. C., 54, *66,* 120, 121(4), 130(22), 136(4), *139, 140*
Chodos, A. A., 300, *322*
Cholak, J., 187(7), *220*
Chopra, D., 352, *416*
Chow, T. J., 31(10), 32(10), *63*
Chrikov, V. I., 385, *417*
Claassen, A., 255(36), *273*
Claassen, I., *182*
Claisse, F., 297, 298(35), 300(35), *321*
Clark, F. G., 107, *114*
Clark, G. L., *323*
Clarke, B. L., 49(87), *65*
Clemency, C., 134(45), *140*
Clift, J., 382, 383, 401, *417*
Cluley, H. J., 54(118), *66*
Codell, M., 134(45), *140*
Cohen, J., 255(42), 256(48), *273*
Cohn, G., 173
Coleman, C. F., 37(65), *65*
Collier, F., 103(77), 107(77), 108(77), *114*
Compton, A. H., 282(3), 285(4), 310(3), *320, 323*
Conrad, A. L., 153(12), *183,* 254(70), *273*
Conrad, F. J., 256(58), *273*
Conte, R. R., 404, *417*
Cook, G. B., 287, *320*
Cooke, F., 97, *113*
Cooke, W. D., 176, 211(99), *222*
Cooper, M. D., 255(45), *273*
Cornet, C., 36(46), *64*
Corning Glass Works, 257(78), *274*
Cosgrove, J. F., 37(62), *65,* 188(55), 189(55), *221*
Costa, P., 404, *417*
Coster, D., 276, *320*
Cotton, F. A., 117(3), *139*
Cottrell, F. G., 156(19), *183*
Cowley, T. G., 213(111), *223*
Cox, J. J., 30
Craig, D., 41(69), *65*
Craig, L. C., 41(69), *65*
Crawford, C. M., 251(12), *272*
Cream, P., 406–408, 409(98), *418*
Crisp, R. S., 346, 376(59), 397, *415, 417*
Crittenden, A. L., 164
Cruickshank, W., 179(56), *184*
Cullen, T. J., 298(40,41), 301, 314, *321–323*
Cunningham, A. F., 192, 209, *221*
Curry, C., 379, 382, 383, 401, *417*
Curry, R. W., 393
Cuthill, J. R., 346, 377, 381, *415, 417*

Dagnall, D. M., 188(41), *221*
Dagnall, R. M., 255(23), *272*
Dallmann, W. E., 83(34), 85(34, 38), 103, *113, 114*
Darken, L S., 393
Das Gupta, K., 405, *418*
Dauncey, L. A., 130(22), *140*
David, D. J., 188(44), *221*
Davidovich, N. K., 271(22), *274*
Davidson, F. D., 366, *416*
Davis, C. M., 134(47), 138, *140*
Davis, D. G., 176, *181*
Davis, H. M., 165(32), *183*
Dean, J. A., 162(27), *183,* 187, 188(16), 192(11), 201(87), 213(10), *220, 222, 223,* 263(83), 274
Delahay, P., 154(16), 156(18), 169(41), *181, 183, 184*
Delaughter, B., 191
DeMars, R. D., 167
Dempsey, E., 404, *417*

Desbarres, J., 177(53), *184*
DeWet, J. F., 58(134), *67*
Diehl, H., 270(91,92), 271(92), *274*
Dodd, C. G., 311, 314, *322*
Doležal, J., 173
Donahue, J. F., 176
Donaldson, E. M., 255(46), *273*
Donovan, P. D., 36(41), *64*
Dowdey, J. E., 307(64), *322*
Drako, O. F., 36(45), *64*
Dryer, H. T., 298(38), *321*
D'Silva, A. P., 188(29), *221*
Dubovenko, L. I., 271(106), *274*
Ducret, L., 36(46), *64*
Duke, J. F., 45, 49(92), *65*
Duncan, L., 178
Duncomb, P., 355, 356, *416*
Dunken, H., 194(67), *222*
Dunn, H. W., 311, 312, *322*
Dyer, J. R., 230, *272*

Eberle, A. R., 34(16), *63,* 256(62), *273*
Eckfeldt, E. L., 175(52), *184*
Eckman, J. R., 70, *112*
Edelhausen, L., 131, *140*
Egin, R. L., 349(33), *416*
Ehlert, R. C., 336, 337, 340, 342–345, 354, 358, 360, 384, 385, *415*
Eichlin, D. W., 149(4), *182*
Eisenhart, C., 14(16), *23*
Elbe, G. v., 195(73), *222*
Elbing, P., 107(80), *114*
Elion, H. A., 292, *321*
Elliott, E., 271(102), *274*
Elliott, J. F., 90, *113*
Ells, V. R., 187(6), *220*
Elving, P. J., 26(1), *63,* 173, *182*
Elwell, W. T., 188(53), *221,* 255(40), 256(55), *273*
Emets, N. P., 268(86), *274*
Eng, K. Y., 133(40), *140*
Engel, C. G., 300(48), *322*
Englesman, J. J., 131, *140*
Englis, D. T., 48(83), *65*
Engstrom, A., 311, 313, *322, 323*
Enke, C. G., 152(8), *183*
Ennos, A. E., 355, *416*
Erdey-Gruz, T., 53(106), *66*
Ermolaev, A. A., 194(68), *222*
Estebaranz, A. A., 36(51), *64*
Etten, N., 35(34), *64*
Evans, C. P., 255(30), *272*
Evans, F. M., 36(42), *64*
Evans, H. B., 287(11), *320*
Evans, J. L., 36(41), *64*
Evcim, N., 271(104, *274*
Evens, F. M., 80(26), 88, 89, *112*
Everett, M. E., 89, *113,* 255(42), 256(48), *273*
Everingham, M. R., 46(73,74), *65*
Ezerskaya, N. A., 164, 173

Faessler, A., 353, 383(72), 399, 405, *416–418*
Fagel, J. E., 96, 97, *113*
Fahlbusch, W. A., 296(27), *321*
Fahlman, A., 404(101), *418*
Faris, J. P., 59(143), 60(144), 62(158), *67*
Fassel, V. A., 36(42,43), 61(151), *64, 67,* 80(26), 83(34), 85(34,38), 88, 89, 103, *112–114,* 188(29,31,36), 206, 213, 217, *221, 223*
Feldman, C., 58(140), *67*
Field, B. D., 54(117), *66*
Fischer, D. W., 286(7), *320,* 346, 352, 377, 382, 384, 397, 400, 401, 403, *415–417*
Fisher, D. J., 154(17), 167(17,35), 169(17), *183,* 215(105), *223*
Fisher, P., 342, 346, *415*
Flanagan, F. J., 298(43), *321*
Flaschen, S. S., 46, 47(78), *65*
Fleischer, K. D., 173
Flugge, S., *323*
Flynt, W. E., 307(64), *322*
Formichev, V. A., 349(32), *416*
Fornwalt, D. E., 411–413, *418*
Foulk, C. W., 172(49), *184*
Fowler, E. W., 45
Fowler, R. M., 129(20), *140*
Frank, C. W., 188(37), *221*
Frank, G., 330
Franklin, J. C., 21(13), *23*
Frazer, J. W., 128(19), 129(21), *140*
Freegarde, M., 34(17), *63*
Freeland, M. Q., 256(64), *273*
Freiser, H., 37(58–60,62–64), 40(48), 41(58), *64, 65*
Friedline, J. E., 74(6b), *112*
Friedrich, K., 96, *113*

Fristrom, R. M., 195(74), 196(77), 211, *222*
Fritz, J. S., 63(163), *67,* 256(64), *273*
Frohmeyer, W., 310, 312, *322*
Fryxell, R. E., 121, *139*
Fugmann, W., 256(54), *273*
Fulton, J. W., 121, *139*
Furman, N. H., 51(105), 53(105), 63(169), *66, 67,* 149(3), 150, 151, 176, *181, 182*
Fuwa, K., 188(30), 199, 217(30), *221*

Gaeke, G. C., 134, *140*
Gahler, A. R., 255(39), *273*
Gale, P., 131(29), *140*
Galway, A. K., 133(35), 135, *140*
Gardner, A. W., 168(39), *183*
Gatehouse, B. M., 199(83), *222*
Gaydon, A. G., 195(72,75), 201, 212(102), *222, 223*
Gaylor, D. W., 299, *322*
Gaylor, V. F., 153(12), *183*
Gegus, E., 45
Geiger, R. A., 271(113), *274*
Geilmann, W., 36(49–51, 53), *64*
Gelberg, A., 271(105), *274*
Geller, K., 194(67), *222*
General Electric Co., 295(23), *321*
Gibbons, R., 355, *416*
Gibbs, G. V., 295(25), 321, *321*
Gibson, J. H., 211, *222*
Gidley, J. A. F., 188(53), *221*
Gilbert, P. T., Jr., 187, 188(14,15), 193(64), 194(69), 195(76), 197(14), 201(88), 208(98), 211, 217, *220, 222, 223*
Gillies, W., 188(25), 205(91), *221, 222*
Glocker, R., 310, 312, *322*
Glodowski, S., 167
Göbbels, P., 176
Goddu, R. F., 252(16), *272*
Goehring, M., 399, *417*
Goldringer, L. S. R., 246(7), *272*
Goleb, J. A., 188(34,35), 199(34,35), *221*
Golightly, D., 217
Goode, G. C., 61(157), *67,* 166
Goodwin, H. B., 79(13), *112*
Gordon, G. M., 298(36), *321*
Gordon, L., 47(82), *65*
Gordon, W. A., 122(10), 124(10), 137, *139*
Gori, A., 53(109), *66*
Gouy, G. L., 187, *220*
Goward, G. W., 87, 89, 90, 97, 107(80), *113, 114*
Graab, J. W., 124(13), *139*
Graham, R. P., 50(100,101), *66*
Graham, T., 70
Grant, J. A., 131, *140*
Grant, J. W., 122(10), 124(10), 137(10), *139*
Grant, N. J., 86(40), 87, *113*
Green, H., 37(66), *65*
Green, I. R., 120, 121(4), 136, *139*
Green, T. E., 45, 305(61), *322*
Gregory, L. J., 122(9), 134(9), 138(9), *139*
Griffin, E. B., 21(13), *23*
Griffin, G. L., 330, *415*
Griffith, C. B., 78(11), 85(37), 86, 88, 97, *112, 113*
Grimes, M. D., 133(44), 134(44), 137(44), *140*
Grossman, W. E. L., 188(36), 206(36), 211(99), *221, 222*
Grubb, W. T., 58(135), *67*
Guilbault, G. G., 271(98), *274*
Guinn, V. P., 31(5), *63*
Guldner, W. G., 36(39), *64,* 72, 80(19), 81(29), *112,* 129(20), *140*
Gunn, E. L., 297, *321*
Gupta, K. P., 379(68), 381, *417*
Guyer, A., 255(38), *273*

Haas, C. S., 46(74), *65*
Hagenah, W. D., 195(71), *222*
Hague, J. L., 129(20), *140*
Hahn, R. B., 256(67), *273*
Hahn, R. L., 133(41), *140*
Hainski, Z., 63(164), *67*
Hair, R. P., 48(86), *65*
Hakkila, E. A., 275, 299(47), 301(50), 302(55), 304(47), 305(50), 311(80,81), 313, 314(85), *322, 323*
Hambly, A. N., 204
Hamrin, K., 404(101), *418*
Hanker, J. S., 271(105), *274*
Hansen, W. R., 80(20), 84(20), 85(20), 91(56,57), 95(56), *112, 113*
Hanson, H. P., 307(64), *322*
Hardy, F. R. F., 255(25), *272*
Hare, G. H., 246(7), *272*
Harrar, J. E., 143, 164, 178
Harrington, W. L., 133(42), *140*

Hartford, W. H., 150
Hartley, A. M., 176
Hartmann, H., 36(44), *64*
Harvey, C. A., 89(51), 90, *113*
Hashitani, H., 45
Hasler, M. F., 291(17), *321*
Hasler, M. G., 332–334, 340–342, 351, *415*
Hatch, W. R., 256(65), *273*
Hawes, R. C., 246(7), *272*
Hazan, I., 271(111), *274*
Hazelby, D., 74(6a), *112*
Headridge, J. B., 253, *272*
Heady, H. H., 107(83), *114*
Heffelfinger, R. E., 35(26), 45, *64*
Heide, H. G., 355, *416*
Heinrich, B. J., 133(44), 134(44), 137(44), *140*
Hell, A., 218(107), *223*
Henderson, D. J., 287(11), *320*
Henke, B. L., 336–339, 349, 358–360, 362, 409, 410, *415, 416*
Henry, R. J., 31(8), *63*
Henry, W. M., 1, 35(26), 45, *64*
Hepp, H., 36(53), *64*
Hercules, D. M., 254(69), *273*
Hermance, H. W., 49(87), *65,* 180(60), *182, 184*
Hermann, J. A., 47(81), *65*
Herringshaw, J. F., 51(103), *66*
Herrington, J., 166
Herrmann, R., 187, 188(11), 199(79,80), *220, 222*
Herzberg, G., 365(50), *416*
Hess, D. C., 36(54), *64*
Hessenbruch, W., 70, *112*
Hettel, H. J., 61(151), *67*
Hevesey, G. v., 276, 286, 310, *320, 322*
Heyrovský, J., 159, *181, 183*
Hibbits, J. O., 49(89), *65*
Hickam, W. M., 79(16), *112*
Hill, C. C., 80(26), 83(34), 85(34), 88, 89, *112, 113*
Hill, U. T., 255(24, 31), *272*
Hirano, S., 45, 51(104), 53(110–112,114), 58(138), 61(112, 138,148–150),63(149), *66, 67*
Hiskey, C. F., 251(11), *272*
Ho, M., 258(81), *274*
Hochgesang, F. P., 313, *323*
Hodecker, J. C., 173
Hoffman, J. I., 33(12), 50(102), *66*
Hoffman, O., 363(49), *416*
Holbrook, W. B., 178
Holliday, J. E., 325, 327(1,2), 330–333, 335, 336, 343, 345, 349, 351, 352, 355, 357, 358(22), 363, 365, 366, 373(2), 374, 375, 376(1), 379, 383(1,22,73,74), 384(74), 388–390, 392, 394–397, 399, 401(1), 402–405, 410–412, 413(9), *414–417*
Holm, V. C. F., 80(22), 91(55), *112, 113*
Holt, B. D., 34(19), *63*
Holzman, R. T., 129(21), *140*
Hood, R. L., 187(5), *220*
Hopkins, P., 131(29), *140*
Horrigan, V. M., 85(38), *113*
Horton, W. S., 79(14), *112*
House, H. P., 201(86), *222*
Howell, J. A., 225, 252(14), *272*
Hoyt, S. L., 80(25), *112*
Hubbard, D. M., 187(7), *220*
Hubbard, G. L., 45
Huff, E. A., 60(145), 61(156), *67*
Hughes, H. K., 308, 313, *322, 323*
Hughes, R. C., 219(118), *223*
Hume, D. N., 252(16), *272*
Hunter, J. A., 131, *140*
Hurley, R. G., 302(55), 311(81), *322, 323*
Hyde, E. K., 293(19), 294(19), *321*

Iden, R. B., 82(33), *112*
Ihida, M., 88, 89, *113*
Iida, Y., 45, 58(138), 61(112,138,148–150), 63(149), *66, 67*
Ilkovič, D., 159, *183*
Imahashi, T., 45
Inczédy, J., 55(124), 59(124), *66*
Ingber, N. M., 34(22), *64*
Ingram, G., 117(2), *139*
Inman, W. R., 255(46), *273*
Iron and Steel Institute, 34(18), 36(18), *63,* 87(42), *113, 114*
Irving, H., 30, 37(61), *64*
Isreeli, J., 193(63), *222*
Issopoulos, P. B., 45

Jackson, H., 45
Jackwerth, E., 45
Jacobson, B. J., 309, *322*

Jaffee, R. I., 79(13), *112*
Jared, R. C., 293(19), 294(19), *321*
Jaundrell-Thomson, F., *323*
Jenkins, I. L., 168(38), *183*
Jenkins, R., 305, *322*
Jensen, R., 270(92), 271(92), *274*
Jenson, R. E., 271(112), *274*
Jessop, G., 168(37), *183*
Johansson, G., 404(101), *418*
Johnson, G. G., Jr., 295(25), *321*
Johnson, J. L., 256(67), *273*
Johnson, J. S., 56(132), *66*
Johnston, J. E., 377(61), *417*
Johnston, W. H., 31(9), *63*
Jones, A. H., 255(32), *272*
Jones, H. C., 167(35), *183,* 215(105), *223*
Jones, J. C. H., 255(27), *272*
Jones, J. L., 291(17), *321*
Jones, J. T., 256(66), *273*
Jones, R. F., 131, *140*
Jones, R. H., 35(32), *64*
Jones, W. G., 188(24), *221*
Jones, W. T., 166
Jordan, L., 70, *112*
Jordon, D. E., 256(63), *273*
Jursik, M. L., 217(114), 218(114), *223*

Kahn, H. L., 188(46), *221, 223*
Kaelble, E. F., *323*
Kallmann, S., 49(89), *65,* 69, 103(77,78), 107(77), 108(77), *114*
Kalnin, I. L., 138, *141*
Kalvoda, R., 169(42), *183*
Kamada, H., 45
Kamin, G. J., 107(81), *114*
Kaminski, J., 138, *141*
Kane, P. F., 167
Kanie, T., 253(17), *272*
Kanzelmeyer, J. H., 45
Karp, H. S., 88(44), *113*
Karplus, R., *150*
Karpovich, C. A., 347(30), *416*
Karttunen, J. O., 287(11), *320*
Kasha, M., 268(87), *274*
Kassir, Z. M., 51(103), *66*
Katsuki, A., 375(58), 376, *417*
Katz, H. L., 256(53), *273*
Keirs, R. J., 268, *274*
Kelley, M. T., 154(17), 167(17,35), 169(17), *183,* 215(105), *223*
Kelly, J. H., 317, *323*
Kemp, J. W., 291(17), 301, *321, 322*
Kemula, W., 166(34), 167, *183*
Kenna, B. T., 256(58), *273*
Kennedy, J. K., 6(4), *23*
Kennicott, P. R., 21(14), *23*
Kertes, S., 255(41), *273*
Kessler, G., 193(63), *222*
Kiessling, R., 404, *417*
Kilday, B. A., 296, *321*
Kirchoff, G., 187, *220*
Kirk, P. L., 41(67), *65*
Kirkbright, G. F., 270(94), *274*
Kirsten, W. J., 194(66), *222*
Kiryushkin, V. V., 194(68), *222*
Kitagawa, H., 255(34), *272*
Klimiova, L. A., 268(88), *274*
Knapp, J. R., 45
Kniseley, R. N., 188(29,36), 206(36), *221*
Koch, O. G., 31(6), 32(6), 37(6), 40(6), 45, *63*
Koch, W., 37(57), *64*
Koch-Dedic, G. A., 31(6), 32(6), 37(6), 40(6), *63*
Koehl, B. G., 82(33), *112*
Koh, P. K., 319, *323*
Kohler, D. F., 82(33), *112*
Koirtyohann, S. R., 212(101), *222*
Kolobkov, V. P., 268(86), *274*
Kolobova, R. M., 385, *417*
Kolthoff, I. M., 26(1), *63,* 151, 164, 171(45), 173, *181, 182, 184*
Kondrat'eva, L. I., 63(168), *67*
Konovalov, E. E., 63(167,168), *67*
Kopa, L., 78, *112*
Korkish, J., 271(111), *274*
Kovacs, E., 255(38), *273*
Kramer, D. N., 271(98), *274*
Kraus, K. A., 55(125), 56(132), 59(125,142), 60(147), *66, 67*
Krefeld, R., 63(164), *67*
Kreingol'd, S. U., 270(96), 271(96), *274*
Kubaschewski, O., 89(51), 90, *113*
Kublik, Z., 166(34), 167, *183*
Kuchmistaya, G. I., 271(99,100), *274*
Kuebler, N. A., 194(70), *222*
Kunin, L. L., 88, 89, *113, 114*
Kunin, R., 55(128,129), 59(128,129), *66*
Kuo, C. W., 133, 136, *140*
Kuta, J., *181*

Kwestroo, W., 30

Laboratory Equipment Corp., 98, *113, 114,* 130, *140*
Laitenen, H. A., 14(9), 20(9), *23,* 152(8), 153(10), 159(22), 164, 171(45), *183, 184*
Lambert, M. C., 297(34), 307, *321, 322*
Lamson, D. W., 149(6), 151(6), *182*
Lan, D. J., 375(58), 376, *417*
Lancaster, W. A., 46(73), *65*
Landerl, J. H., 153(12), *183*
Lane, R. O., 363(48), *416*
Lang, W., 199(79,80), *222*
Langer, A., *173*
Laqua, K., 195(71), *222*
Larionova, I. E., 63(168), *67*
Larsen, R. P., 34(22), *64*
Lastovskii, R. P., 270(96), 271(96), *274*
Laug, E. P., 31(7), *63*
Lawrence, G. L., 153(14), *183*
Ledingham, R. B., 349(33), *416*
Lemm, H., 98, *113*
Lent, R. E., 349(33), *416*
Lerner, M. W., 34(16), *63*
Leslie, W. D., 256(63), *273*
Levich, B., 171(47), *184*
Lewin, S. Z., 256, *273*
Lewis, B., 195(73), *222*
Lewis, G. N., 268(87,89,90), *274*
Lewis, L. L., 74(7), 88(44), *112, 113,* 133(34), *140,* 218(117), *223*
Liden, K., 287(13), *320*
Liebhafsky, H. A., 307(63,67), *322, 323*
Liefeld, R., 352, 377, 383, *416, 417*
Lindh, A. E., 316(89), *323*
Lindsey, A. J., *182*
Lingane, J. J., 50(95), 55(95), *65,* 150, 152(8,9), 159(22), 170(44), 176, 178, *181, 183, 184*
Lipkin, D., 268(89,90), *274*
Litvinova, N. F., 79(17), 83(17), *112*
Liu, R., 103(78), *114*
Llewellyn-Jones, F., 205(89), *222*
Lockyer, R., 188(45), *221*
Logothetis, E. M., 404, *417*
Longobucco, R. J., 313(87), *323*
Lott, P. F., 263, *274*
Lounamaa, N., 256(47,54), *273*
Loveland, J. W., *182*
Lowe, B. J., 319, *323*
Lowy, S. L., 271(109), *274*
Lublin, P., 317, *323*
Lucas-Tooth, J., 304, *322*
Luescher, W., 255(38), *273*
Luke, C. L., 34(15,20,21), 35(27,29), 45, 46, 47(78), 48(84,85), 49(90,91), 58(136), *63–65, 67,* 255(26,44), 256(61), *272, 273,* 298(39), 305, *321, 322*
Lukirskii, A. P., 346, 347, 349, 377, 385, *415–417*
Lundberg, B., 309, *322*
Lundegardh, H., 187, *220*
Lundell, G. E. F., 33(12), 50(102), *63, 66*
L'vov, B. V., 199(84), *222*
Lye, R. G., 404, *417*
Lyssy, G. H., 133(37), *140*

McAlister, A. J., 377(67), 381, *417*
McAllster, R. A., 346(26), *415*
McClellan, B. E., 255(35), *272*
MacColl, R. S., 298(37), *321*
MacDonald, A. M. G., 126(14), *139*
McDuffie, B., 51(105), 53(105), *66*
McGarvey, F. X., 55(129), 59(129), *66*
McIntyre, D. B., 298(37), *321*
Mack, D. L., 35(30,31), 45, *64*
McKinley, T. D., 93(60), 95(63), *113*
McKinney, C. R., 31(10), 32(10), *63*
MacMillan, H. R., 256(65), *273*
McNely, D. J., 298(36), *321*
Magel, T. T., 268(89), *274*
Maienthal, E. J., 166
Malissa, H., 135, *140,* 191
Malkina, E. N., 256(49), *273*
Mallett, M. W., 36(40), *64,* 69, 78(11), 79(18), 80(20,23), 82(18,33), 84(20,35), 85(20,37), 86, 88, 91(56,57), 95(56), 97, *112, 113*
Malmstadt, H. V., 206, *222*
Maltby, J. G., 29(3), *63*
Mandelstam, S., 36(47), *64*
Manning, D. C., 206(94), 217, *222*
Manning, D. L., 167
Mansfield, C. T., 196(78), *222*
Manzione, A. V., 411–413, *418*
Mareček, J., 45
Margoshes, M., 187, *220*
Marinenko, G., 175(51), *184*
Mark, H. B., Jr., 55(121,122), *66*
Markovich, P. J., 287(11), *320*

Marshall, R. R., 36(54), *64*
Martin, E. L., 178
Martin, J. F., 74(6b), 92, *112, 113*
Massey, E. M., 136, *140*
Massie, W. H. S., 131, *140*
Masson, C. R., 88, *113*
Matheson, L. A., 164(28), 165(28), *183*
Mathre, O. B., 40
Mattson, R. A., 286(9), *320,* 336, 337, 340, 343–345, 354, 358–361, 384, 385, *415, 416*
Matyska, B., 173
Mavrodineanu, R., 187, 188(13,17,18), 193, 197(13), 199(81), 201, 214(104), 219(118), *220, 222, 223*
Maxwell, J. A., 50(100,101), *66*
Maykuth, D. J., 79(13), *112*
Meggers, W. F., 205(92), *222*
Mehalchick, E. J., 50(93), *65*
Meinke, W. W., 6(1), *23*
Meisel, A., 385, *417*
Meites, L., 30, 55(120), *66,* 159(21), 160(24), 161(25,26), 162(25), 163, *181, 183,* 238(5), *272*
Melford, D. A., 355, 356, *416*
Mellish, C. E., 287(12), *320*
Mellon, M. G., 248(9), 254(68), *272, 273*
Melnick, L. M., 74(6b,7), 88(44), 92(58), 97, *112, 113,* 150
Meloan, C. E., 188(37), *221*
Menis, O., 167, 201(86), *222*
Merkle, E. J., 124(13), *139*
Mero, J. L., 298(36), *321*
Merritt, L. L., 178, 263(83), *274*
Metz, C. F., 150
Meyer, A., 131(26), *140*
Meyer, R. A., 133(40), *140*
Michaelis, R. E., 296, *321*
Migliore, J. C., 31(5), *63*
Mikkeleit, W., 194(67), *222*
Miller, F. J., 153(14,15), *183*
Mislan, J. P., 188(32), *221*
Mitchell, A. C. G., 188, *220*
Mitchell, B. J., 302, *322*
Mitchell, J., 173(50), *184*
Mizuike, A., 25, 45, 51(104), 53(110–112, 114,115), 58(138), 61(112,138,148–150), 63(149), *66, 67*
Moak, W. D., 135, *140*
Monnier, D., 45
Mooney, C. F., 330(4), *415*
Mooney, Y., 330
Morgan, E., 164
Morris, C. J., 128(19), *140*
Morrison, G. H., 6, *23,* 37(58–60,62,63), 40(58), 41(58), 46, 60(146), 62(160,161), *64, 65, 67,* 133(42), *140*
Moskal, J. F., 406–409, *418*
Mossotti, V. G., 188(36), 195, 206(36), *221, 222*
Mostyn, R. A., 192, 209, *221*
Motojima, K., 45
Müller, O. H., 164
Mueller, T. R., 153(13), *183*
Mulford, C. E., 191
Murase, T., 60(147), *67*
Murphy, J. E., 107(83), *114*
Murphy, T. J., 179(58), *184*
Muschaweck, J., 35(34), *64*
Muzzarelli, R. A. A., 62(162), *67*
Myers, R. J., 176

Nakashima, F., 61(155), 62(159), *67*
Nardozzi, M. J., 133(34), *140*
National Academy of Sciences, 15(15), *23*
Neal, T. E., 55(122), *66*
Neeb, K. H., 36(50,52), 37(55,56), *64*
Nefedow, W., 385, *417*
Nelson, F., 59(142), 60(147), *67*
Nelson, L. S., 194, *222*
Nelson, W. F., 316(91), *323*
Nemnonov, S. A., 385, *417*
Newman, E. J., 48(86), *65*
Newton, D. C., 46(77), *65*
New York Academy of Sciences, 6(3), *23*
Nichols, N., 164(28), 165(28), *183*
Nicholson, J. B., 330, 332–334, 340–342, 351, *415*
Nicholson, M. M., 167
Nicholson, R. S., 165(30), 166(30), 169(30), *183*
Nielsch, W., 255(33), *272*
Niemann, R. C., 287(11), *320*
Nightingale, C. F., 134(38), *140*
Nishimura, K., 35(37), *64*
Nishiya, T., 45
Nordling, C., 404(101), *418*
Norman, V. J., 256(50), *273*
Norwitz, G., 134(45), *140,* 255(42), 256(48), *273*

Oberhoffer, P., 70, *112*
Oberthin, H., 49(89), *65,* 103(78), *114*
O'Bryan, H. M., 384, *417*
O'Connell, R. F., 36(48), *64*
Oelsen, W., 176
Ogilvie, R. E., 292, 303(57), 309, 319, *321–323*
O'Laughlin, J. W., 107(81), *114*
Oldfield, J. H., 35(30), 36(38), 45, *64*
Olsen, R., 270(92), 271(92), *274*
Olson, C. E., 173
Ong, P. S., 337, 339, 340, 347, 351, 352, 406–408, 409(98), *415, 418*
Ordoveza, F., 126(16), *139*
Orgel, L. E., 229, *272*
Ott, W. L., 256(65), *273*
Overbeck, R. C., 53(116), *66*
Owens, E. B., 45

Page, J. A., 50(101), *66,* 178, 179(57), *184*
Palilla, F., 251(13), *272*
Pan, Y. D., 173
Parissakis, G., 45
Parker, A., 78, 88, *112, 113*
Parker, C. A., 47(79), *65,* 254(71,72), 258(80), 271(71), *273, 274*
Parker, W. G., 197
Parks, T. D., 173
Parratt, L. G., 326, 373, 376(56), *414, 416*
Patchornik, A., 131, *140*
Patser, G. V., 295(26), *321*
Pauling, L., 401, 404, *417, 418*
Payne, J. A., 287(12), *320*
Payne, S. T., 45
Pearce, M. L., 88, *113*
Peed, W. F., 311, 312, *322*
Peekema, R. M., 164, 178
Pehlke, R. D., 90, *113*
Peĭzulayev, S. I., 63(168), *67*
Pelavin, M., 193(63), *222*
Pellin, R. A., 46(74), *65*
Pellisser, G. E., 74(6b), *112*
Pellowe, E. F., 255(25), *272*
Pepkowitz, L. P., 135, *140*
Perkin-Elmer Corp., 188(56), *221*
Perkins, O. E., 35(26), *64*
Peters, D. G., 152(9), *183*
Peters, E. D., 126(15), *139*
Peters, T., 35(28), *64*
Peterson, D. T., 81(28), *112*
Peterson, J. I., 97, *113*
Petrikova, M. N., 41(68) 51(68), *65*
Pfann, W. G., 63(165, 166), *67*
Pfeiffer, H. G., *323*
Pflaum, R. T., 271(112), *274*
Pforr, G., 194(67), *222*
Phillips, D. S., 45
Phillips, H. O., 56(132), *66*
Phillips, J. P., 149(7), *183*
Pickering, W. T., 256(59), *273*
Pickett, E. E., 212(101), *222*
Pietri, C. E., 61(152, 153), *67*
Pikulik, L. G., 268(86), *274*
Pinchuk, G. P., 63(168), *67*
Poellmitz, G. S., 307(63), *322*
Pohl, F. A., 35(24,25), 45, *64*
Poluektov, N. S., 187, 188(12,47), 207(12), *220, 221*
Porterfield, W. W., 149(5), 176, *182*
Potter, J. L., 107(83), *114*
Powell, L. N., 131(29), *140*
Powell, W. A., 271(107), *274*
Powers, M. C., 295(24), *321*
Proctor, K. L., 256(53), *273*
Prod'hom, G., 45
Profitt, P. M. C., 54(118), *66*
Przybylowicz, E. P., *151*
Pshenitsyn, N. K., 164, 173
Puckett, J. E., 133(44), 134(44), 137(44), *140*
Pyne, C., 304, *322*

Radley, J. A., 271(102), *274*
Rains, T. C., 58(140), 67, 201(86), *222*
Ramelow, J., 188(33), 199(33), *221*
Ramirez-Muñoz, J., 214(113), 216, 218, *223*
Ramquist, L., 404, *418*
Randles, J. E. B., 164(28), 165, *183*
Rands, J., 35(36), *64*
Rann, C. S., 204, *222*
Ranzetta, G. V. T., 408, *418*
Rapp, R., 92(58), *113*
Rasmuson, J. O., 213(111), *223*
Ratnikova, V. D., 164
Razumova, L. S., 271(99), *274*
Reber, L., 271(104), *274*
Rechnitz, G. A., *182,* 270(95), *274*
Rees, W. T., 30, 35(23), *64,* 254(71,72), 258(80), 271(71), *273, 274*
Reeve, L., 80(24), *112*

Reilley, C. N., 149, 151(6), 176, *182,* 251(12), *272*
Rein, J. E., 178
Rengstorff, G. W. P., 45
Répás, P., 45
Reynolds, G. F., 165(32), 166, *183*
Rheinhart, R. C., 149(4), *182*
Rhodes, J. R., 319, *323*
Ricci, E., 133(41), *140*
Rice-Jones, W. G., 121, *139*
Richardson, D., 187(5), *220,* 331, *415*
Riddiford, A. C., 171(47), *184*
Rieman, W., III, 55(126), 59(126), *66*
Ringbom, A., 244, *272*
Roberts, E. D., 218(116), *223*
Roberts, K. H., 54(117), *66*
Robins, D. A., 404, *417*
Robinson, J. W., 188(54), *221, 223*
Rodden, C. J., 150
Rogers, L B., 50(98), *65*
Rohde, R. K., 255(37), *273*
Rooney, R. C., 45, 54(119), *66,* 121, *139,* 165(31), 166, *183*
Rose, H. J., Jr., 298(43), *321, 323*
Rossi, G., 53(164), *67*
Roubalová, D., 173
Rubeska, I., 188(48), 217, *221*
Ruch, R. R., 31(5), 60(146), 62(160,161), *63, 67*
Ruediger, K., 199(80), *222*
Rumsk, M. A., 347(30), *416*
Rupp, R. L., 46, *65*
Ružička, J., 42(70–72), 59(70,141), *65, 67*
Ryan, J. W., 46(75), *65*
Rynasiewicz, J., 46(75), *65*

Saeki, M., 51(104), *66*
Sagawa, T., 372(55), 376(55), 397, *416, 417*
Sajo, I., 45
Salamon, M., 42(71), *65*
Salmon, L., 58(137), *67*
Salpeter, E. E., 366, 368, *416*
Salutsky, M. L., 47(82,82), *65*
Sambucetti, C. J., 53(109), *66*
Samson, C., 297, *321*
Samuelson, O., 55(123), 59(123), *66*
Sand, H. J. S., 170, *184*
Sandell, E. B., 33(13), 37(13), 40, 49(13), *63,* 253, 271(113), *272, 274*
Sanders, J., 46(77), *65*
Sandstrom, A. E., 279(2), *320,* 327, 364, 365, *415*
Sarian, S., 298(46), *321*
Sattur, T. W., 218, *223*
Savvin, S. B., 256(49), *273*
Saylor, J. H., 271(107), *274*
Scapp, H. A., 255(30), *272*
Schaffer, E. W., Jr., 175(52), *184*
Scheil, 80(25), *112*
Schleser, H., 188(49), *221*
Schmid, R. W., 149(6), 151(6), *182*
Schmidt, H., *182*
Schmidt, W. E., 53(107), *66*
Schmidts, W., 135, *140*
Schmitt, D. H., 63(163), *67*
Schmitt, H. J., 129(20), *140*
Schmuckler, G., 56(131), *66*
Schneider, E.-L., 45
Schoffman, E., 191
Scholes, I. R., 255(29), *272*
Scholes, P. H., 166
Schreiber, E., 35(35), *64*
Schrenk, W. G., 188(37), *221*
Schüller, H., 188, *220*
Scott, D. V., 408, *418*
Scott, F., 121, *139*
Scott, F. A., 164, 178
Scott, T. C., 218, *223*
Scribner, B. F., 6(1), 21(12), *23,* 187, 188(17,18), *220,* 320(98), *323*
Seaborn, J. E., 165(32), *183*
Seal, R. T., 289, *320*
Sebens, C., 188(27), 217, *221*
Segatto, P. R., 178
Selmer-Olson, A. R., 256(57), *273*
Şen Gupta, J. G., 139(61), *141*
Ševčik, A., 165, *183*
Shaffer, P. T. B., 393
Shain, I., 59(99), *66,* 165(30), 166(30,33), 167, 169(30), *183*
Shalgosky, H. I., 166
Shalitin, Y., 131, *140*
Shanahan, C. E. A., 97, *113*
Sherman, J., 302, *322*
Shifrin, N., 218(107), *223*
Shimizu, M., 375(58), 376, *417*
Shipitsyn, S. A., 194, *222*
Shpol'ski, E. V., 268(88), *274*
Shults, W. D., 177(55), 178, *182, 184*
Shuvaev, A. T., 385, *417*

Sidorenko, V. V., 270(96), 271(96), *274*
Siegbahn, M., 286, *320*
Siegel, I., 316(91), *323*
Sierer, P. D., Jr., 319, *323*
Siggia, S., 149(4), *182*
Sill, C. W., 271(110), *274*
Simmler, J. R., 54(117), *66*
Simon, W., 133(37), *140*
Singer, E., 45
Singer, L., 97, *113*
Skinner, H. W. B., 326, 372, 375–377, 384, *414, 416*
Skogerboe, R., 6, 85(38), *113,* 133(42), *140,* 206(95), *222*
Slavin, W., 188(27,42,52), 206(94), 217(27), *221–223,* 246(8), *272*
Sleeper, M. P., 46(75), *65*
Sloman, H. A., 78, 79, 80(15), 81(9,15), 82(15, 83(15), 89, 90, *112, 113*
Smales, A. A., 58(137), *67*
Smiley, W. G., 97, *113*
Smith, D. E., *182*
Smith, D. M., 173(50), *184*
Smith, E. C., 31(8), *63*
Smith, H. A., 96, 97, *113*
Smith, R., 189(61), *221*
Smith, S. W., 53(113), *66,* 175(52), *184*
Smith, W. H., 82(32), *112*
Snell, C. T., 253, *272*
Snell, F. D., 253, *272*
Solet, I. S., 137, *140*
Solon, E., 177(53), *184*
Solovev, E. A., 271(108), *274*
Sommer, P. F., 133(37), *140*
Southern Analytical Ltd., 166
Southworth, B. C., 173
Spano, E. F., 305(61), *322*
Spielberg, N., 306(62), *322*
Spielholtz, G. I., 270(92), 271(92), *274*
Spigarelli, J., 137(53), *140*
Spitz, E. W., 54(117), *66*
Srehla, G., 251(10), *272*
Staab, R. A., 188(39,40), *221*
Stackelberg, M. v., *182*
Starfelt, N., 287(13), *320*
Starý, J., 41(70–72), 59(72,141), *65, 67,* 191
Steers, J. E., 97, *113*
Stegemann, H., 30
Stephens, F. B., 126(17), 127(17), 132(31), *139, 140,* 178
Sternglass, E. J., 363, *416*
Stickney, M. E., 246(7), *272*
Still, J. E., 120, 121(4), 130, 136(4), *139, 140*
Stobart, J. A., 255(43), *273*
Stock, J. T., 172(48), *182, 184*
Strahm, R. D., 150
Strock, L. W., 35(28), *64*
Ströhl, G., 36(44), *64*
Stuckey, W. K., 133(36), 134, *140*
Stupar, J., 217
Sturton, J. M., 256(56), *273*
Sullivan, J. V., 188(26), 205(26), 214(26), *221*
Sunderland, W., 122(9), 126(17), 127(17), 132(31,32), 134(9), 138(9,56), *139, 140*
Suttle, J. F., 47(81), *65*
Swift, E. H., 176
Sylvania Electric Products, 128(18), *140*

Tait, P. C., 126(15), *139*
Takos, R. C., 92(58), *113*
Tanaka, N., 50(96), *65, 182*
Tardon, S., 188(50), *221*
Taylor, J. K., 53(113), *66,* 166, 175(51,52), 179(58), *184*
Temyanko, V. S., 164
Tera, F., 60(146), 62(160,161), *67*
TeSelle, L. D., 126(15), *139*
Thatcher, J. W., 286(8), 298(45), 305, *320–322*
Theuerer, H. C., 63(166), *67*
Thiers, R. E., 30, 31(4), 32(4,11), 33, *63,* 210, *223*
Thilliez, G., 188(51), *221*
Thomas, A. M., 61(154), *67*
Thompson, B. J., 382, 383(71), 401, *417*
Thompson, C., 201(87), *222*
Thompson, J. G., 80(21,22), 91(55), *112, 113*
Thompson, S. G., 293(19), 294(19), *321*
Thurmond, C. D., 80(19), *112*
Tipler, G. A., 133, *140*
Tomboulian, D. H., 326, 371, 372(53), 376(53), *414, 416*
Townsend, J. E., 298(44), *321*
Trotter, J., 346, 382, 389(70), *415, 417*
Trzeciak, M. J., 80(20), 84(20), 85(20), *112*

Tsuchibuchi, A., 35(37), *64*
Tumney, Z. T., 122(10), 124(10, 137(10), *139*
Turovtseva, Z. M., 79(17), 83(17), 88, 89, 98(75), *112–114*
Tuthill, S. M., 54(117), *66*
Tyrrell, A. C., 46(77), *65*

Udenfriend, S., 254(73), 263, *273*
Ulrich, W. F., 218(107), *223*
U.S. Atomic Energy Commission, 28(2), *63*

Vacher, H. C., 80(21), *112*
Vainshtein, E. E., 385, *417*
Vallee, B. L., 188(30), 199, 210, 217(30), *221, 223*
Van Aman, R. E., 45
Van Camp, M. A., 79(12), 81(12), *112*
Van Niekerk, J. N., 58(134), *67*
Van Nordstrand, R. A., 316, *323*
Van Ostenburg, D. O., 375, 376, *416*
Van Reuth, E. C., 379(68), 381, *417*
Varian Techtron, *223*
Vasileva, L. N., 167
Vassos, B. H., 55(122), *66*
Vazsonyi-Zilahy, A., 53(106), *66*
Veale, C. R., 50(94), *65*
Veillon, C., 189(60), *221*
Veleker, T. J., 45, 50(93), *65*
Velicka, I., 188(48), *221*
Vernadsky, V. I., 98(75), *114*
Vetter, K. J., 164
Vickers, T. J., 188(38), 189(59), 196(78), *221, 222*
Vidale, G. L., 199(82), *222*
Vinogradova, E. N., 167
Violante, E. J., 130, 131, *140*
Visser, J., 30, 131(26), *140*
Vladimirova, V. M., 271(99,100), *274*
Volkova, A. I., 36(45), *64*
Vollmer, J., 205(90), *222*
Vorlíček, J., 138, *141*
Vose, G. P., 307(65), 309, *322*

Wadlow, H. V., 180(60), *182, 184*
Wagner, R., 316(91), *323*
Wahlin, E., 169(40), *183*
Waldron, H. F., 95, *113*
Walker, J. M., 133, 134, 136(52), 137, *140*
Walsh, A., 188, 199(83), 205(26), 214, *220–222*
Walter, D. I., 73(5), *112*
Walton, H. F., 33(14), 55(127,130), 59(127, 130), *63, 66*
Wasserman, A. M., 98(75), *114*
Waterbury, G. R., 150, 299(47), 301(50), 302(55), 304(47), 305(50), 311(80,81), 313, *322, 323*
Waterman, W. R., 255(29), *272*
Watson, R. E., 377(67), 381, *417*
Weart, H. W., 298(46), *321*
Weberling, R. P., 188(55), 189(55), *221*, 251(13), *272*
Weimer, E. R., 258(81), *274*
Weissler, A., 254(74,75), 256(60), 271(97, 101), 272(114), *273, 274*
Welcher, F. J., 179(59), *184*
Weldrick, G. J., 61(154), *67*
Wendt, R. H., 188(31), *221*
Wenzel, A. W., 61(152,153), *67*
West, P., 126(16), *139*
West, P. W., 134, *140*
West, T. S., 49(88), *65*, 188(41), *221*, 255(23), 270(94), *272, 274*
Westenberg, A. A., 195(74), 196(77), 211(100), *222*
Westfall, F. O., 205(92), *222*
White, C. E., 254(74,75), 258(79,81), 271(97,101), 272(114), *273, 274*
White, D. C., 124(12), 131, *139*
White, E. W., 295(25), *321*
White, J. C., 45
White, T. T., 126(15), *139*
Wickbold, R., 30
Wiech, G., 373, *416*
Wilczewski, J. W., 308, *322*
Wilkinson, G., 117(3), *139*
Willard, H. H., 47(82), *65*, 162(27), *183*, 263, *274*
Williams, M. L., 346(26), 377(67), 381, *415, 417*
Williams, R. J. P., 37(61), *64*
Williams, S. E., 346, 376(59), 397, *415, 417*
Williams, W. S., 404, *418*
Willis, C. P., 271(110), *274*
Willis, J. B., 188(28), 204, 207(97), 209, 211, 217(97), *221, 222*
Wilson, H. N., 256(55), *273*
Wilson, R., 348, *416*
Wilson, R. E., 153(11), *183*

Winefordner, J. D., 188(38–40), 189, 196, *221, 222*
Winge, R. K., 36(43), *64*
Winslow, E. H., 307(63), *322, 323*
Wise, E. N., 177(53), *184*
Witbeck, F. F., 96, 97, *113*
Witt, E., 53(109), *66*
Wittern, B., 271(105), *274*
Wolff, A. K., 136(51), *140*
Wolfhard, H. G., 195(72), 197, *222*
Wollaston, W. H., 188, *220*
Wood, D. F., 256(66), *273*
Wood, R. G., 50(94), *65*
Woodriff, R., 188(33), 199(33), 206(95), *221, 222*
Woodward, C., 270(94), *274*
Wybenga, F. T., 58(134), *67*
Wyckoff, R. W. G., 366, *416*

Yamada, K., 45, 58(138), 61(138), *67*
Yamasaki, G. K., 188(25), 205(91), *221, 222*
Yokoyama, Y., 188(35), 199(35), *221*
Youden, W. J., 14(8,10), 20(8,10), *23*
Young, J. R., 363(47), *416*
Young, P., 188(41), *221,* 255(23), *272*
Youtz, M. A., 153(11), *183*

Zaffarano, D. J., 363(48), *416*
Zamochnick, S. B., 270(95), *274*
Zechman, G. R., Jr., 295(25), *321*
Zeegers, P. J. T., 189(61), *221*
Zeitz, L., 291(17), *321*
Zelinskii, V. V., 268(86), *274*
Zelyukova, Yu. V., 188(47), *221*
Zemansky, M. W., 188, *220*

Subject Index

Absorbent, 125
for carbon, 125
manganese dioxide, use of, 126
Accuracy, 18
cross check, 19
homogeneity, 19
realistic requirements, 18
"spiking," 19
standards, 19
synthetic standards, 19
Aluminum, impurities in, 53, 60
in iron, 54
Analysis of thin films, *see* Thin films, analysis of
Analytical flame spectroscopy, 186
Antimony, alkali metals in, 37
in lead, 49
Apparatus, cleaning of glass, silica and plastic ware, 31
Applications of absorption spectroscopy, 306
absorbtiometry with monochromatic X-rays, 308
absorbtiometry with polychromatic X-rays, 306
analysis by absorbtion edge fine structure, 314
X-ray absorbtion edge fine structure, 314
Applications of X-ray emission and fluorescence, 295
quantitative analysis, 298
sample preparation, 296
selection of a line, 295
trace analysis, 304
Arsenic, alkali metals in, 37
impurities in, 43
Atomic absorbtion flame spectroscopy, 188
sources, 204
spectrometer, 212
Auger process, 367
Automation and data handling, 21
automatic analyzers, 22
automatic data readout, 21
computer programs, 21
Automatic titrators, 149

Back-extraction, 42
Back-washing, 42
Batch operation, 58
in ion exchange, 58
Be *K*, 340
BeO, 344
Bismuth, impurities in, 44
in lead, 48
lead in, 35
B *K*, 340
B *K* bands from boron, 397
of TiB_2, 397
of VB_2, 397
of ZrB_2, 397
"Blank," 123
cold, 123
correction for, 123
hot furnace, 123
routine, 123
Blazed replica grating, 331
blaze angle, 332
selective effect, 332
total reflected beam, 332
Bond strength, 405
heats of formation, 405
melting temperatures, 405
Boron, in germanium, impurities in, 43
in nickel, 54
in plutonium, 61
in silicon, 46, 48
in silicon tetrachloride, 46, 50
in sodium, 46
in uranium, 61
in various metals, 34
Breakthrough curves, 58
Burner and flame for flame spectroscopy, 195

Cadmium, impurities in, 43
in throium compounds, 61
Calcium, in zirconium, 37
Carbon, in germanium, 36
in silicon, 36
Carbon determination in various metals, 135–137
Carbon dioxide, 119
detection of, 129
formation and release, 120
gas chromatography for, 133
infrared detection of, 133
titrimetric method for, 132
Carbides, 117
Cell, 144
Galvanic, 144
Polarographic, 160
Change in shape and wavelength of the emission bands, of elements when chemically combined, 383
with excitation voltage, 351
Changes in peak intensity with chemical combination, 413
Chromatic elution, 57
Chromatography, 77, 99
Chromium, impurities in, 35, 43
Chronoamperometry, 151
apparatus for, 155
current-potential relationship, 157
current-time relationship, 156
potential-sweep, 164
analytical determinations by differential cathode ray polarography, 166
current-potential curve, 165
Chronopotentiometry, 151, 169
apparatus for, 170
C *K* band, 351, 356
C *K* bands of diamond, 390
C *K* emission bands from carbides, 388
C *K* band from carbon in solution in austenite and martensite, 394
C *K* band from graphitized iron, 392
C *K* band transition metal carbides, 389
Cobalt, in nickel, 61, 63
in titanium, 53
in zirconium, 53
Colorimetric methods for sulfur dioxide, 134
Column operation, 56
Combustion method for carbon and sulphur, 115, 116
description, 116
general problems, 116
Combustion train, 125
construction practices, 126
schematic for, 127
Concentration factor, 28
Conductometric methods for CO_2, 120
commercial apparatus, 130
high frequency method, 131
sensitivity of, 131
steels and cast iron, 130
Conductometry, 180
high frequency, 181
low frequency, 180
Contamination, 29
airborne contamination, 31
due to apparatus, 29
due to reagents, 29
Copper, gold in, 53, 61
impurities in, 43
iron in, 48
lead in, 48
selenium in, 49
silver in, 53
tellurium in, 49
Co-precipitation, 47
Cottrell equation, 156
Coulometry, 173
constant current, 174
apparatus for, 175
examples of determinations, 176
controlled-potential, 175
examples of determinations, 178
selectivity in, 177
types of electronic integrators, 177
Crystals for soft X-ray region, 336
lead lignocerate, 340
lead stearate crystals, 337, 340, 343, 384
phthalic acid salts, 336
potassium acid phthalate (KAP), 336, 343, 384
Crystal spectrometer, 342, 349
Current, background, 154
charging, 154
diffusion, 156
Faradaic, 154
limiting, 171
residual, 154

Degassing, 71
Density of states N(E) curve, 370

Fermi edge, 373, 375
partial density of states, 372
Depth from which X-rays come, 363
Desiccant, 125
Detectability and sensitivity, 7
"absolute detectability," 12
chemical, 8
confidence limits, 15
electron microprobe, 9
flame (atomic absorption and emission), 9
matrix interferences, 11
minimum theoretical mass requirement, 13
minute samples, 13
optical emission spectrography, 9
paired observations, 15
spark source mass spectrography, 9
standard deviation, 14
thermal (combustion and vacuum fusion), 9
various parameters, effects of, 12
Detector for soft X-rays, 345
CuBe photomultipliers, 346
gas counter, 345
geiger counter, 347
photoelectric multiplier, 345
photosurface, 346
scintillation counter, 345
cellulose nitrate, 349
counter voltage, 347
formvar films, 349
gas pressure, 347
proportional counter, 347
thin windows, 346
type of gas, 347
Diffraction crystals, 288
Dispersion, 343
Distribution coefficient, 56
in ion exchange, 56
Distribution ratio, 37
in extraction, 37
Double layer electrode, 154

Electroanalytical techniques, 143, 144
Electrode, 144
counter, 156
double layer, 154
dropping mercury (DME), 151
current-time characteristics, 161
glass, 146
hanging mercury drop (HMDE), 166
indicator, 144
mercury, range of applicability, 153
normal hydrogen (NHE), 145
oxide film on noble metal, 152
pH, 146
pM, 149
platinum, range of applicability, 153
polarographic, 160
reference, 144
rotating disc, 171
rotating Pt, 171
saturated calomel (SCE), 144
silver, 149
silver-silver chloride, 146
specific ion, 146
working, 151
Electrodeposition, 50, 179
of matrix or interfering elements, 53
of trace elements, 52
on a mercury cathode, 50
Electrode process, 157
irreversible, 157
reversible, 157
Electrography and electrospot testing, 179
Electrogravimetry, 179
Electrolysis, controlled potential electrolysis, 50
electrolysis cells, 51
internal electrolysis, 50
mercury cathode electrolysis, 51
Electrolysis apparatus, 155
constant current, apparatus for, 155, 170, 175
constant or controlled potential, 155
three electrode, 155
two electrode, 155
Electron distribution and bonding from soft X-rays, 398
changes in emission bands with chemical composition, 398
intensity distribution changes, 398
peak energy shift, 399
electronegativity difference, 401
electroneutrality principle, 404
electron transfer, 401
ionic bonding, 401
metallic bonding, 404
Electron excitation of soft X-rays, 349
Electron microprobes, 351, 355
take-off angle, 351

Electron penetration depth, 363
Electroseparations, 156, 179
Emission bands from oxides, 384
Energy level diagram, 364
  emission bands from alloys, 379
Enrichment factor, 28
Equilibrium pressure method, 95
  nitrogen determination, 97
  oxygen determination, 95–96
Errors, sources of in combustion methods, 122
Evaporation chambers, 32
Extraction, applications of, 42
  continuous extraction technique, 41
  discontinuous countercurrent distribution extraction technique, 41
  extraction systems, 37
  liquid-liquid, 37
  of matrix elements, 44
  techniques of, 41
  of trace impurities, 43

Fe L_{III} bands, 379
Fire assay, 63
Falme emission, 199
  band emission, 201
  continuous emission, 202
  line emission, 199
Flame emission spectroscopy, 187
Fluxes, 120
  for carbon analysis, 121
  choice of, 120
  factors for use of, 120
  lead for use with molybdenum and silicon, 137
  for sulfur analysis, 122
  types of, 121
Furnace blank, 75
Furnaces for combustion methods, 128
  containers for, 129
  induction, 128
  resistance, 128

Gallium arsenide, impurities in, 35, 44
Gas chromatography for CO_2, 133
  column retention, 133
  for ferrous type alloys, 133
  liquid nitrogen or oxygen traps, 133
  sensitivity of for ferrous metals, 133
Gas chromatography for SO_2, 134
Gases, in metals, 36
Gathering precipitate, 49
Germanium, boron in, 48
  impurities in, 35, 43
Gold, in copper, 53
  impurities in, 44
Graphite, 117
Graphitic compounds, 117
Gratings, 328
  critical angle, 329
  cut-off wavelength, 330
  zero order, 332
Grating spectrometer, 327, 342, 349
  Rowland circle, 327, 342
  slit, 327
  slit widths, 339

Half-wave potential, E 1/2, 157
  tables of data (tables 3 and 4), 163
High purity acids, impurities in, 36
High purity reagent, 30
  purification and preparation of, 30
Hot vacuum extraction method, 91
Hydrogen determination in various metals, 85, 87, 91
Hydrogen peroxide, 134
  absorption of sulfur dioxide, 134
Hydrogen sulfide evolution method for sulfur dioxide, 134

Identification parameters and techniques, 2
  activation analysis, 3
  atomic absorbtion, 3
  carbon determination, 4
  chemical, 3
  electron microprobe, 2
  electron microscopy, 3
  emission spectrography, 2
  flame emission, 2
  light microscopy, 3
  metallography, 3
  nitrogen determination, 4
  scanning electron microscopy, 3
  spark source mass spectroscopy, 2
  sulphur determination, 4
  vacuum fusion, 4
  X-ray diffraction, 2
  X-ray spectrography, 2
Infrared detection for CO_2, 133
Igniter, 121
Ilkovic equation, 159
Index of symmetry, 370

Inert gas fusion method, 36, 97
blank, 104
conductometric procedure, 103
furnace, 103
furnace and reaction tube, 99
instrument calibration, 105
oxidation of CO to CO_2, 99
oxygen determination in various metals, 108–111
sample preparation, 103, 104
thermal conductivity, 99
Inner level width, 366
Instrumentation for X-ray Spectroscopy, 286
diffraction crystals, 287
pulse height analysis, 294
spectrometer construction, 290
X-ray detectors, 292
X-ray sources, 286
Internal standards, 300
Interstitial carbides, 118
Iodine titration of sulfur dioxide, 134
Ion bombardment cleaning, 357
Ion exchange, 55
application of, 60
complexing agents, 59
effect of solution composition, 59
ion exchange paper, 56
ion exchange resins, 55
loss and contamination in, 59
techniques of, 56
Ion exchanger, 55
cellulose base, 56
liquid, 37
solid, 55
synthetic inorganic, 56
Iron, aluminum in, 54
in copper, 48
impurities in, 43
in tin, 35

L_{III} bands for TiB_2, $TiC_{0.95}$ and $VC_{1.095}$, 40
Lead, antimony in, 49
bismuth in, 48
in high purity bismuth, 35
iron in, 48
selenium in, 49
in telluric acid, 50
tellurium in, 49
thallium in, 49
Liquid junction, 144

M_V band of Mo, 373
M_V band of Nb, 373
Magnesium, impurities in, 53, 60
Manganese dioxide, 126
preparation of, 126
use of, 126
Manometry, 75
Masking agents, 39
in extraction, 39
Mass spectrography, 76
Metals and other materials, extraction of matrix elements, 44
of trace impurities, 43
Microcrystalline graphite, 117
Mo M_V N_{III} line, 365

N K emission bands from nitrides, 386
NiL_{III} band, 343, 352, 377
Nickel, boron in, 54
cobalt in, 48, 61
in copper, 55
impurities in, 43
Nitrogen, in metals, 34
Nitrogen determination in various metals, 87–91
equilibrium nitrogen contents, 91
thermodynamic considerations, 90

O K bands, 343, 384
Oscillometry, 180
Oxidation-reduction (redox) buffers, 50
Oxygen, checking purity of, 125
purifying, 125
Oxygen determination in various metals, 78–85

Peak wavelength shift, 385, 398, 406
Percent extraction, 39
Precision, 14
Phosphorous, alkali metals in, 37
in silicon tetrachloride, 47
"Plate theory," 57
Plutonium, impurities in, 43, 61
silicon in, 61
Pneumatic nebulizer, 193
Polarography, 151
alternating current, 168
cathode ray, 164
classical, 159

derivative, 167
pulse, 168
oscillographic, 169
single-sweep oscillographic, 164
square wave, 168
tast or strobe, 169
Postprecipitation, 47
Potentiometry, 145
Potentiostat, 156
Precipitation, 47
carrier in, 48
collector, 48
fractional precipitation, 48
from homogeneous solutions, 47
of ordinary amounts of elements, 47
of trace elements, 48
Preconcentration, 25
evaluation of preconcentration techniques, 27
selection of preconcentration techniques, 27
Pulse height analysis, 294

Quantitative analysis, 350
of light elements, 406
accuracy, 410
C *K* band, 407
C *K* band intensities from carbides, 410
lowest detectable limit (L.D.) wt%, 410
X-ray excitation, 409

Radiative processes, 228
Rare earths, 49
in metals, alloys and oxides, 49
in stainless steel, 54
in zirconium, 61
Randles-Sevcik equation, 165
Recovery, of desired element, 27
Refractory metals, impurities in, 36
Resolution, 343
Resolving power, 342

Salting-out agents, 37
Sample form and preparation, 20
amount available, 20
cleaning, 21
contamination, 21
pretreatment, 20
sample conversion, 20
surfaces, 21
Sample preparation, for flame spectroscopy, 189
for fusion techniques, 74
Sand equation, 170
Satellite lines, 367, 383, 386
Scope and sensitivity of Analytical flame spectroscopy, 215
Selenium, impurities in, 43
impurities in selenium oxide, 35
Self-absorbtion, 352
Sensitivity, defined, 135
Separation, 25
by distillation from solutions, 33
by electrodeposition, 50
evaluation of separation techniques, 27
by ion exchange, 55
by liquid-liquid extraction, 37
by precipitation, 47
selection of separation techniques, 27
by selective dissolution, 46
by volatalization, 33
by volatilization at high temperatures, 36
Separation factor, 28
Silicon, boron in, 48
impurities in, 35, 43
in plutonium, 34, 61
in uranium, 34
Silicon tetrachloride, phosporous and boron in, 47
Sodium, boron in, 46
Soft X-ray emission bands, 370
auger transition, 376
changes in shape, 385
Mo emission edge, 375
temperature correction, 375, 377
width of the band, 376
Soft X-ray emission lines, 365
Soft X-ray spectroscopy, 326
Specific heat measurements, 375, 379
Spectrofluorimetry, 226
Spectrofluorimetric analysis, 254
applications, 269
instrumentation, 254
methodology, 263
scope, 254
Spectrophotometry, 226
Spectrophotometric analysis, 236
applications, 253
instrumentation, 236
methodology, 243
scope, 236
Specular reflection, 339
Standard potential, 145
Standard sample, 124

ferrous type alloys, 125
NBS standard sample 166 B, 125
problem of, 125
Standardization, 123
by combustion, 124
direct, 123
gas chromatography, 124
standard substance, 124
Statistics of X-ray measurements, 293
Steel, rare earths in, 54
Stripping, 42
in extraction technique, 42
Stripping analysis, 50, 166
examples of determination, 167
Substochiometric technique, 42, 59
Sulphur, in nickel, 34
Sulphur compounds, 119
metallic, 119
oxides, 119
Sulfur determination, 137
in solid lubricants, 138
in various metals, 137
Sulphur, elemental, 118
Sulphur dioxide, 121
colorimetric methods for, 134
detection of, 133
formation and release, 121
gas chromatography for, 134
hydrogen sulphide evolution method for, 134
reaction with oxygen, 121
removal of, 126
Sulfur trioxide, 119
adsorption of, 121
correction factor, 122
Supporting electrolyte, 151

Target, contamination, 354
carbon combination, 356, 363
cold chamber, 355
hydrocarbon layer, 355
Tellurium, in copper, 49
in lead, 49
Thallium, in lead, 49
Theory of X-ray Spectrography, 277
absorption of X-rays, 281
diffraction of X-rays, 280
intensity of X-ray lines, 284
origin of X-rays, 277
scattering of X-rays, 280
Thin films, analysis of, 316
absorbtion analysis, 316
fluorescence analysis, 319
Ti L_{III} band, 377
Ti L_{III} band from TiNi, 404
Ti $L_{II,III}$ emission band, 385
Titanium, cobalt in, 53
Titration, 145
amperometric, 172
examples of determinations, 173
types of curves, 172
biamperometric, 172
conductometric, 180
"dead stop," 172
EDTA pM, 149
Karl Fischer, 173
potentiometric, 145
Titrimetric methods for CO_2, 131
acetone, 131
benzylamine methyl formamide, 131
sensitivity of, 132
standard acid-base titration, 131
Tracer technique, 27
Transition metal emission bands, 373
Transition metals, 361
Transition probability, 365, 371
Transition time, in chronopotentiometry, 170

Unresolved problems, 22
background impurities, 22
clean operational conditions, 23
composition of surfaces, 22
interstitial elements, 22
primary metal standards, 23
stoichiometric analyses, 22
Uranium, beryllium in, 62
boron in, 34, 61
cadmium in, 62
impurities in, 43, 53, 61, 62
rare earths in, 61
silicon in, 34

Vacuum fusion method, 36, 69, 70
furnace, 72
oxidation of CO to CO_2, 73
power supply, 71
pumping system, 73
Voltametry, 151
cyclic, 169
forced convection, 152

with forced convection, 170
working electrode materials, 152

Wave, current-potential, 157

X-ray detectors, 292
X-ray excitation for soft x-rays, 358
X-ray fluorescence analysis, 360
X-ray versus electron excitation, 353
X-ray intensity, P/B or resolution, 339
X-ray intensity techniques, 299
X-ray sources, 286
X-ray spectrography, 275

Yield, of desired element, 27

Zinc, in cadmium, 55
in copper, 55
impurities in, 43
Zirconium, calcium in, 37
cobalt in, 53
Zone melting, 63